Electronics for the Physicist with Applications

Electronics for the Physicist with Applications

C. F. G. DELANEY
Professor of Experimental Physics
Trinity College, University of Dublin

ELLIS HORWOOD LIMITED
Publishers Chichester

Halsted Press: a division of
JOHN WILEY & SONS
New York - Chichester - Brisbane - Toronto

First published in 1980 by

ELLIS HORWOOD LIMITED
Market Cross House, Cooper Street, Chichester, West Sussex, PO19 1EB, England

The publisher's colophon is reproduced from James Gillison's drawing of the ancient Market Cross, Chichester.

Distributors:

Australia, New Zealand, South-east Asia:
Jacaranda-Wiley Ltd., Jacaranda Press,
JOHN WILEY & SONS INC.,
G.P.O. Box 859, Brisbane, Queensland 40001, Australia.

Canada:
JOHN WILEY & SONS CANADA LIMITED
22 Worcester Road, Rexdale, Ontario, Canada.

Europe, Africa:
JOHN WILEY & SONS LIMITED
Baffins Lane, Chichester, West Sussex, England.

North and South America and the rest of the world:
Halsted Press, a division of
JOHN WILEY & SONS
605 Third Avenue, New York, N.Y. 10016, U.S.A.

British Library Cataloguing in Publication Data
Delaney, Cyril
Electronics for the physicist with applications.
1. Electronic apparatus and appliances
2. Physical instruments
I. Title
621.381'02'453 TK7870 79–42737

ISBN 0–85312–149–4 (Ellis Horwood Ltd., Publishers, Library Edition)
ISBN 0–470–26902–2 (Halsted Press, Library Edition)
ISBN 0–470–26903–0 (Halsted Press, Student Edition)

Typeset in Press Roman by Ellis Horwood Ltd.
Printed in Great Britain by W. & J. Mackay Ltd., Chatham

Contents

Acknowledgements

I am indebted to the following organizations for permission to use circuits and data from their official publications.

General Electric Company, U.S.A.
Institute of Electrical and Electronic Engineers, Inc.
Mullard, Ltd
Texas Instruments, Ltd
United States Department of Defense.

I would also like to express my thanks to my wife, and to my colleagues in the Physics Department at Trinity College, who helped with correction of the proofs.

CYRIL DELANEY

Foreword

There are a large number of texts on electronics specifically for students of electrical and electronic engineering, but not so many aimed at students in the science area and particuarly those in the physical sciences. It was with this in mind that the present text was prepared. It had its origins in lectures originally given to students in the third year of a four year experimental physics course but which were later extended downwards to second year students and upwards to final year students. However, since the mathematics required for understanding the text (apart from an acquaintance with calculus and elementary a.c. theory) is developed as required, the book should also be useful to those in a wider range of scientific and allied fields. The present work is a revised version of an earlier edition with the addition of substantial new material, and it is hoped that it will meet with the same acceptance as did the earlier volume.

C.F.G. Delaney

Department of Pure and Applied Physics
University of Dublin
Trinity College
Dublin 2

January 1980

CHAPTER 1

Introduction to detectors and basic semiconductor processes

1·1 Signals and detectors

The physicist is concerned with the amplification, measurement and processing of two main types of signal. The first, common in telecommunications, is the repetitive signal, which is usually sinusoidal or near sinusoidal. Such signals may be either genuine radio- or audio-frequency signals carrying information, or timing signals in measuring equipment. The other signals with which he will be concerned – indeed even largely concerned – are very different indeed. They are much closer to discrete, sudden steps of voltage, which may occur at random times. One source of such signals is a nuclear radiation detector, the simplest example of which is the air ionization chamber, shown schematically in Figure 1. In its most elementary form, this consists of a pair of parallel plates, a centimetre or so apart in air, with a potential of about 100 volts applied between them. When an ionizing particle, such as an alpha particle enters, a trail of positive and negative ions is produced in the air of the chamber. The positive ions are drawn to the left-hand plate by the negative bias applied by the battery, while the negative ions go to the other plate. When the collection process is finished, the result is the deposition of a charge $Q = -ne$ (where n is the number of negative ions produced and e is the electronic charge) on the right-hand plate, or, more exactly, on its capacity to ground C. This produces a voltage $V = Q/C$ at the amplifier input. The capacity C represents the sum of the capacitance of the chamber plus any stray capacitance at the amplifier input. Reversing the polarity of the external battery will reverse the sign of the charge collected and hence that of the voltage.

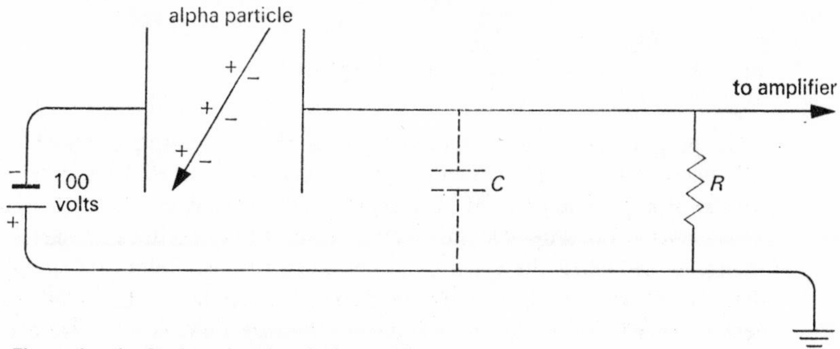

Figure 1. Ionization chamber (schematic).

The appearance of this voltage signal is not instantaneous due to the finite time taken for the ions to reach the plates. During this time the voltage on the amplifier input is building up by induction to its final value of Q/C. For a chamber such as we are discussing, this time might be of the order of a millisecond. Thus, while the signal from the detector may be approximated by the voltage step shown in Figure 2(a), in practice we may have to take account of the more precise representation shown in Figure 2(b). Here the rise, for the reason mentioned above, is not infinitely steep; in addition the top is not perfectly flat because of the subsequent slow leakage of the charge on C through the large resistor R.

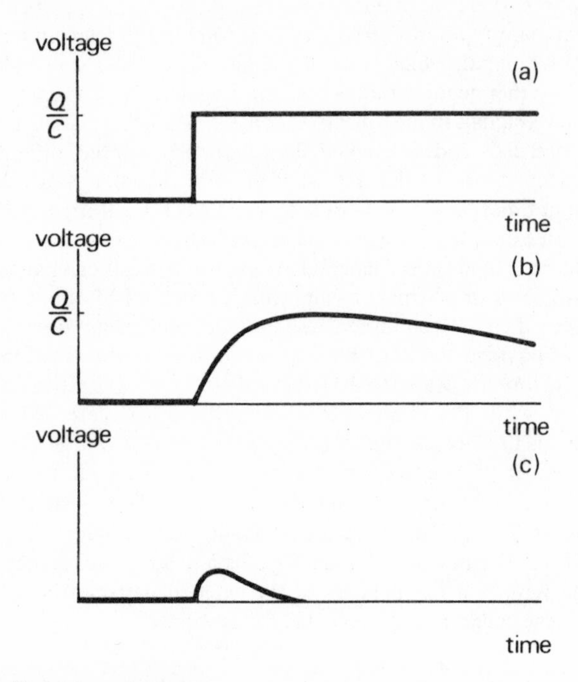

Figure 2. Signal from detector.
(a) idealized (b) in practice (c) with small value of R.

The proportional counter is a development of the ionization chamber in which the electrodes, instead of being two parallel plates as we have drawn them, consist of a metallic cylinder of diameter a few centimeters, down the centre of which runs a thin wire held positive with respect to the cylinder. Because of this change in geometry the electric field between the electrodes rises sharply near the wire. This, combined with the increased potential usually applied to the device, means that as electrons approach the wire they have enough energy to produce other secondary electrons in the gas, and these in turn can produce further electrons. The charge collected by the wire is now very much larger (typically by a factor of several hundred) than originally produced by the ionizing particle, and this built-in amplification makes subsequent recording of the event simpler. The output charge (and the voltage) are still proportional to the original charge Q deposited, and hence the name 'proportional' counter. In a further development, the Geiger counter, the output voltage is even larger but the device has lost its proportional character, that is, one obtains a large output signal but its size is independent of the nature or energy of the incoming particle. Indeed we can think of the Geiger counter as producing a standard signal if triggered by any nuclear particle.

The semiconductor or solid state detector is the solid state analogue of the ionization chamber. As we shall see in Section 1.3 a semiconductor diode consists of two regions of reasonable electrical conductivity with a normally non-conducting layer between – in fact something resembling in principle the chamber of Figure 1. Its action is also similar, although in this case the current produced by the ionizing particle consists of negative electrons and positive 'holes' of which we shall have more to say later. In spite of its small physical size as compared with an air ionization chamber, and lack of built-in amplification as in a proportional counter, the semiconductor detector has its own peculiar virtues. It can, for example, distinguish between two ionizing particles of nearly identical energies producing nearly identical amounts of charge, and it is widely used in atomic and nuclear physics.

The scintillation counter operates rather differently. Basically it depends on the fact that certain materials (for example anthracene, or a sodium iodide crystal activated with small amounts of the correct impurity) produce a tiny flash of light (or scintillation) when traversed by a nuclear particle. In some cases the flash may be large enough to be just seen by the eye, but normally it is converted into a burst of electrons by a photocathode, and the resulting electrical pulse of charge then amplified. The conversion and amplification are in practice done by a single device known as a photomultiplier. In spite of the very different physical principles on which the scintillation counter operates, the final electrical signal is similar to that of Figure 2(b). In the present case the finite rise time of the signal is due to the fact that the flash in the scintillating material does not decay immediately, and the electrical output signal can obviously not reach its maximum until all the light has been converted into electrical charge. The subsequent fall of the output pulse is, as before, due to the leakage of the charge on the photomultiplier output stray capacitance C through the photomultiplier load resistance R.

Although the output pulses from all the devices we have discussed are basically similar, due to the different mechanisms involved in each case, the actual rise time may vary enormously: in a scintillation counter, for example, it could be only a small fraction of a microsecond, which on the millisecond time scale for the air ionization chamber would represent an almost perfect step. The fall time of course is at our disposal through our choice of R, and we will later develop criteria for this choice. One possible criterion is that the fall should be fast enough for the signal to have effectively disappeared before the next particle arrives: however, the matter is not quite so straightforward as this, and we will discuss it further in section 3·7.

We conclude this discussion with two general remarks. Firstly, although we shall normally be dealing with voltage signals produced as just described, it is clear that it is the arriving charge which is the quantity giving us the basic information about the energy of the nuclear particle which entered the detector. It might therefore be more logical to deal directly with this charge rather than introduce the stray capacitance C, which is not always easily determinable. We shall be discussing in Chapter 6 the 'charge sensitive amplifier' which produces a voltage output proportional to Q but independent of C.

Secondly, in all we have said up to now we have assumed a two stage process in the production of the signal from a nuclear radiation detector – the initial growth of the voltage on the capacitance as the charge is collected from the detector (or for a scintillation counter as the light flash decays) and the subsequent slow decay as the charge leaks away. What we are really saying is that provided R is large enough, the main process is the charging of the capacitance C and that the later slow decay of the charge through R is subsidiary. This indeed is the normal case we shall have to deal with, yielding as it does a voltage Q/C which is a measure of the total energy of the incoming particle. However if R is made sufficiently small then the current, instead of going to charge the capacitance C, flows instead largely through R. The voltage output (equal to the current multiplied by R) now follows the current in the radiation detector, and we expect it to look as in Figure 2(c). Note that the voltage falls rapidly to zero as the charge is collected from the detector (or in the case of the scintillation counter as the light flash decays away). Note also that the time of disappearance of the signal in Figure 2(c) corresponds to the maximum of the signal in Figure 2(b), since the effective end of the detector current in Figure 2(c) means that for Figure 2(b) the total available charge has arrived on the capacitance. Although the current mode signals we have been discussing have the advantage of being fast, they have the corresponding disadvantage of being much smaller in size than their charge signal counterparts. The choice of which mode to employ depends very much on the particular experiment in hand, but for the applications in this book we will usually tend to think in terms of the charge type of signal.

Whatever type of detector we use and whatever mode we employ we will normally need some form of external amplification before we can process or record our signals, since they may well be in the millivolt or even microvolt range. For the construction of amplifiers we have a choice of thermionic valves, unipolar transistors (that is, devices whose operation depends on charge carriers of a single polarity), or bipolar transistors, the conventional variety, whose operation depends on two polarities of carrier. We shall see that valves and unipolar transistors can be classed as 'voltage-controlled devices'; bipolar transistors are 'current-controlled devices'. The main roles for thermionic valves nowadays are in high-frequency, high-power applications which are largely beyond the scope of this book: in the field we are concerned with here, they are superior to bipolar transistors in low-noise, high-input-impedance applications only. Even in this area their performance has been surpassed by the unipolar field effect transistor.

A brief account of the basic physical principles underlying the action of the semiconductor diode will allow us to understand the characteristics of both the unipolar and bipolar transistor. References (3 and 4) are among the many books which treat the physical background in greater detail.

1·2 Basic semiconductor processes

The semiconductor materials most commonly used for diodes and transistors are germanium and silicon. Both are elements of valence four – that is, in the crystalline state an atom is held by bonds formed through sharing its four outer electrons with four other atoms. At absolute zero all these bonds will be intact, but at room temperature a few of them will be fractured due to thermal vibration, with the electrons set free to move through the crystal, thus making it slightly conducting. (In another way of looking at the process the electron is considered to be lifted from the valence band, across a forbidden band of energies, to the conduction band.) The 'hole', or absence of negative charge, left by the electron, can also appear to move: for example, if an electron in an atom immediately to the left of the hole moves to the right under the action of an electric field, and fills the hole, then the hole appears to have moved to the left. Thus, in addition to electrons moving through the crystal, there are holes which act as if they possessed positive charges, since they tend to move in the opposite direction to the negatively charged electrons. The energy required to produce such a 'hole–electron pair' is 0·7 eV for germanium, and 1·1 eV for silicon. (1 eV = 1 electron volt = energy acquired by an electron in falling through a potential difference of 1 volt = $1{\cdot}6 \times 10^{-19}$ joules.)

The 'intrinsic conductivity' due to these holes and electrons is small: for germanium it corresponds to a resistivity of about 50 ohm-cm at room temperature, which should be compared with values for the good metallic conductors of less than 10^{-5} ohm-cm (and with those for the good insulators of greater than 10^{14} ohm-cm). The intrinsic resistivity for silicon, of about 2×10^{5} ohm-cm, is very much higher than for germanium, due to the exponential dependence of this quantity on forbidden energy gap width.

'Extrinsic conductivity' is of more importance. This is a very much larger effect obtained by 'doping', that is by adding controlled amounts of a selected impurity to the pure semiconductor. Suppose, for example, we add a small amount (say one part in 10^{8}) of an element of valence five, e.g., phosphorus or arsenic, to germanium. The impurity atoms will try to fit into the germanium lattice, but at each impurity site there will be one electron over, which will be only very loosely bound. At room temperature nearly all these weak bonds will be fractured, and the electrons made available for conduction. For the figure of one part in 10^8, the germanium conductivity would be increased by a factor of over ten. Since an electron has left the impurity atom, this is now positively charged. It is not however a free hole, since it has little attraction for an electron, and therefore cannot be made to 'move' in the manner previously described: it is instead a fixed ionized atom.

Such a semiconductor is known as an *n*- (for negative) type semiconductor, since the current is largely carried by electrons produced by impurity atoms. These electrons are also known as the 'majority' carriers, but there are also a few 'minority' carriers – the holes produced spontaneously during the formation of electron-hole pairs, as in the intrinsic case. A further point of terminology to be noted is that the phosphorus or arsenic is known as a 'donor' impurity, for the obvious reason that it 'donates' an electron to the conduction process. On the other hand it is possible to add an 'acceptor' impurity. This would be an element of valence three, indium, for example, or aluminium. Again its atoms would try to fit into the silicon or germanium lattice, but in this case there would be one electron per atom too few. An electron from a nearby atom will be attracted to the site, thus producing a hole where this electron came from, and producing a fixed, negatively charged, ionized acceptor atom at the impurity site. The hole will be available for conduction. A semiconductor of this type, where the majority carriers of current are holes, is known as a *p*- (for positive) type semiconductor. Again, there will be a few minority carriers present, electrons from thermally generated hole–electron pairs.

1·3 The junction diode

A semiconductor junction diode is formed, as its name implies, at the junction between a *p*-type and an *n*-type region. It cannot be made simply by placing two such regions in contact, but may, for example, be produced by growing first a crystal with a *p*-type impurity, and then changing over to a *n*-type impurity. Let us imagine however for the purpose of argument, that we have been able to place a piece of *p*-type material in sufficiently intimate contact with a piece of *n*-type material, and investigate in a simple way what happens. We might imagine that there would be an initial flow of electrons from the *n*-type material over to the *p*-type material, where they would combine with the holes. Similarly there would be a diffusion of holes over to the *n*-type material, these two flows constituting a current in the same direction because of the difference in sign of the carriers. This current will naturally not be a continuous one, and will be stopped by the creation of a potential barrier. This is formed because the electrons which migrate from the *n* to the *p* region leave behind them the fixed positive charge of the donor atoms, which tend to prevent further electrons leaving. Similarly in the *p*-type material, a negative charge will be built up at the acceptor sites, which will inhibit the flow of holes. A barrier layer is thus produced, and no net current flows across it. This is known as the 'depletion layer', since although there are fixed positive and negative charges in it, due to the presence of ionized donor and acceptor atoms, there are no mobile charge carriers, either holes or electrons. The width of this layer W (Figure 3), is typically of the order of microns. Its height, V_0, is typically 0·5 volts for germanium, 0·8 volts for silicon.

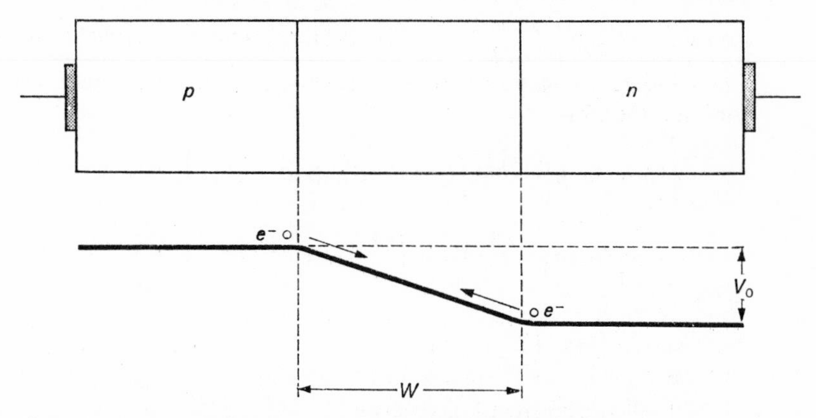

Figure 3. Potential barrier and depletion layer for a junction diode.

The equilibrium at the barrier is a dynamic rather than a static one – that is the net current flow is zero, not because no currents flow, but because equal currents flow in opposite directions. Let us consider the electron flow first. There is, to begin with, a flow from left to right in Figure 3. Electrons formed by the thermal breaking of bonds in the p-type material may drift to the barrier and 'fall' rapidly down it, producing what is called the 'thermal' current. On the other hand electrons in the n-type material have a range of energies, and some of them will have sufficient to climb the barrier from right to left, and combine with holes in the p-type region. This 'recombination' current is in the opposite direction to the previous one. Similar remarks apply to the two hole currents. The picture of the barrier for the holes will be just the bottom part of Figure 3 turned upside down. In equilibrium the barrier automatically sets itself so that the net current across it is zero.

We can derive a quantitative expression for the diode current as follows. In accordance with Boltzmann's law, the number of current carriers able to climb the barrier of height V_0 volts, that is the number with energy greater than eV_0 electron volts, is proportional to

$$\exp\left(-\frac{eV_0}{kT}\right) = C\exp\left(-\frac{eV_0}{kT}\right)$$

say, where T is the absolute temperature, k Boltzmann's constant, e the electronic charge, and C some constant. (Fermi–Dirac statistics are ideally necessary but the classical Boltzmann formula given above is a good approximation for semiconductors.) The expression we have just written down is the recombination current, which in equilibrium is equal to the thermal current. If an external voltage V is applied to the diode so as to help the flow of carriers up the potential 'hill' (that is if a battery is connected across the diode with its positive terminal to the p-type material), then the recombination current is increased to $C\exp\left\{-\frac{e(V_0 - V)}{kT}\right\}$. The thermal current remains unaltered being a fundamental spontaneous process depending only on temperature. The net current is therefore

$$I = C\exp\left[-\frac{e(V_0 - V)}{kT}\right] - C\exp\left(-\frac{eV_0}{kT}\right)$$

$$= C\exp\left(-\frac{eV_0}{kT}\right)\left[\exp\left(\frac{eV}{kT}\right) - 1\right]$$

or

$$I = I_0\left[\exp\left(\frac{eV}{kT}\right) - 1\right] \qquad \textbf{1.1}$$

where the thermal current I_0 is given by

$$I_0 = C\exp\left(-\frac{eV_0}{kT}\right) \qquad \textbf{1.2}$$

A number of important consequences follow from the exponential form of this equation. First of all suppose we put a reverse voltage on our diode, that is make V negative. The exponential term in equation **1.1** will eventually become negligible and we will be left only with the thermal current I_0, which for a small silicon diode might typically be 0·1 microamps, although diodes with reverse currents of the order of one hundredth of this figure are commercially available. For germanium the figure will be orders of magnitude higher, because of its higher intrinsic conductivity as previously noted. In both cases the reverse current increases rapidly with temperature as can be seen from equation **1.2.**

We do not need to reverse bias the diode very much to bring its current to the value I_0, since the quantity kT/e is only about $\frac{1}{40}$ volt. If for example we put on a reverse bias of only 0·1 volt the current is $I_0[\exp(-4) - 1]$ or over 98 per cent of I_0. Increasing the reverse bias then beyond a small fraction of a volt leaves the diode current substantially constant at I_0 (Figure 4). What does happen with increasing reverse bias is that the mobile charges are moved back from the junction, thus increasing the width of the depletion layer. A simple theory shows that the width is in fact proportional to the square root of the reverse bias, provided this latter is large compared with kT/e, that is for all but the very smallest voltages.

1·3·1 *Junction capacitance*

The junction diode is composed of two pieces of semiconducting material of reasonable conductivity, separated by an insulating depletion layer. It thus resembles a parallel-plate capacitor consisting of two conducting plates with dielectric between. The capacity, which might typically be of the order of one picofarad (10^{-12} farad) for a small diode, is usually a nuisance, since it represents a path for a varying signal even when the diode is 'cut off' under reverse bias. However it has been turned to advantage in the use of the diode as a variable capacitor, whose capacity value can be controlled by an applied voltage. This is possible since the width of the depletion layer is a function of the reverse bias, and therefore so is the capacity – in fact we would expect the capacity to be inversely proportional to the square root of the bias. A very convenient device is thus obtained.

1·3·2 *Forward bias*

Now consider the characteristics of the diode under forward-bias conditions. Because of the smallness of the quantity kT/e quite a small forward bias will produce a large current. For example a forward bias of only $\frac{1}{20}$ volt will produce a current of $I_0\{\exp(2) - 1\}$ or over $6I_0$. This sharp rise can be seen in Figure 4.

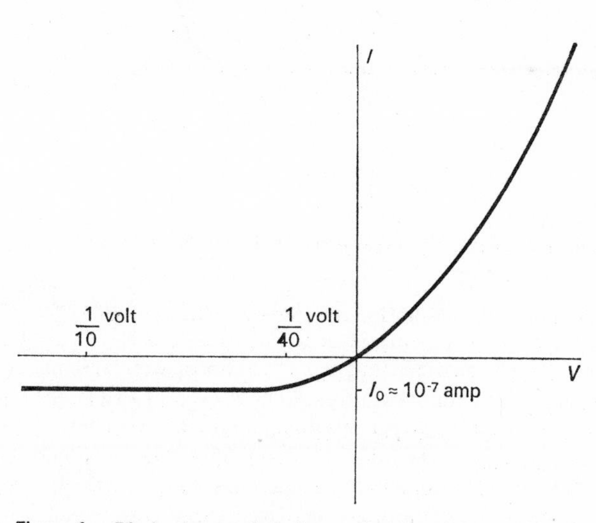

Figure 4. Diode characteristic for small voltages.

Even at values of $6I_0$ however we are still only talking in terms of microampere currents and we are usually interested in currents of the order of milliamperes. Let us redraw Figure 4 to show larger quantities on both axes (Figure 5). For a forward bias of 0·25 volt we expect a current of

$$I_0[\exp(10) - 1] \approx I_0 \exp(10) = I_0(2{\cdot}2 \times 10^4)$$

This equals 2·2 mA using a typical value of 10^{-7} for I_0. With 0·35 volt forward bias we expect $I_0 \exp(14)$ or $I_0(1{\cdot}2 \times 10^6) = 120$ mA. The forward part of the characteristic is thus as shown in Figure 5. (For various reasons – one being the appreciable voltage drop occurring due to the non-zero resistance of the body of the semiconducting material – the current will in practice rise more slowly than our simple theory would predict, particularly at the larger current values. In other words, we will have to apply a voltage larger than expected to obtain a particular current. For example, it might require something like 0·8 volts to produce 120 mA. Nevertheless, even with this reservation, we are still obtaining a very rapid current rise.)

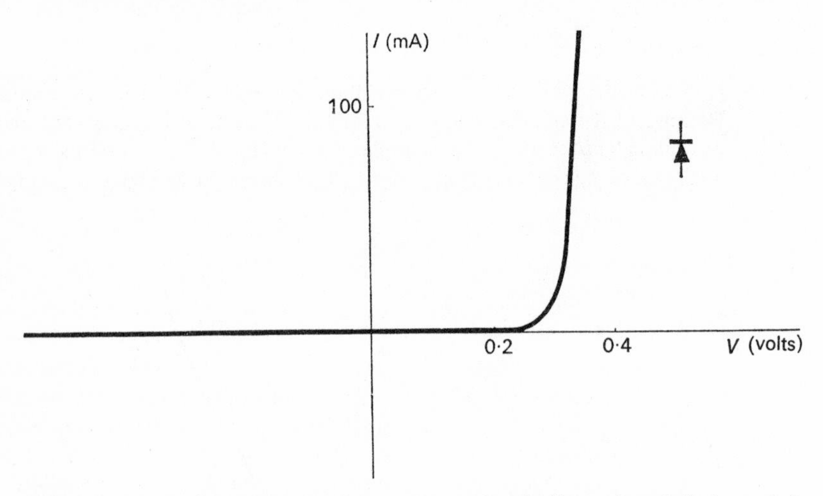

Figure 5. Diode characteristic for larger voltages than in Figure 4, and diode symbol.

The curve giving the reverse part of the characteristic in Figure 5 runs almost exactly along the axis, since a current of 10^{-7} A cannot be shown on the scale now being used. The silicon diode thus acts as an almost perfect *rectifier*, that is, allows only a negligible amount of current to flow, even when a large voltage is applied to it in one direction, yet permits a very large current to flow when quite a small voltage is applied in the other direction. Shown beside Figure 5 is the usual symbol for a diode (although it is sometimes shown enclosed in a small circle). The arrowhead shows the direction of easy current flow, and it is to here that the positive pole of the battery should be connected to bias the diode 'on'.

The manufacturer will usually indicate physically which end is which by a small dot placed near the 'cathode' end of the diode – that is the opposite end to the arrowhead.

The characteristics of a germanium diode are generally similar, but as we have noted, the reverse current I_0 will be very much larger than that for silicon. Because of this higher value of I_0 the forward characteristics will lie above the corresponding one for silicon.

We can describe the forward characteristics of a diode numerically by giving a value which in some way defines its resistance. Since the characteristic is so markedly non-linear an expression of the form V/I, as for a resistor, is not appropriate. Instead we define the 'dynamic', 'incremental' or 'differential' resistance, as it is variously called, as $r = dV/dI$, which gives the inverse of the slope of the characteristic at the point in question. Since

$$I = I_0 \left\{ \exp\left(\frac{eV}{kT}\right) - 1 \right\} \approx I_0 \exp\left(\frac{eV}{kT}\right)$$

$$\frac{dI}{dV} = \frac{e}{kT} I_0 \exp\left(\frac{eV}{kT}\right) = \frac{e}{kT} I$$

So

$$r = \frac{dV}{dI} = \frac{kT}{eI} = \frac{\frac{1}{40}}{I} \text{ ohms} \qquad \mathbf{1.3}$$

So, even if I is only 1 mA the dynamic resistance is still only 25 ohms.

1·4 Zener breakdown

Figure 6 shows the characteristics of the diode when the voltages are higher than in Figure 5. Note that at a voltage V_Z, which may vary from a few volts to many hundreds, depending on the resistivity of the semiconducting materials, the diode breaks down, that is the current magnitude rises extremely rapidly (and destructively if not limited by external means) for a small change in voltage. This is known as 'avalanche' or Zener breakdown after two processes which have been suggested to explain it. In the former, which is predominant for diodes with large values of V_Z, the effect is due to electrons passing through the depletion layer acquiring sufficient energy to generate further electron–hole pairs, and thus a drastic increase in current. In the latter, which predominates for diodes with low breakdown voltages, the increase is due to the passage of electrons *through*, rather than over the potential barrier, by means of a quantum mechanical effect known as 'tunnelling'. (We shall refer again, in a little more detail, to tunnelling for a diode biased in the forward direction in section 7·5.) In practice no distinction is drawn between these two modes of breakdown. They normally represent the practical limit of operation of a diode under reverse bias although, as will be discussed in Chapter 9, the phenomenon has been turned to good account in voltage stabilizing circuits.

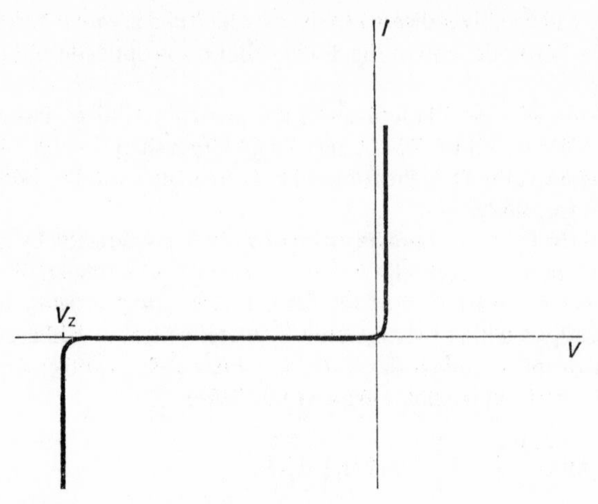

Figure 6. Diode characteristic for larger voltages than in Figure 5.

1·5 Asymmetrical junctions

In discussing the physical basis for diode action we implied that the diode junction was 'symmetrical', that is, the current was equally contributed to by hole and electron flow. This is not necessarily nor indeed usually so. The diode will work quite satisfactorily if, say, the *n*-type material is highly doped with impurities, and the *p*-type weakly doped, in which case the current will be largely carried by electrons. In the same way diodes in which the current is largely carried by holes can be produced with strong *p*-type material. This is of importance when considering bipolar transistors.

References and further reading

1. W. J. PRICE, *Nuclear Radiation Detection,* McGraw-Hill, 1964.
2. GLENN F. KNOLL, *Radiation Detection and Measurement,* Wiley, 1979.
3. D. A. FRASER, *The Physics of Semiconductor Devices,* O.U.P., 1977.
4. J. E. CARROLL, *Physical Models for Semiconductor Devices,* Edward Arnold, 1974.

CHAPTER 2

The unipolar, or field effect transistor, and its properties as an amplifier

2·1 Principles of operation

We discuss unipolar rather than bipolar transistors first, because as we shall see their characteristics can be described mathematically in a rather simpler way. The unipolar transistor, or field effect transistor as it is also known because of the dependence of its operation on electric fields in the semiconductor, is shown schematically in Figure 7. For a picture of what the device looks like in

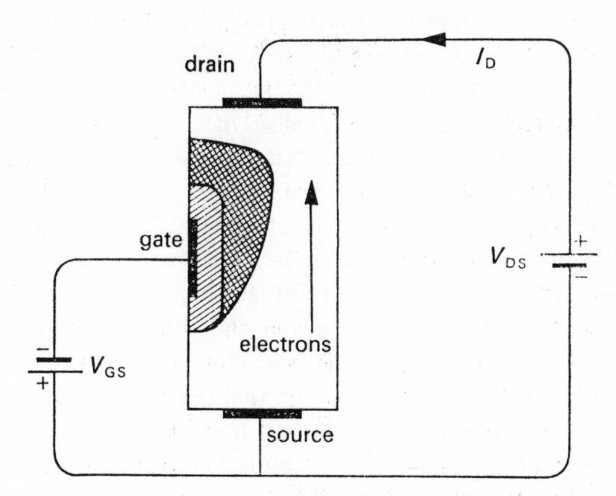

Figure 7. Field effect transistor (schematic).

reality and a description of how this and other semiconductor devices are fabricated see reference 1. In an idealized picture a bar of *n*-type silicon has an ohmic, that is a conventional, contact at either end, with a voltage between them that can be varied from zero to say 30 volts. In Figure 7, the upper contact is known as the *drain*; the lower contact as the *source*. On one side of the bar we make a rectifying contact – that is we diffuse in some *p*-type material (shown solid in the diagram) so as to make a *p–n* junction. This junction is the *gate*, and it will be held at a negative voltage with respect to the source, as shown. In this investigation of the field effect transistor (which we shall in future refer to as the FET) this gate to source voltage, V_{GS}, will be varied from zero to a few volts negative. With V_{GS} initially fixed at zero, we can investigate how the current flowing between drain and source (the drain current I_D) depends on the drain to source voltage V_{DS}. When we start with V_{DS} equal to zero (that is all voltages in the circuit zero) there is nonetheless a depletion layer at the *p-n* junction at the gate, as we discussed in Chapter 1 for a junction diode with no applied bias. Furthermore if this junction is asymmetrical, with the *p*-type material much more heavily doped than the *n*-type, it is clear that the depletion region will lie largely in the *n*-type material, since a much greater depth will have to be depleted here to match the number of carriers which can be obtained from a small depth in the rich *p*-type material. This depletion layer is shown in single hatching in the diagram.

Now let us gradually increase the voltage V_{DS}. To begin with, the bar of silicon will act simply as a resistor, and we will obtain a linear relation between the current I_D and the voltage V_{DS}, as shown by the part A of the curve in Figure 8. However as the voltage V_{DS} is increased the *p–n* junction at the gate becomes more reverse biased, and the depletion region spreads out into the body of the silicon, as shown by the double-hatched region in Figure 7. (This spreading does not occur at the source end of the depletion layer, since the gate to source voltage is fixed at zero. The asymmetrical shape of the depletion layer is thus explained.) The current is now forced to flow in a channel which becomes increasingly constricted, with the result that the characteristic begins to turn over. Eventually the channel will be closed or 'pinched off'. This does not mean, of course, that the current ceases, but merely that the electrons forming the drain current now have to be swept across a depletion layer by part of the drain voltage. What happens, in fact, is that as the drain voltage is increased beyond that necessary for pinch off, both the length of the depletion layer to be traversed by the electrons and the voltage across it increase, but the voltage between that part of the pinched off region nearest the source and the source itself remains unchanged. Since it is this latter voltage that determines the current coming from the source (and going on to the drain) we find that the characteristic curve now becomes practically horizontal as shown in part B of the curve in Figure 8. The FET is then said to be in the 'pinch-off' or 'saturation' region of its characteristics. In practice we find that the curve still has a slight positive slope, and this small increase in current with applied voltage can be attributed (see for example reference 2) to the shortening of the effective channel length due to the growth in the pinched off region.

If V_{GS} is fixed at some small negative voltage, and again V_{DS} is gradually increased, then, since the depletion layer will initially be larger, the slope of the characteristic will be initially smaller and the curve will turn over, and pinch off will occur, at a lower current value. We thus see how the family of curves shown in Figure 8 arises. Typical values of current and voltage are also shown. If the characteristics are continued to even higher values of V_{DS} internal breakdown occurs and the curves start to rise sharply again.

This FET, working as it does with electrons flowing in a channel in *n*-type silicon, is known as an n-channel field effect transistor. In the case of p-channel field effect transistors, the polarities of V_{DS} and V_{GS} will have to be reversed, since it will be holes rather than electrons which will be acting as carriers. Properly the nomenclature for both these FETs is *junction-gate* field effect transistors or JFETs. *Insulated-gate* field effect transistors (IGFETs) also exist in which, as their name implies, an insulating layer separates the semiconductor channel from the gate (an evaporated metal layer). As the insulating layer is normally oxide grown on the silicon substrate, this device is more commonly known as a metal-oxide-semiconductor FET or MOSFET. For a detailed account of the basic workings of MOSFETs and in particular the distinction between 'enhancement' and 'depletion' operation, references 2, 3 or 4 should be consulted. All MOSFET characteristics in general closely resemble those of JFETs, and the depletion type in particular operates in a similar manner to a JFET, except that, since the gate is insulated, either polarity or voltage may be used for V_{GS}. Thus in Figure 8 the characteristic curves could be extended to values of $V_{GS} > 0$ (and similarly to negative values for a p-channel device). Although MOSFETs now play a very important role in electronics (particularly in complex integrated circuitry, where their small size and simplicity of manufacture is advantageous) we will concentrate on the junction-gate FET, since it has the property, very important for may applications in physics, of being a 'low-noise' device, in a sense to be discussed later.

Two salient points should be noted. Firstly FETs are *voltage-controlled* devices, that is, it is the voltage on the gate electrode which is responsible for the basic mechanism of the FET. Secondly they are *majority-carrier* devices, that is in n-channel devices it is the electrons which carry the current, and similarly in p-channel devices it is the holes. The latter may seem such an obvious remark that it is worth adding that in bipolar transistors it is precisely the presence of holes injected into n-type material (or of electrons into p-type material) that make their operation possible. Majority-carrier devices are much less sensitive to nuclear radiations than minority-carrier devices, which makes them of special value for certain space and nuclear physics applications. Voltage-controlled devices are suitable for high-input-impedance, low-noise applications; that is applications which are usually associated with the early stages of amplifiers for nuclear physics experiments.

2·2 FET parameters

The area of the FET characteristics in which we shall be most interested is the 'pinch-off' region, where the characteristic curves are almost horizontal. To describe the operation of the FET in this region mathematically, we take as independent variables the gate to source voltage V_{GS}, and the drain to source voltage V_{DS}, with I_D, the drain current, as dependent variable. Then

$$\delta I_D = \left(\frac{\partial I_D}{\partial V_{GS}}\right)_{V_{DS}} \delta V_{GS} + \left(\frac{\partial I_D}{\partial V_{DS}}\right)_{V_{GS}} \delta V_{DS} \qquad \textbf{2.1}$$

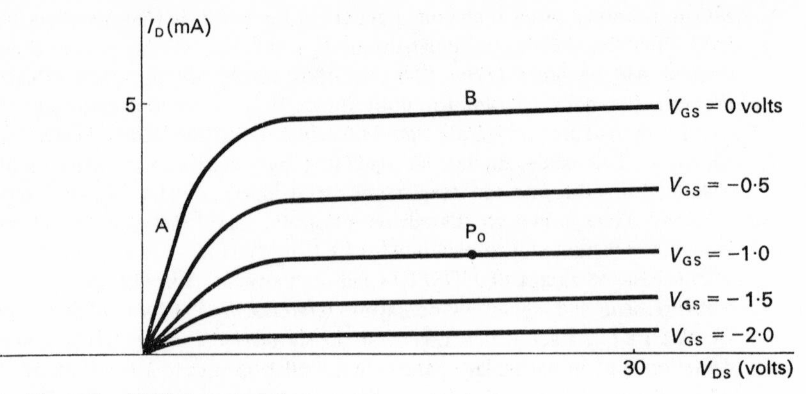

Figure 8. FET characteristics.

This gives the small change in drain current, δI_D consequent on small changes in gate voltage δV_{GS}, and drain voltage δV_{DS}, from some selected operating point such as P_0 in Figure 8. The quantity $(\partial I_D/\partial V_{DS})$ gives the slope of the characteristic curve at the point P_0 for the particular gate–source voltage in question, while $(\partial I_D/\partial V_{GS})$ gives the spacing of the characteristic curves near the point P_0 for the particular drain–source voltage at P_0. An examination of Figure 8 will show that $(\partial I_D/\partial V_{GS})$, that is the spacing of the curves, is by no means constant, and the same is true of $(\partial I_D/\partial V_{DS})$ though this is not so easily seen on the scale to which the diagram is drawn. However if we restrict ourselves to small variations around a specified operating point these quantities will vary very little, and we shall be able to quote definite values for them – remembering at the same time to specify the position of the operating point itself. In fact, later we shall be discussing methods of 'feedback' whereby the effect of the variation of these parameters with operating point can be minimized.

Equation **2.1** can be written as

$$i_d = g_m v_{gs} + g_d v_{ds} \qquad \textbf{2.2}$$

where we have written i_d for δI_D, the small *change* in drain current; similarly v_{gs} for δV_{GS} and v_{ds} for δV_{DS}. Furthermore we have written $(\partial I_D/\partial V_{GS})$ as g_m the mutual conductance (or transconductance as it is also known) and g_d (the drain or output conductance) for $(\partial I_D/\partial V_{DS})$. Both g_m and g_d have dimensions ohms^{-1}. Sometimes g_d is replaced by $1/r_d$ where r_d is the 'drain resistance' (it is of course an incremental resistance, just as in the case of the diode). So

$$i_d = g_m v_{gs} + \frac{1}{r_d} v_{ds} \qquad \textbf{2.3}$$

Finally, in yet another notation, the reason for which will be clear at a later stage, g_m is written as y_{fs}, and $g_d(=1/r_d)$ as y_{os}. We will normally make use of the notation in equations **2.2** and **2.3**. A typical value for g_m for an FET at an operating point of P_0 given by $V_{DS} = 15$ volts, $V_{GS} = -1$ volt would be 3×10^{-3} ohms^{-1} (often written 3 millimhos or 3 mA volt^{-1}). The corresponding value for g_d would typically be $10\,\mu$mhos (10^{-5} ohms^{-1}) corresponding to a value of $100{,}000\,\Omega$ for r_d. The small value of g_d means that it can often be neglected, leaving us with just one parameter, g_m, to describe the operation of the FET.

An equation like **2.3** also describes the operation of a triode or pentode valve, although g_m for these devices is usually considerably larger than for the FET. The value of g_d for a FET typically lies between the corresponding values for triode and pentode, that for the triode being the largest. In triode theory we often meet the quantity μ, the amplification factor, and we can define it similarly for the FET. It is given by $\mu = (\partial V_{DS}/\partial V_{GS})_{I_D}$, and from equation **2.1**, by putting $\delta I_D = 0$ (that is I_D = constant), we can show that $\mu = g_m/g_d = g_m r_d$. However for the FET we shall have more occasion to define its performance simply in terms of g_m as indicated earlier.

One quantity has been missing from this discussion – the current I_G, which might be expected in the gate lead. However, for a reverse biased silicon junction, which the gate contact is, this will be a tiny fraction of a microamp at most, and is thus negligible. For bipolar transistors a very different situation exists.

2·3 The FET as an amplifier: setting the operating point

Figures 9(a), 9(b) and 9(c) show various stages in the development of an FET amplifier stage. The basic arrangement and the standard symbol for an *n*-channel FET is shown in Figure 9(a): for a *p*-channel device the direction of the arrow would be reversed. (Note however that this arrow is purely an identifying symbol, and does not indicate the direction of flow of the gate current I_G under normal biasing conditions, as can be easily seen.)

The batteries V_{DD} and V_{GG} supply the voltages necessary for the operation of the device: note however that a resistor R_d, the drain load resistor, is inserted between V_{DD} and the drain, so V_{DS} is not equal to V_{DD} because of the potential drop $I_D R_d$ across the resistor R_d. We have shown the input as a signal generator producing a small voltage signal v_{in}. When this is applied to the FET, it changes the voltage on the gate, which changes I_D. This in turn produces an *increase* in the voltage across R_d, and a corresponding *decrease* in V_{DS}, thus giving v_{out}, the small change in output. We shall be deriving the exact relation between v_{out} and v_{in} shortly. It has been assumed in the preceding treatment that the d.c. resistance of the input signal generator is low (ideally zero), so that it does not interfere with the application of the biasing voltage V_{GG} to the gate. The arrangement we have just described is known as a 'common-source' amplifier stage for the following reason. One side of the input is connected to the gate,

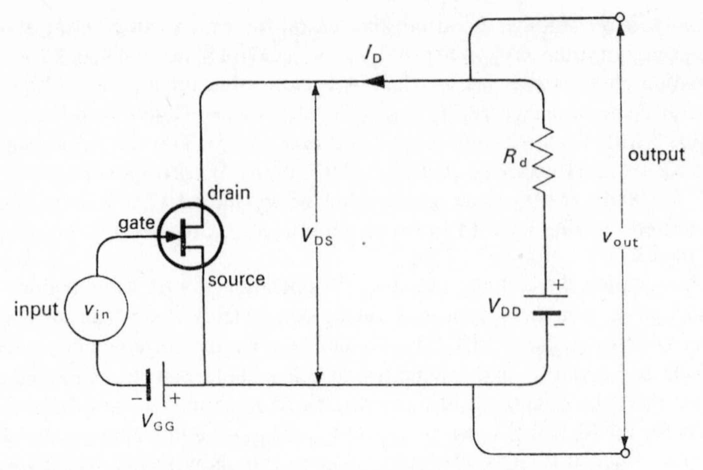

Figure 9. Various stages in the development of an FET amplifier.
(a) Basic arrangement.

and the other (via the battery V_{GG}) to the source; one side of the output is connected to the drain, and the other to the source. The input and output have one terminal in common – the source; hence the terminology. Since it is usual, but not essential to put this common terminal to ground, the arrangement is also known as a 'grounded-source' stage.

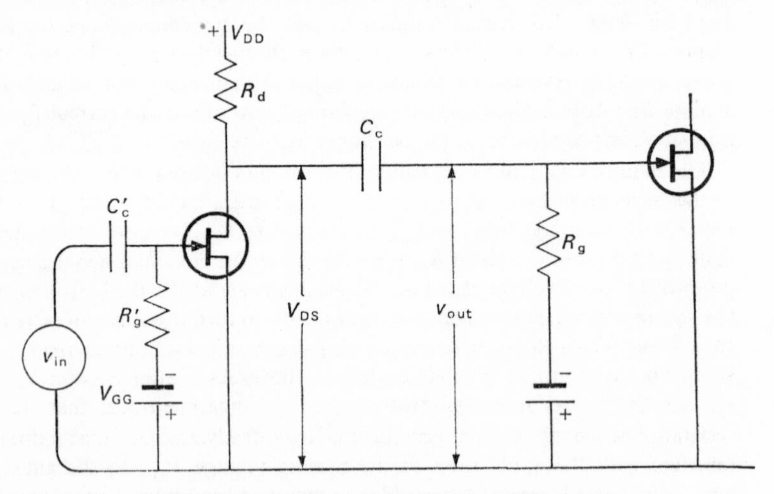

Figure 9. Various stages in the development of an FET amplifier.
(b) FET with following stage.

We have seen that v_{out} is given by $-R_d \times$ (change in I_D) or $-R_d i_d$. However there is in addition at the output, a constant voltage V_{DS}. Figure 9(b), which shows not only the f.e.t. stage but also the succeeding stage to which its output may be connected to provide further amplification, indicates how this unwanted voltage may be eliminated. Before discussing the question of the output, it should be noted that we do not in these diagrams show the battery V_{DD} of Figure 9(a) explicitly. Instead we use the symbol V_{DD} at the end of R_d, as in Figure 9(b), and imply that between that point and the bottom line of the diagram, a battery of voltage V_{DD} is connected. Returning to the discussion of the output, we note the presence of a capacitor C_c which has the effect of blocking off from the output the steady voltage V_{DS}. On the other hand the capacitor can present quite a small impedance to the changing signal voltage. The impedance of a capacitor of size C to a sinusoidal signal of frequency f is $1/(2\pi fC)$, so by making the capacitor size large we can make its impedance very small. The coupling capacitor C_c can then be thought of as blocking completely the d.c. voltage while providing almost unimpeded passage for the signal voltage. This aspect will be discussed more critically at a later stage. Because it is possible that the input voltage to the stage may also contain a d.c. component along with the signal - indeed if it happens to be a preceding FET stage it certainly will – a capacitor C_c' has also been included in the input lead. This in turn raises the question of how we are to apply the gate bias voltage necessary for correct functioning of the FET. The solution is as shown in Figure 9(b): it is applied through a resistor R_g' – the gate resistor. If this resistor were not there and the battery V_{GG} connected directly from source to gate, then V_{GG} would obviously hold the gate at a fixed voltage relative to the source, and there would be no possibility of its being varied by the incoming signal. The resistor then can be thought of as providing an element across which the input signal can be produced. The only current flowing in R_g' is the very small gate current I_G. R_g' then can be very large (say in the megohm region) without causing any appreciable voltage drop, $R_g' I_G$, in the voltage applied by V_{GG} to the gate. It is an advantage to have R_g' large because it then represents only a small load on the input (remember that C_c' can be thought of as 'not being there' for signals). R_g similarly acts as the gate resistor of the next stage. We refer to the complete arrangement as 'resistor–capacitor' (RC) coupling.

It is inconvenient to have to provide a separate voltage supply V_{GG} for the gate bias. Figure 9(c) shows how this may be avoided, and the final arrangement for this amplifying stage. A resistor R_s is introduced between source and the 'ground' line. Due to the flow of I_D through this resistor, the source is held at a positive potential of amount $R_s I_D$ with respect to ground. The gate is now at ground potential, so it is in fact negative with respect to the source which is just the situation provided by the battery V_{GG} in the previous figure. By a suitable choice of R_s we can provide the correct amount of bias previously provided by the battery. There is now, however, an important change from the basic circuits of Figures 9(a) and 9(b), in so far as the source is no longer at the join of one input and one output terminal, but separated from them by R_s. To overcome this difficulty we place across R_s a large capacitor C_s. Behaving as a small impedance to signals, this capacitor effectively short circuits R_s from the point of view of signals, and places the source once again at ground in this respect. Since from the point of view of d.c. the capacitor is an open circuit, the bias voltage necessary to set the operating point is not affected. Such an arrangement is known as 'self-biasing' or 'automatic biasing'.

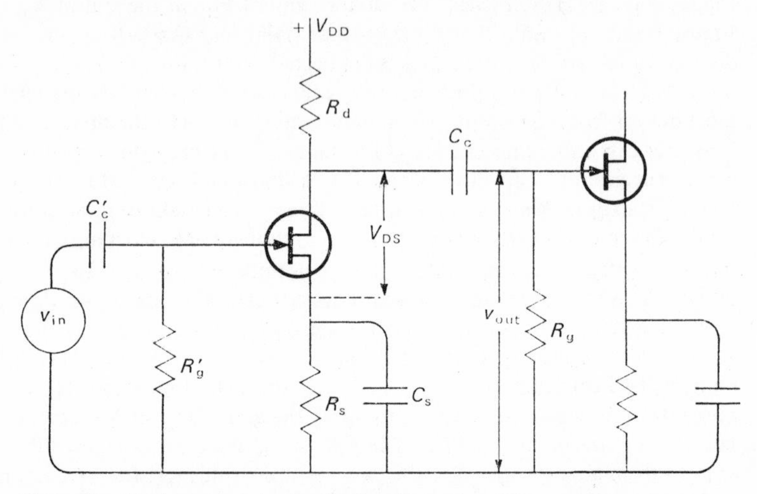

Figure 9. Various stages in the development of an FET amplifier. (c) Final arrangement.

2·4 A numerical calculation on the operating point

Let us now attempt a typical computation of an operating point for a common-source, *RC*-coupled amplifier stage. We intend to select a 'quiescent' operating point (that is the point at which the FET will rest in the absence of a signal) somewhere near the centre of the pinch-off region of the FET characteristics, where the variations in g_m and g_d will not be too great. Such a point P_0 is shown in Figures 8 and 10. The quiescent operating conditions are thus $V_{DS} = 15$ volts, $I_D = 2$ mA and $V_{GS} = -1$ volt. To provide the necessary gate–source bias we need a source resistor R_s given by $R_s(2 \times 10^{-3}\text{ A}) = 1$ volt or $R_s = 500$ ohms. The voltage between drain and source has been set at 15 volts, and since the voltage across R_s is 1 volt, the total voltage from drain to ground is 16 volts. To calculate V_{DD} we must know R_d in order to calculate the voltage drop across it. We have as yet no criterion for selecting R_d: its value is in fact related to the performance we require from the stage at high frequencies. Let us arbitrarily take R_d to be 10 kohm ($=10^4$ ohms), a value we shall later find to be reasonable. The drop across R_d is thus $10^4(2 \times 10^{-3})$ volts $= 20$ volts, and the total value of V_{DD} needed is $1 + 15 + 20 = 36$ volts. In fairness it should be added that the calculations may not always be as simple as this, either because we may have to work to a drain supply voltage V_{DD} already specified, or because the value of R_s may work out to be a non-standard size. In either event it may be necessary to make a number of empirical adjustments to the position of P_0 before a point satisfying all the conditions can be found.

It remains to specify a value for C_s. For the range of frequencies to be amplified, the impedance of C_s must be very much less than R_s, that is $1/(2\pi f C_s) \ll R_s$. Clearly this will be hardest to satisfy for the lowest frequency we have to deal with: suppose this is 20 Hz (that is, 20 cycles per second) then $C_s \gg 1/(2\pi \times 20 \times R_s)$ or 16 μF. If C_s is 250 μF, we shall handsomely fulfil these conditions. We have already suggested a value of about 1 megohm for R_g (and R_g'): we shall be dealing with suitable values for C_c (and C_c'), and their relation to the values for R_g and R_g' in some detail later.

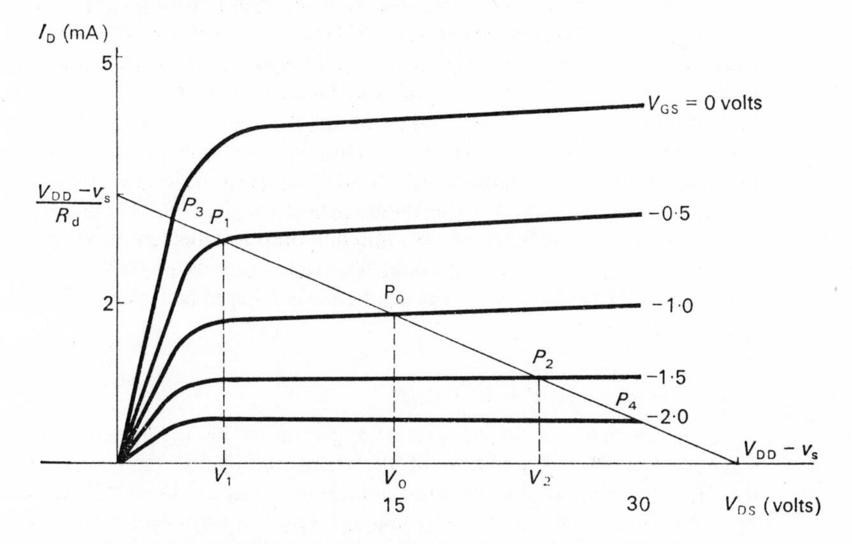

Figure 10. FET characteristics illustrating the choice of operating point.

Also shown in Figure 10 is a 'load line' which indicates how the operating point moves when a signal is applied to the input of the FET. To understand how it arises we note that at any instant the total drain current i_D (composed of the quiescent value I_D and the varying signal component i_d) is related to the corresponding voltage v_{DS} as follows

$$v_{DS} = V_{DD} - i_D R_d - v_S$$

where as usual V_{DD} is the supply voltage, and v_S is the voltage across R_s. Now v_S will in fact be held constant at 1 volt, since we have arranged for the changing component of the drain current to be shunted to ground through C_s. Therefore

$$v_{DS} = (V_{DD} - v_S) - i_D R_d \qquad \textbf{2.4}$$

or

$$i_D = \frac{V_{DD} - v_S}{R_d} - \frac{v_{DS}}{R_d} \qquad \mathbf{2.5}$$

This is of the form $y = A - Bx$ and represents a straight line of slope $-1/R_d$, intercepting the current axis at a value of $(V_{DD} - v_S)/R_d \approx V_{DD}/R_d$ and the voltage axis at a value of $V_{DD} - v_S \approx V_{DD}$ (Figure 10). The load line can be used to show the sort of variations in input and output signals we may reasonably expect to obtain. Certainly we cannot go much farther than the point P_3 in one direction, as here we are on the characteristic $V_{GS} = 0$, where the gate junction ceases to be reverse biased, nor can we go farther than P_4 in the other direction, because here we have reached the maximum permitted value of V_{DS}. Even restricting ourselves to the range P_1 to P_2 (that is with the input voltage swinging ±0·5 volts from its quiescent value of −1 volt) may not be adequate as can be seen from the corresponding output voltages – marked V_1 and V_2 on the bottom axis. Clearly $V_0 V_1$ is appreciably larger than $V_0 V_2$ which means that the output from a positive half-volt input signal is appreciably larger than from a negative half-volt input signal. The situation arises because of the non-uniform spacing of the characteristic curves, that is from the non-constancy of g_m (and to a lesser extent from the non-constancy of g_d). Unless such distortion is acceptable the magnitude of the input and output signals must be considerably reduced. Later we shall look at feedback methods whereby the amplification of FETs and other devices can be made independent of such variations of their parameters.

2·5 Midband voltage gain of an FET stage

We now calculate the voltage gain or amplification for the *RC*-coupled FET stage discussed. The addition of the word 'midband' (sometimes 'mid-frequency') indicates that we are thinking in terms of sinusoidal signals of moderate frequencies, neither very high nor very low, say in the 1000 to 10,000 Hz region. The more complex treatment required if we go outside this band is dealt with in sections 2·8 and 2·9. Because of the coupling capacitor C_c on the output of the FET stage, we shall be interested only in the varying signal voltages in the circuit. Consequently the relevant equations are equation **2.2**, which we repeat here for convenience,

$$i_d = g_m v_{gs} + g_d v_{ds} \qquad \mathbf{2.2}$$

and

$$v_{ds} = -i_d R_d \qquad \mathbf{2.6}$$

Equation **2.6** states that when the drain current increases by a small amount i_d, the drain to source voltage *decreases* by an amount $i_d R_d$. (This can also be seen directly from equation **2.4**, on removing the quiescent components.) Eliminating i_d

$$v_{ds} = \frac{-v_{gs} g_m R_d}{1 + g_d R_d}$$

But v_{gs} is equal to the signal voltage v_{in} applied to the stage, while v_{ds} is the output signal appearing at the drain (that is across the load resistor R_d). Hence the voltage gain is given by

$$\frac{v_{out}}{v_{in}} = \frac{-g_m R_d}{1 + g_d R_d} = -A_0 \text{ (say)} \tag{2.7}$$

the minus sign showing that the FET stage inverts the polarity of the applied signal. A typical value for A_0, using the values previously indicated for g_m, g_d and R_d, would be thirty. As the term $g_d R_d$ is small, a good approximation to equation **2.7** would be

$$A_0 = g_m R_d \tag{2.8}$$

2·6 Equivalent circuits for the FET stage (output side)

It is useful to think of the FET, from the point of view of the output it produces, as being replaced by an *equivalent circuit*. The FET will be shown to act as a voltage generator of size $(g_m/g_d)v_{in}$ in series with a resistor of size $1/g_d = r_d$,

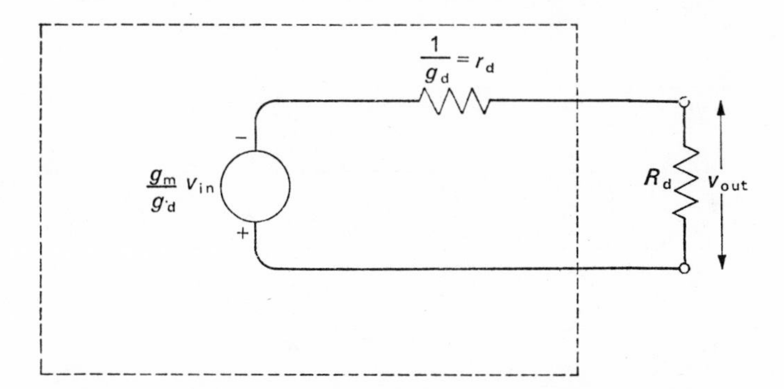

Figure 11. Equivalent circuit for the FET common-source amplifier.
(a)

that is, a voltage generator with an internal resistance (Figure 11(a)). This equivalent circuit is shown inside the dashed box: the usual drain resistor R_d is shown connected externally. The whole picture is a very simplified one; for example comparing it with Figure 9(a) we see that the battery V_{DD} has been left out. This is quite in order because we are only interested in changes of voltage, not quiescent values. Similarly we have not shown in Figure 11(a) the components R_s and C_s of Figure 9(c) as they also are only concerned with setting up the quiescent conditions.

It is easy to show that this equivalent circuit gives the same result as the previous theory: the output voltage appearing across R_d is clearly the voltage of the generator $(g_m/g_d)v_{in}$ multiplied by the fraction $R_d/(R_d + 1/g_d)$ or

$$v_{out} = \frac{g_m v_{in} R_d}{g_d\left(R_d + \dfrac{1}{g_d}\right)} \qquad \textbf{2.9}$$

which on simplification is the same as equation **2.7** (apart from the minus sign). We did not consider the polarity of the generator in deriving equation **2.9,** and when we do, complete accord is found between equivalent circuit and theory.

An alternative equivalent circuit involves a current generator, which is perhaps not so easy to understand instinctively as a voltage generator. A perfect voltage generator is one which gives a fixed voltage no matter how small a load resistor is placed across it. Real voltage generators have an internal series resistance, so that the voltage available drops as the magnitude of the external resistor is reduced. (For the d.c. case an ordinary dry cell is a good example of this.) A perfect current generator, on the other hand, is one which will drive a constant *current* through an external resistor, no matter how *large* it is. To indicate the divergence of a real current generator from this ideal, a series resistance cannot be used as in the case of a voltage generator, because the total output current of the generator is, by definition, constant. Instead a resistance is placed in parallel with the generator (R_{int} of Figure 11(b)), and as a division of

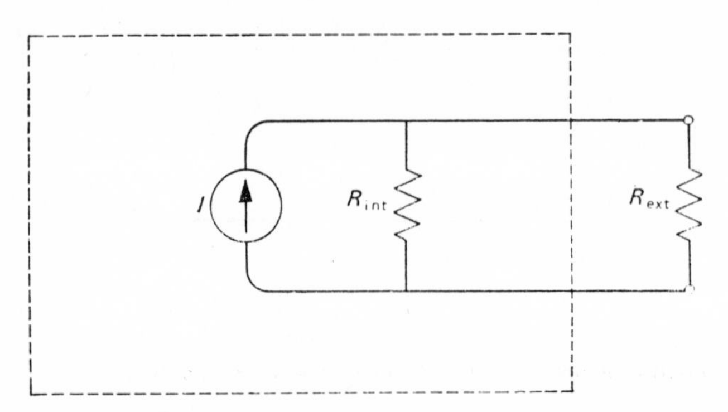

Figure 11. Current generator equivalent circuit.
(b)

the total constant current I now takes place between the external resistor R_{ext} and the internal resistance of the generator, once again we have a variation of the external current with external load.

The equivalent circuit for the FET is shown in the dashed box in Figure 11(c).

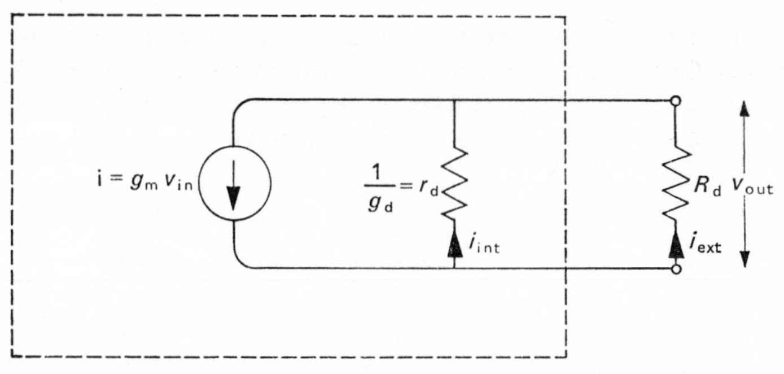

Figure 11. Equivalent circuit for the FET common-source amplifier.
(c)

Note the direction of the current arrow to ensure polarity agreement with theory. The current through R_d is given by

$$i_{ext} = g_m v_{in} \frac{\frac{1}{R_d}}{\frac{1}{R_d} + g_d}$$

and the output voltage $v_{out} = -i_{ext} R_d$. Solving for v_{out} gives the same result as before, which indicates that this circuit with a current generator is equally good for describing the FET action.

So far we have been considering the equivalent circuit for the FET, with the drain load R_d as the external load. Let us now consider the FET plus drain load, that is the complete stage, as the basic element, and ask how will it act when a load, say the next FET stage, is placed upon it. Figure 11(d) shows the situation,

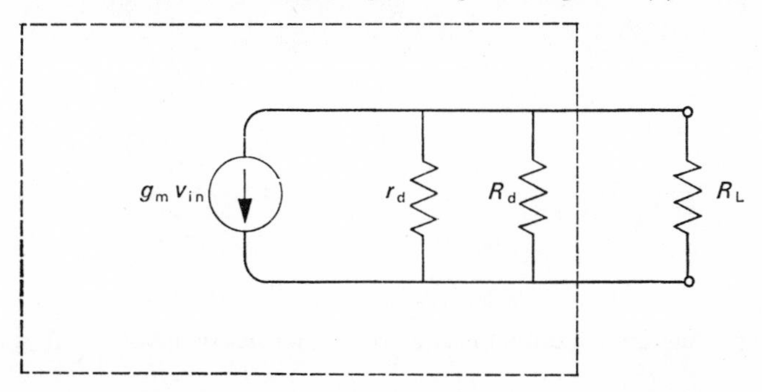

Figure 11. Equivalent circuit for the FET common-source amplifier.
(d)

with the external load R_L in place. Clearly r_d and R_d can be combined into a single resistance R_d^* as shown in Figure 11(e). The complete FET stage is thus equivalent to a current generator of size $g_m v_{in}$ with a shunt resistance R_d^*

corresponding to r_d and R_d in parallel. It has already been stated, as yet without proof, that the values of R_d will be very much less than r_d, therefore in R_d^*, the large resistance r_d in parallel with the smaller one R_d, can be neglected to give $R_d^* \approx R_d$. This is the approximation used in equation **2.8.**

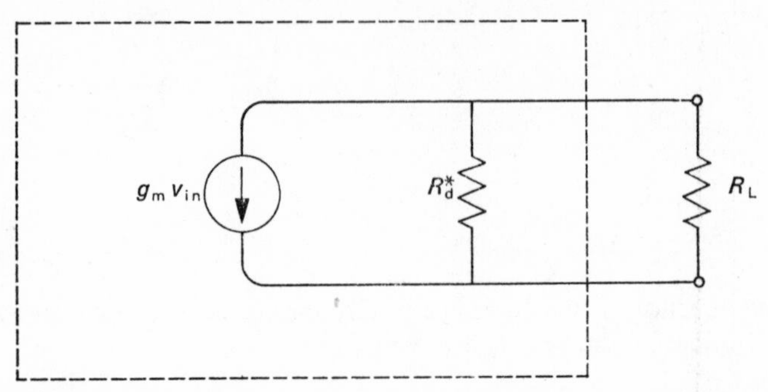

Figure 11. Equivalent circuit for the FET common-source amplifier.
(e)

A few further general points about equivalent circuits should be made. Any voltage generator like the one in Figure 12(a) can be transformed into an equivalent current generator as shown in Figure 12(b). This can be seen by checking the current which each will drive through an arbitrary external resistor R_L. This fact would have enabled us to move directly from Figure 11(a) to 11(c). It will also allow us to move from Figure 11(e) to 11(f). This last diagram shows an equivalent circuit using a voltage generator which may equally well be employed to represent the complete FET stage.

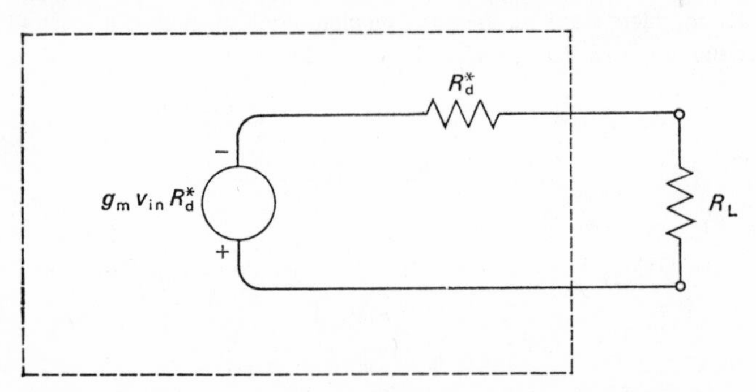

Figure 11. Equivalent circuit for the FET common-source amplifier.
(f)

Finally we discuss an important method which will prove useful in certain computations on equivalent circuits. Suppose for example the circuit of Figure 12(a) were enclosed in a box, and v and R were to be determined. To determine v we measure the voltage at the terminals with a voltmeter of such high resistance that it does not load the circuit appreciably, that is, we determine the 'open-circuit voltage'. To determine R we can, of course, load up the circuit with a known resistor R_L, note the new output voltage, and hence compute R. It would be quicker (although not really feasible in practice) to short circuit the terminals (that is, make $R_L = 0$), because this would give the 'short-circuit current' $= v/R$, which when divided into the open-circuit voltage gives R, and completes the description of the circuit. Suppose however that in the box there was not just one voltage generator and one resistance, but a whole network of resistances and voltage generators. *Thévenin's theorem* (which we quote without proof) tells us that nonetheless the circuit can be characterized by making the same two measurements, the open-circuit voltage and the short-circuit current, and obtaining from these values for v and R. The complicated circuit will behave, from the point of view of external loads, as if it were a single voltage generator of size v in series with a resistance R. (It can then be turned into its current generator dual of Figure 12(b), if necessary.) The term 'output impedance' is often used to denote the internal resistance R.

A very simple illustration of the preceding method is obtained from its application to the complete circuit of Figure 11(a) (and not just the part in the dashed box). The open-circuit-voltage is

$$\frac{g_m v_{in}}{g_d} \frac{R_d}{R_d + r_d} = g_m v_{in} \frac{r_d R_d}{R_d + r_d} = g_m v_{in} R_d^*$$

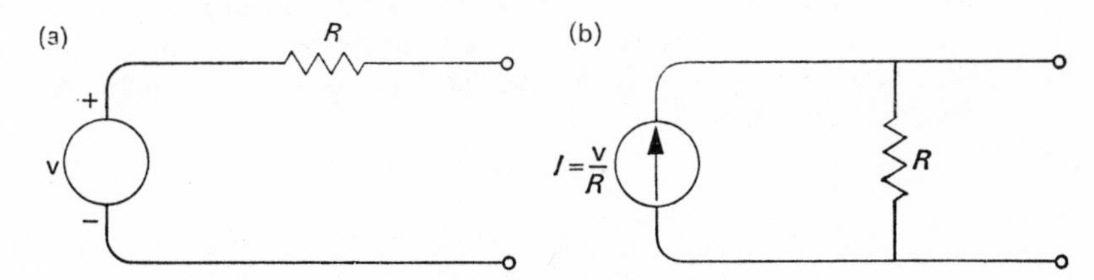

Figure 12. Equivalent voltage and current generator circuits.

The short-circuit current, obtained by putting a short across R_d and measuring the current there, is $g_m v_{in}/g_d$ divided by r_d, or simply $g_m v_{in}$. The output impedence of this FET stage is thus the open-circuit voltage divided by the short-circuit current $= g_m v_{in} R_d^*/g_m v_{in} = R_d^*$ (or R_d approximately for the usual case where $R_d \ll r_d$). Figure 11(a) can thus be replaced by a voltage source of size $g_m v_{in} R_d^*$ in series with a resistance R_d^*, which is what is in the dashed box in Figure 11(f), and which we previously obtained in a more roundabout way. (There is one other way in which the output impedance R could have been deduced for the circuit of Figure 11(a) and similar circuits, and that is by imagining the voltage generator shorted out, and considering a direct resistance measurement between the output terminals. This will clearly give the correct result, but, in general, the previous method for the determination of R is more useful.)

2·7 Complete equivalent circuit for the FET stage

The discussion so far has been concerned with the output side only of the FET equivalent circuit - that is how it appears from the point of view of the circuit into which it is feeding. To complete the picture we must inquire how an FET appears from the point of view of the circuit from which it is being fed, and construct the corresponding equivalent circuit. It is in fact extremely simple.

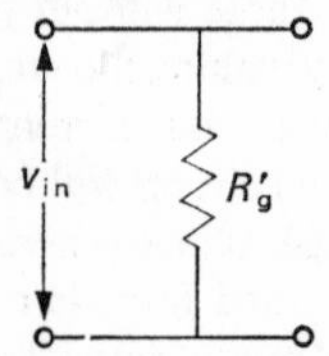

Figure 13. Equivalent circuit for input of FET common-source stage.

The current flowing in the gate junction of an FET is so small as to be negligible for most purposes, so this does not represent any load on whatever is feeding the FET. The only load is therefore the gate resistor [R_g' or R_g of Figure 9(c)] which is shown in the equivalent circuit for the input of Figure 13. As this resistor itself will normally be in the megohm region, it, too, can be neglected in some calculations. Figure 14 shows a complete equivalent circuit for a FET stage, incorporating on the output side the circuit of Figure 11(f) – although of course the circuit of Figure 11(e) using a current generator could equally well have been used. As in Figure 9(c) the input and output circuits have one terminal in common.

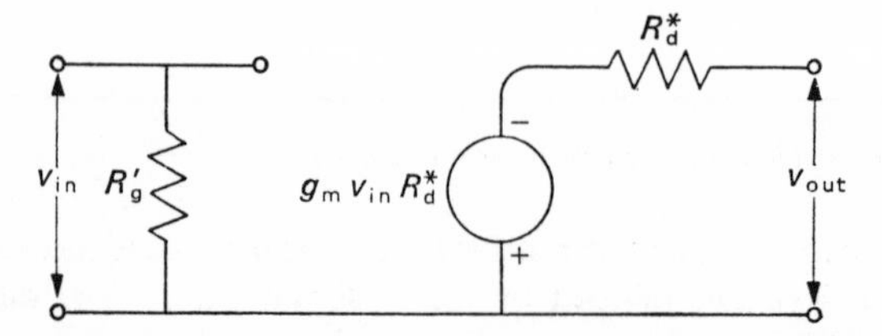

Figure 14. Complete equivalent circuit for FET common-source stage.

This equivalent circuit has been developed in connexion with the equations of paragraph 2·5, and is subject to the same limitation imposed on them – it is only applicable in the midband region. We shall now discuss the modifications necessary at higher and lower frequencies.

2·8 Gain at high frequencies

Working with signals of higher and higher frequencies, small 'stray' capacities become increasingly important. This is because the impedance of a capacitor of size C to a sinusoidal signal of frequency f is $1/(2\pi fC)$, that is, inversely proportional to frequency. At one megahertz the impedance of a one picofarad capacitor is approximately 160 kohm, and thus even a few picofarads of stray capacity provide a relatively easy alternative path for a signal, with considerable resultant change in the characteristics of an amplifier stage. We are most concerned with the capacities that exist across the gate junction in the FET, that is the gate to drain capacity C_{gd}, and the gate to source capacity C_{gs}, which might be typically two and four picofarads respectively. (Sometimes the total input capacity $C_{gd} + C_{gs}$ is given and written as C_{iss} the 'common-source, short-circuit input capacity'. C_{gd} is then given separately, the notation C_{rss} being used, indicating 'common-source, short-circuit, reverse-transfer capacity'. The 'reverse-transfer' part of the terminology implies a connexion between input and output, which will be seen to be a role this capacity plays, while the 'short-circuit' part refers to a condition of measurement which ensures that we really do measure C_{rss} and not an augmented value of this quantity, which can happen in certain circumstances.)

Figure 15 shows an FET transistor stage fed from a source of signals of internal resistance r_s. This could be a signal generator with internal resistance of the amount indicated, or it could represent a preceding FET stage, in which case r_s, in accordance with the equivalent circuit of Figure 11(f) would be equal

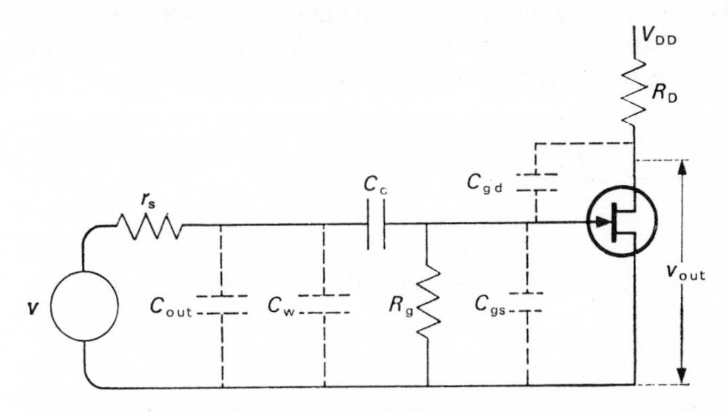

Figure 15. FET showing stray capacities.

to R_d^*. The stray capacities C_{gd} and C_{gs} are shown, as well as C_{out}, which represents whatever stray capacity is associated with the source of signals, and C_w, which represents the stray capacity to ground of the connecting wiring. C_c is the usual coupling capacitor and R_g the gate resistor. Neither the bias resistor R_s nor its decoupling capacitor C_s are shown in the circuit because they are associated with the setting of the operating point only, and not with signals. Figure 16 is a simplified form of Figure 15. The FET has been replaced by its equivalent circuit; the three capacities C_{out}, C_w and C_{gs}, which are in parallel, have been replaced by a single capacity C_1. The large resistor R_g, now in parallel with the relatively low impedance of C_1, has been omitted, while the coupling capacitor C_c, already of small impedance at moderate frequencies is entirely negligible at higher frequencies and can be replaced by a short.

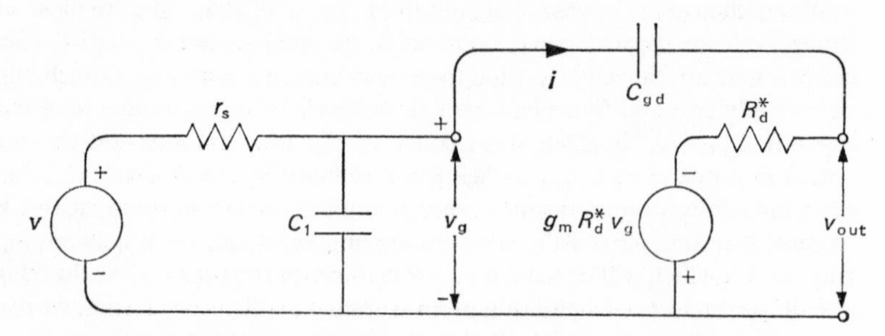

Figure 16. FET common-source equivalent circuit at high frequencies.

The calculations will be carried out in two parts. In the first we shall concentrate on C_1 and neglect the effect of C_{gd} (even though this is usually the most important) while in the second we shall discuss the role of C_{gd} and neglect C_1. We start by imagining the connexion to C_{gd} to be open circuited: in that case the voltage v_g at the gate of the FET is determined simply by the voltage-dividing action of C_1 and r_s and is given by

$$v_g = v \frac{\dfrac{1}{j\omega C_1}}{r_s + \dfrac{1}{j\omega C_1}}$$

where $\omega = 2\pi f$, f being the frequency of the signal, and $j = \sqrt{-1}$. Hence

$$v_g = \frac{v}{1 + j\omega r_s C_1} \qquad 2.10$$

To find the output voltage we note that as no current flows in R_d^* (the connexions to C_{gd} were open circuited) we have $v_{out} = -g_m v_g R_d^*$. Putting in the value for v_g from equation 2.10 and re-arranging to obtain the gain v_{out}/v we obtain

$$\text{gain} = \frac{v_{out}}{v} = -\frac{g_m R_d^*}{1 + j\omega r_s C_1}$$

$$= -\frac{A_0}{1 + j\omega r_s C_1} \qquad 2.11$$

where A_0 is the midband gain previously defined.

The performance at high frequency is specified by noting that at $\omega = 1/(r_s C_1)$ the gain becomes $-A_0/(1+j)$, that is its magnitude has dropped to $1/\sqrt{2}$ of its midband value. This is known as the 'upper half-power point' (and the corresponding frequency the upper half-power frequency) because, as power is proportional to (voltage)^2 there would be a drop of half in power here. It is also referred to as the 3 db (decibel) point, as $10^{-0\cdot 3} \simeq 1/2$. Note that the phase angle has shifted by 45° at this point.

For good high frequency performance $1/(r_s C_1)$ should be large, that is $r_s C_1$ small. To reduce C_1 the wiring capacity C_w should be reduced to a minimum, as should C_{out} (if we have any control over it), and an FET chosen with a low value of C_{gs}. If it is possible to reduce r_s this should be done as well, and if r_s represents the R_d^* of a previous FET stage it clearly is possible. R_d^* consists of R_d and r_d in parallel, so for good high-frequency performance R_d must be reduced because r_d is, by the nature of the FET, large. This justifies the earlier observation that FETs will normally be operated with $R_d \ll r_d$. However it is not worth making a numerical calculation on this point until the role of C_{gd} has been discussed, but it is worth observing that as R_d is reduced to improve the frequency response, the midband gain A_0 of the stage in question ($= g_m R_d^* \approx g_m R_d$) is simultaneously reduced, a point we shall return to later.

We now consider the effect of C_{gd}. When the bottom end of C_{gd} (in Figure 15) has a voltage of say 1 volt applied to it the upper end has a voltage of $-A_0$ applied to it, because of the gain $-A_0$ of the stage. C_{gd} thus becomes charged to a voltage $A_0 + 1$ times as large as C_1 would be by the same signal. This might indicate that C_{gd} would act as a capacity $(A_0 + 1)$ times as large as it actually is as we shall now demonstrate more exactly. Referring to Figure 16 (and this time forgetting about C_1) we have

$$v - v_g = ir_s \qquad \textbf{2.12}$$

which is a statement about the voltage drop across r_s and

$$g_m R_d^* v_g + v = i(r_s + Z_{gd} + R_d^*) \qquad \textbf{2.13}$$

which is a statement about the total e.m.f. around the circuit, and in which we have written Z_{gd} for $\dfrac{1}{j\omega C_{gd}}$ the impedance of C_{gd}. Solving for v_g

$$v_g = \frac{v(Z_{gd} + R_d^*)}{Z_{gd} + R_d^* + r_s(1 + g_m R_d^*)}$$

or

$$v_g = \frac{v}{1 + r_s(1 + g_m R_d^*)(Z_{gd} + R_d^*)^{-1}} \qquad \textbf{2.14}$$

Equation **2.14** can be simplified because Z_{gd} is large. At midband frequencies Z_{gd} was ignored altogether – that is, it was considered of infinite impedance – so, provided we do not go to really high frequencies it will certainly be much larger than R_d^*, and consequently R_d^* can be ignored in the term $Z_{gd} + R_d^*$ above.

Hence 2.14 becomes

$$v_g = \frac{v}{1 + j\omega r_s(1 + A_0)\, C_{gd}} \qquad 2.15$$

where the value for Z_{gd} has been substituted and also $g_m R_d^*$ replaced by A_0. The value for v_{out} in this case is not given by $-g_m v_g R_d^*$ because there is a voltage drop across R_d^* due to the current i flowing in it. But R_d^* will be small compared with Z_d and hence most of the voltage drop will be across Z_{gd} and not across R_d^*. The answer $v_{out} = -g_m v_g R_d^*$ is thus after all a reasonable approximation, so finally we have

$$\text{gain} = \frac{v_{out}}{v} = \frac{-g_m R_d^*}{1 + j\omega r_s(1 + A_0)\, C_{gd}} \qquad 2.16$$

Once again, C_{gd} has been effectively multiplied by the factor $1 + A_0$. This Miller effect, as it is known, was first noted for triode valves, and was the reason for the introduction of tetrodes and pentodes, where an additional grid is used to reduce the capacity corresponding to C_{gd} to a very small amount, and thus allow a large value of A_0 without unmanageable value of $(1 + A_0)\, C_{gd}$. No such device can be employed with FETs, but later we shall discuss the use of two FETs in a '*cascode*' circuit which produces the same result.

An expression for the gain in the upper frequency range taking into account the effect of both C_1 and C_{gd} can now be written as follows

$$\text{gain} = A_U = \frac{-g_m R_d^*}{1 + j\omega r_s\{C_1 + (1 + A_0)\, C_{gd}\}} \qquad 2.17$$

The upper half-power frequency f_U is thus

$$f_U = \frac{1}{2\pi r_s \{C_1 + (1 + A_0)\, C_{gd}\}} \qquad 2.18$$

The value of reducing the various stray capacities, but particularly C_{gd} is obvious. We must also reduce r_s and A_0 for best results. The reduction of r_s has already been discussed, and implies the reduction of the gain of the previous stage, if this is what the voltage source and r_s represent. In addition A_0, the gain of the stage we are mainly concerned with, must also be reduced, for f_U to be large. For a given FET, we are thus faced with a choice: we can have good high frequency response at the expense of midband gain, or vice versa, but not both. A 'figure of merit' F (also known as the 'gain–bandwidth product') has some value for comparing various FETs. It is defined as the product of the midband gain and the upper half-power frequency, and hence

$$F = A_0\, f_U = g_m\, R_d^* f_u = \frac{g_m R_d^*}{2\pi r_s\{C_1 + (1 + A_0)\, C_{gd}\}} \qquad 2.19$$

(F may also be interpreted as the frequency at which the magnitude of the gain A_U becomes approximately unity. This can be seen from **2.17** by putting $A_U = 1$, and neglecting at the high values of ω involved, the first term in the denominator.)

If the FET stage in question is being driven by a similar stage, $r_s = R_d^*$, the resistance terms in equation **2.19** disappear. In the most favourable case where the wiring capacity is negligible,

$$F = \frac{g_m}{2\pi\{C_1 + (1 + A_0)C_{gd}\}} = \frac{g_m}{2\pi\{C_{out} + C_{gs} + (1 + A_0)C_{gd}\}}$$

$$= \frac{g_m}{2\pi(C_{out} + C_{in})} \qquad \textbf{2.20}$$

where C_{in} (sometimes C_{eq}) has been written for the 'equivalent input capacity' $C_{gs} + (1 + A_0)C_{gd}$. Unlike the pentode case, where no Miller effect exists, the present value for C_{in} is not an unambiguous function of the FET's own characteristics, because it depends on A_0, and thus the particular operating conditions selected by the choice of R_d. In this respect the figure of merit for the FET (or triode) is of much less value than for the pentode. All we can say in general is that for good high frequency performance we need a high value of g_m, and a low value of C_{gs} and C_{gd}, that of C_{gd} being particularly important.

Let us examine the sort of performance we are likely to obtain in practice. If $r_s = R_d = 10$ kilohms, $C_{gd} = 2$ pF, $C_{gs} = 4$ pF, $C_{out} + C_w = 6$ pF, $g_m = 3$ millimhos, we obtain $A_0 = g_m R_d^* \approx g_m R_d = 30$. The total effective capacity is $6 + 4 + (31 \times 2) = 72$ pF and hence $f_U = 220$ kHz. If we reduce r_s and R_d to 2 kohm we obtain $A_0 = 6$ and $f_U = 3{\cdot}3$ MHz. This sort of performance would be inadequate for many applications in physics, and underlines the need for the use of configurations such as the cascode arrangement previously mentioned, where the Miller effect is virtually eliminated.

2·9 Gain at low frequencies

At the lower part of the frequency spectrum, very different considerations arise. The impedances of the stray capacities of the previous section are now extremely high, and their role as alternative and unwanted current paths is entirely negligible. What now becomes important is the impedance of coupling capacitors like C_c in Figure 9(c). Figure 17 shows a circuit equivalent to Figure 9(c) with this taken into account. Note that the bias resistor R_s does not appear in the diagram. This assumes that even at low frequencies, C_s is still large enough to be considered a short circuit across R_s for signals: it may not in fact be possible or economic to provide such a large value of C_s in all circumstances. In line with our high frequency considerations of the last section, R_d^* has been replaced by R_d, to which it is approximately equal for small values of R_d.

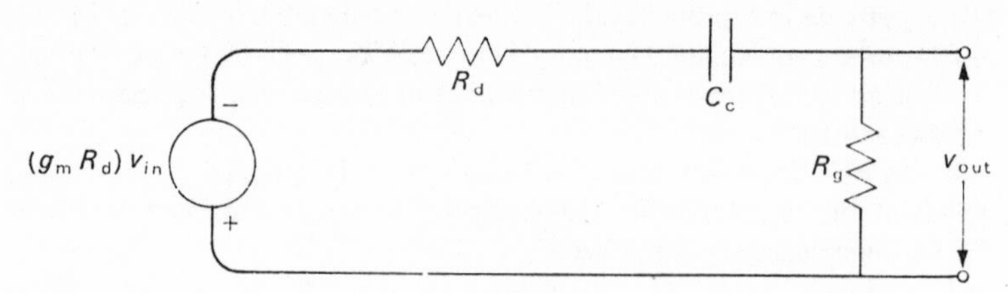

Figure 17. FET common-source equivalent circuit and coupling network at low frequencies.

The voltage appearing across R_g, that is the stage output, is thus

$$v_{out} = \frac{-(g_m R_d v_{in}) R_g}{R_g + Z_c + R_d}$$

where as usual we have written Z_c for the impedance of the capacitor C_c. Solving for v_{out}/v_{in}, dividing above and below by R_g, and putting in the value for Z_c, we obtain a value for the gain in the lower frequency range as follows

$$\text{gain} = A_L = \frac{-g_m R_d}{1 + \dfrac{R_d}{R_g} + \dfrac{1}{j\omega R_g C_c}}$$

Remembering the relative values of R_g and R_d we see that the middle term in the denominator can be neglected, and putting $g_m R_d = A_0$, the midband gain, we have

$$A_L = \frac{-A_0}{1 + \dfrac{1}{j\omega R_g C_c}} \qquad \textbf{2.21}$$

The magnitude of the gain thus drops from its midband value A_0 as ω is decreased, and becomes equal to $A_0/\sqrt{2}$ (that is the half-power point is reached) when $\omega = \dfrac{1}{R_g C_c}$. The lower half-power frequency is thus given by

$$f_L = \frac{1}{2\pi R_g C_c} \qquad \textbf{2.22}$$

For good low-frequency performance f_L should be small, that is R_g and/or C_c should be large. For a value of R_g of 1 MΩ and a lower half-power frequency of 20 Hz, C_c would be 0·004 μF, which is physically quite convenient. We shall find more difficulty in this respect with bipolar transistors. A lower half-power point at 20 Hz does not necessarily mean that the stage is usable for a particular purpose at this frequency, as we have a drop of $1/\sqrt{2}$ in gain and a phase shift of 45° at this point. If the demands were more stringent, this stage could not be used below 100 hertz. The phase shift would be particularly troublesome in the case of a complex waveform, which could be thought of as made up of a number of sinusoidal components of different frequencies. Unless we are on the flat part of the response curve for all the components, not only would the amplitude of each component be a function of its frequency, but the phase relationships between them would be upset. So the moral is to keep well away from the half-power frequencies (both upper and lower) for faithful amplification.

It is not always desirable to extend the low-frequency response (and indeed the high-frequency response) downwards (or upwards) as far as possible. If for example extremely good low-frequency response is not required for the problem in hand, it is sometimes positively harmful to provide it, as we may unnecessarily amplify unwanted stray signals, such as pick-up from the mains. The coupling capacitor C_c may often be chosen to *limit* the low-frequency response of the stage – a point we shall return to in another form in the next chapter. Similar remarks apply to the upper half-power frequency.

2·10 Overall frequency response

A plot of the magnitude of voltage gain versus frequency for the amplifier stage we have been discussing, would show (for $R_d = 10\ k\Omega$), a flat portion with a value of the gain of 30, extending from below 100 Hz right up beyond 100 kHz, the half-power points being, as we saw, at 20 Hz and 220 kHz. (See Figure 18.)

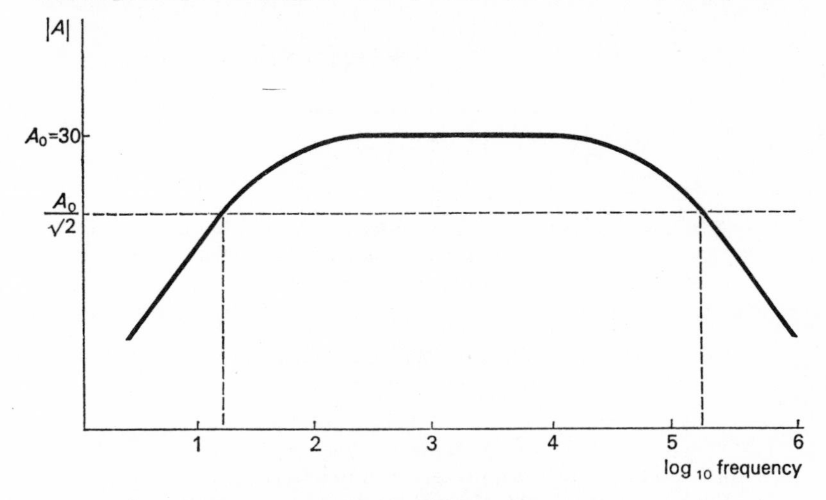

Figure 18. Frequency response of FET amplifier stage.

We talk about the *bandwidth* of the stage as being (220,000 – 20) Hz which of course is very closely 220 kHz, the upper half-power frequency.

It has been very convenient to be able to think of the stage as having this large midband region of constant gain, because it enables us to treat quite separately the regions at the high- and low-frequency ends of the spectrum where the response fell off. We shall not attempt to derive the complete expression for the total frequency response. However, the result for the magnitude of the gain can be written down almost by inspection, and as we shall require this in a later chapter, we shall now obtain it. Equation **2.17** can be written

$$A_U = \frac{-A_0}{1 + \dfrac{j\omega}{\omega_U}} \qquad \textbf{2.22a}$$

where ω_U is related to f_U in an obvious way. The magnitude of A_U is thus

$$|A_U| = \frac{A_0}{\left\{1 + \left(\dfrac{\omega}{\omega_U}\right)^2\right\}^{\frac{1}{2}}} \qquad \textbf{2.23}$$

Similarly from equation **2.21** we have, for the low frequency end,

$$|A_L| = \frac{A_0}{\left\{1 + \left(\dfrac{\omega_L}{\omega}\right)^2\right\}^{\frac{1}{2}}} \qquad \textbf{2.24}$$

An expression which embodies these two particular solutions is

$$|A| = \frac{A_0}{\left\{1 + \left(\dfrac{\omega}{\omega_U}\right)^2\right\}^{\frac{1}{2}} \left\{1 + \left(\dfrac{\omega_L}{\omega}\right)^2\right\}^{\frac{1}{2}}} \qquad \textbf{2.25}$$

as it reduces to equation **2.24** for small values of ω, to equation **2.23** for large values of ω, and also to A_0 for $\omega_L \ll \omega \ll \omega_U$. By a slight manipulation of the contents of the second bracket in equation **2.25** we obtain

$$|A| = \frac{A_0\,\omega}{\omega_L \left\{1 + \left(\dfrac{\omega}{\omega_U}\right)^2\right\}^{\frac{1}{2}} \left\{1 + \left(\dfrac{\omega}{\omega_L}\right)^2\right\}^{\frac{1}{2}}} \qquad \textbf{2.26}$$

which is the form in which we shall refer to it later.

Equation **2.26** has been derived for the case of a single stage with one low-frequency, and one high-frequency constant, but it is reasonable to suppose that the corresponding expressions for a complete amplifier will have the same general form. We shall discuss this problem again, from a rather different point of view in Chapter 3.

2·11 Other forms of load and coupling

Until now we have been discussing amplifier stages with resistive loads and RC coupling, and it is this type with its flat frequency response over a wide range which will be most useful. Many other varieties of load and coupling are available, and for example, an inductor, or an inductor in parallel with a capacitor could be used as the drain load. One could also use the primary of a transformer as the drain load, and the secondary could then be directly connected to the gate of the next stage without any coupling capacitor, as there would be no d.c. voltage appearing in the secondary winding. Because of the limited use of many of these arrangements for our purposes, we shall describe briefly only one of these possibilities – the 'tuned-drain' amplifier. This device allows the amplification of essentially one particular frequency and the rejection of all others.

It is obtained by replacing the load resistor R_d of Figure 9(c) by a capacitor C in parallel with an inductor L (see Figure 19(a)), the rest of the circuit being left unaltered. The impedance of these two elements in parallel is given by

$$\frac{1}{Z} = (jL\omega)^{-1} + \left(\frac{1}{jC\omega}\right)^{-1}$$

$$= \frac{1 - \omega^2 LC}{jL\omega}$$

Hence

$$Z = \frac{jL\omega}{1 - \omega^2 LC} \qquad \textbf{2.27}$$

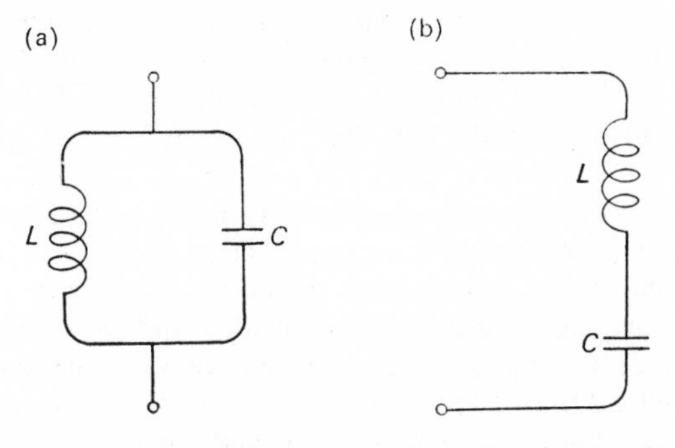

Figure 19. (a) Parallel LC circuit. (b) Series LC circuit.

Clearly Z becomes infinite for $\omega = \omega_0 = 1/\sqrt{(LC)}$, where $f_0 = \omega_0/2\pi$ is known as the resonant frequency. In practice this will not occur because of the small but finite resistance of the inductor. Let us call this R, then,

$$\frac{1}{Z} = (jL\omega + R)^{-1} + \left(\frac{1}{jC\omega}\right)^{-1}$$

or

$$Z = \frac{jL\omega + R}{1 - \omega^2 LC + j\omega RC} \qquad \textbf{2.28}$$

If, as is normally the case, R is small compared with the impedance of L and C, Z will have a maximum close to the previous value of ω_0. At this point the term $1 - \omega^2 LC$ in the denominator will be small, and may be neglected, as may the term R in the numerator. The maximum value of Z is thus given by

$$Z = \frac{jL\omega}{j\omega RC}$$

or

$$Z = \frac{L}{RC} \qquad \textbf{2.29}$$

The circuit thus acts as if it were a purely resistive load of size L/RC, for the resonant frequency f_0. For frequencies far from f_0 the impedance will be very small. Consequently there is appreciable gain only for frequencies around f_0 and this gives the selective property characteristic of this type of amplifier stage. Taking as an example values of L, C, and R of 0·1 millihenrys (=10^{-4} henrys), 1000 picofarads (10^{-9} farads), and 10 ohms respectively, we find $\omega_0 = 3 \times 10^6$ approximately corresponding to a frequency of about 0·5 MHz. The impedance of the circuit at resonance, L/RC, turns out to be 10 kΩ. From equation **2.28** the impedances at other frequencies can be computed, but a simple approximation will show the orders of magnitude. At a frequency of one third of the resonant frequency the impedance of the capacitor is of magnitude $1/C\omega = 1/(10^{-9} \times 10^6) = 1000$ ohms, while the corresponding value for the inductor is $L\omega = 10^{-4} \times 10^6 = 100$ ohms. The total impedance is thus due almost entirely to the inductor (as the two elements are in parallel) and is negligible compared with the resonance value; the gain is thus also negligible. For a frequency of $3f_0$ the roles of capacitor and inductor are reversed but the result is similar – the gain is down by the same large factor.

The matter can be expressed in another way: the FET load of Figure 19(a) consists of two elements of quite low impedance, in parallel, and yet because of the different nature of inductance and capacitance we have been able to construct a high impedance load from them. For example $L\omega$ has a value of 300 ohms at resonance (and $1/C\omega$ has a similar value) and yet the impedance presented to the FET is, as we have seen, $L/RC = 10$ kilohms. We refer to the Q of the circuit as a measure of how much greater the effective resistance L/RC is than either of its components. These ratios are $(L/RC)/L\omega_0$ and $(L/RC)/(1/C\omega_0)$ and are of course equal. So $Q = 1/RC\omega_0$ or $L\omega_0/R$ and from the second of these expressions we see that its value depends on our ability to design a coil with a good inductance-to-resistance ratio. The value of Q of about thirty for the present numerical example would be typical, or a little on the low side. The Q value is also a measure of the sharpness of the resonance – that is the rate at which the impedance presented by the circuit falls as we move away from f_0. Large values of Q give rapid rates of fall and hence an amplifier with good selectivity – that is one which discriminates well against all frequencies except f_0.

The combination of an inductor and capacitor in parallel in the way we have been discussing is known as a 'parallel-resonance circuit'. An inductor and capacitor in series as in Figure 19(b) is similarly known as a 'series-resonance circuit'. Its impedance is $Z = jL\omega + 1/jC\omega$ which for a value of $\omega = 1/\sqrt{(LC)}$ is zero, or for the non-ideal case, just R, the coil resistance. Thus even a small alternating voltage v at the resonant frequency would drive a very large (ideally infinite) current round the circuit. We shall refer to the series-resonance circuit again in connexion with crystal oscillators.

References and further reading

1. A. SCHMITZ, 'Solid Circuits', *Phillips Technical Review*, vol. 27, no. 7, 1966, p. 192.
2. A. BAR-LEV, *Semiconductor and Electronic Devices*, Prentice Hall, 1979.
3. J. MILLMAN and C. C. HALKIAS, *Integrated Electronics*, McGraw-Hill, 1972.
4. R. KING, *Electronic Circuits and Systems*, Nelson, 1975.

CHAPTER 3

Amplification of step voltages and pulses

3·1 Some important networks

We now turn to the problem of step and pulse amplification, and begin by studying the passage of a step signal of voltage through an amplifier stage. Although the initial treatment here is quite different from that in the sinusoidal case we shall find that the same parameters as before do in fact control the performance and we shall eventually draw the two treatments together.

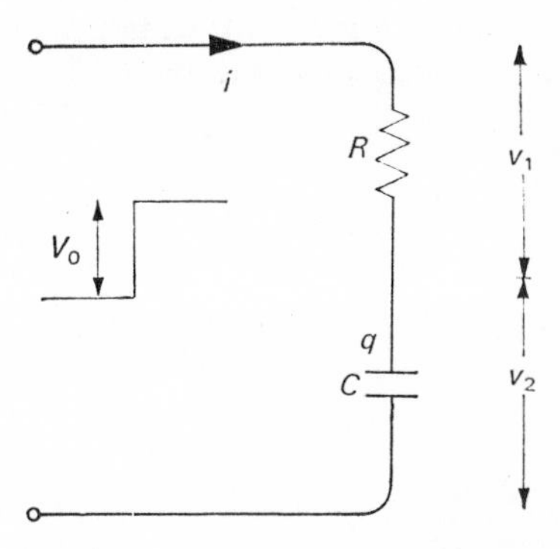

Figure 20. Series *RC* circuit fed with voltage step.

Let us begin by seeing the effect of certain simple networks on voltage step signals. Consider the circuit of Figure 20. Suppose we suddenly apply a voltage V_0 to the resistor R and capacitor C in series – by, for example, connecting a battery of the appropriate size across the input terminals – and inquire what are the voltages appearing across these two components. This of course is just the problem of charging a capacitor from a battery through a resistor. Our basic statement is clearly,

$$v_1 + v_2 = V_0 \qquad \mathbf{3.1}$$

in which the voltages v_1 and v_2 are given by

$$v_1 = iR \qquad \mathbf{3.2}$$

and

$$v_2 = \frac{q}{C} = \int \frac{i\,dt}{C} \tag{3.3}$$

where q is the instantaneous value of the charge on the capacitor, and i the current flowing. To find the voltage appearing across the capacitor, we can eliminate v_1 using the relation $v_1 = iR = RC\frac{dv_2}{dt}$ to give us (from equation **3.1**)

$$RC\frac{dv_2}{dt} + v_2 = V_0 \tag{3.4}$$

The solution of this is

$$v_2 = V_0\left\{1 - \exp\left(\frac{-t}{RC}\right)\right\} \tag{3.5}$$

where the constant in the integration has been determined by remembering that the capacitor is originally uncharged, and $v_2 = 0$ at $t = 0$. If we take an output across C (see Figure 21) we obtain the result shown in Figure 22. The output

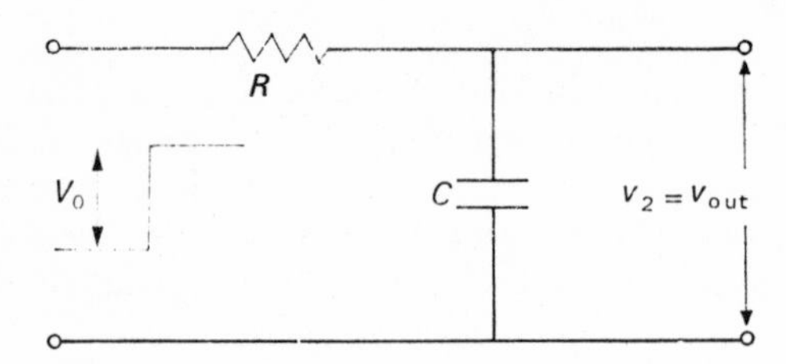

Figure 21. Integrating circuit.

voltage rises exponentially to a maximum value of V_0, with a characteristic 'rise time' given by RC. By this we mean that when $t = RC$ the voltage has risen to within exp(−1) = 1/2·7 of its final value (see Figure 22). This rise time is also defined by the point at which the line giving the initial slope of the curve intersects the final voltage level at V_0. This can be seen by either differentiating equation **3.5** or simply expanding it for small values of t to give

$$v_2 = V_0\left\{1 - \left(1 - \frac{t}{RC} + \ldots\right)\right\} = \frac{V_0}{RC}t \tag{3.6}$$

When $v_2 = V_0$, $t = RC$ as stated. (In order to avoid confusion it should be noted here that there is another definition of rise time – which we shall use on at least one occasion – as the time taken for the output to rise from 10 per cent to 90 per cent of its final value. For the exponentially rising curve here, it is easy to show that this new rise time is just 2·2 RC. We shall refer to this second definition as

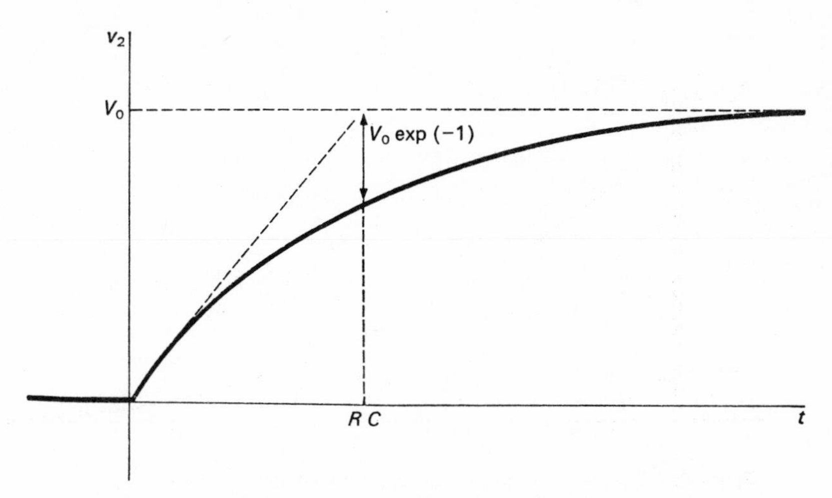

Figure 22. Variation of voltage across capacitor with time.

the '10 to 90 per cent rise time' and retain the simple term 'rise time' for the constant RC in an exponential rise.)

The circuit of Figure 21 is sometimes called an 'integrating circuit' because it performs a crude kind of integration. What we put in at the input was a constant voltage V_0; what appears at the output for small values of t (that is small values of v_2 also) is, from equation **3.6**, $(V_0/RC)t$ or $(1/RC) \int V_0\, dt$. So apart from the multiplying constant $1/RC$ the output is the integral of the input, provided v_2 is small compared to its final value V_0. That this is true for quite general inputs, and not for a step only, can be seen from equation **3.4** if v_2 is neglected with respect to V_0. We shall later discuss circuits which will accurately integrate signals without this restriction on amplitude.

Let us now look at the voltage v_1 across the resistor. We know that as the capacitor charges, the charging current will drop exponentially to zero, so we expect that the voltage across the resistor will do likewise. This we shall now show formally. Using equation **3.3** to eliminate v_2, we have

$$v_2 = \int \frac{i\, dt}{C} = \int \frac{v_1\, dt}{RC}$$

and subsituting in equation **3.1** we obtain

$$v_1 + \int \frac{v_1\, dt}{RC} = V_0 \qquad \textbf{3.7}$$

On differentiating

$$\frac{dv_1}{dt} + \frac{v_1}{RC} = \frac{dV_0}{dt} = 0 \qquad \textbf{3.8}$$

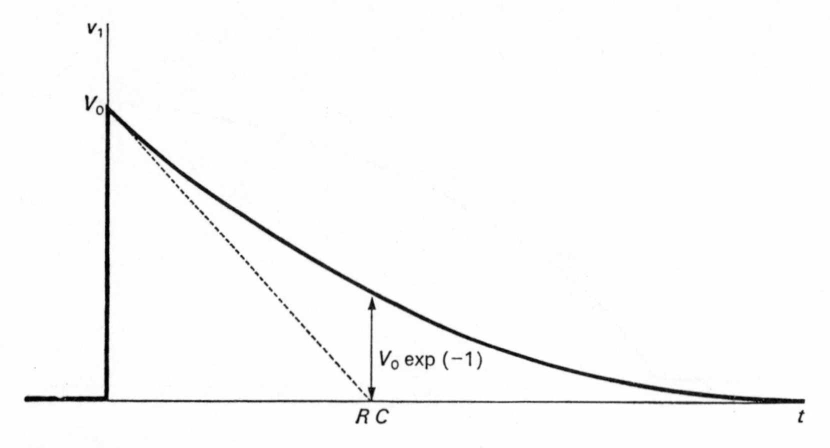

Figure 23. Variation of voltage across resistor with time.

The solution of this is

$$v_1 = V_0 \exp\left(\frac{-t}{RC}\right) \qquad 3.9$$

where the constant in the integration has been determined by remembering that the capacitor is initially uncharged, and therefore the full voltage V_0 appears across R; that is $v_1 = V_0$ at $t = 0$. Here the characteristic 'fall time' is given by RC, which is the time for the voltage v_1 to fall to $\exp(-1) = 1/2{\cdot}7$ of its initial value (see Figure 23). A similar relationship with the initial slope also exists.

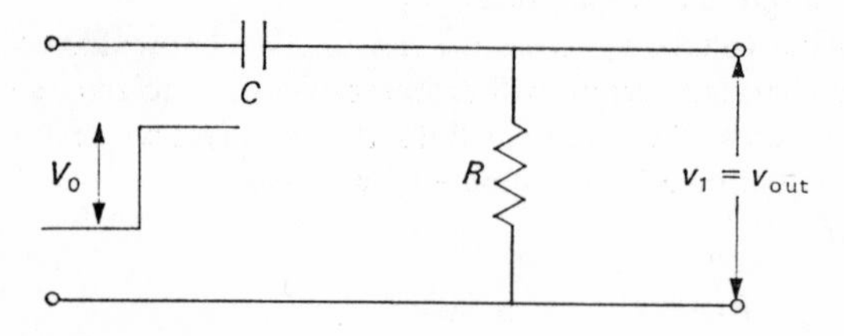

Figure 24. Differentiating circuit.

Figure 24 shows the circuit redrawn in a way we shall more often meet. (The previous result, of course, still applies.) It is referred to as a 'differentiating' circuit since in this case the output approximates to RC times the differential of the input, provided v_1 is small with respect to V_0 (equation 3.7). For example, when we put in a constant voltage V_0 we get out RC times its differential (which is zero!) provided we wait until v_1 has become small with respect to V_0.

Later we shall meet differentiating circuits which perform this function much better.

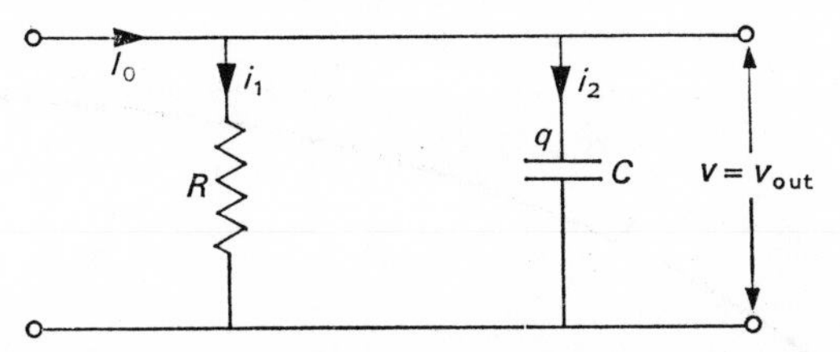

Figure 25. Parallel RC circuit fed with current step.

Since we have been using current as well as voltage sources in our equivalent circuits in the previous chapter, we should discuss now one further circuit (Figure 25). Here a step of *current* I_0 is applied to a resistor and a capacitor in parallel. The equation describing the situation is

$$i_1 + i_2 = I_0$$

or

$$\frac{v}{R} + \frac{dq}{dt} = I_0$$

or

$$\frac{v}{R} + C\frac{dv}{dt} = I_0$$

which is

$$\frac{dv}{dt} + \frac{v}{RC} = \frac{I_0}{C}$$

The solution of this, assuming the capacitor was originally uncharged is

$$v = v_{out} = I_0 R\left\{1 - \exp\left(\frac{-t}{RC}\right)\right\} \qquad 3.10$$

Figure 26 shows a plot of this with the rise time RC indicated.

Finally we refer to a circuit which, while not connected with the present discussion has some formal relationship with the circuits in question. It will be required in another context later. It is shown in Figure 27. The equation describing it is

$$v_1 + v_2 = V_0$$

or

$$iR + L\frac{di}{dt} = V_0$$

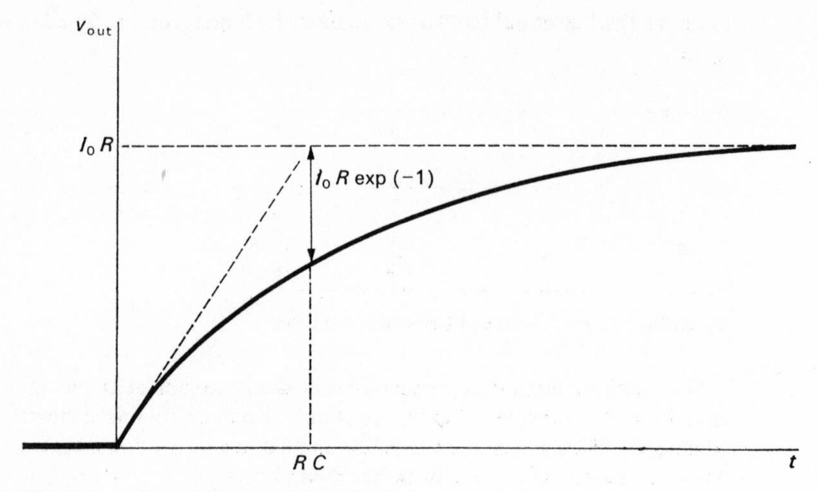

Figure 26. Output from parallel RC circuit.

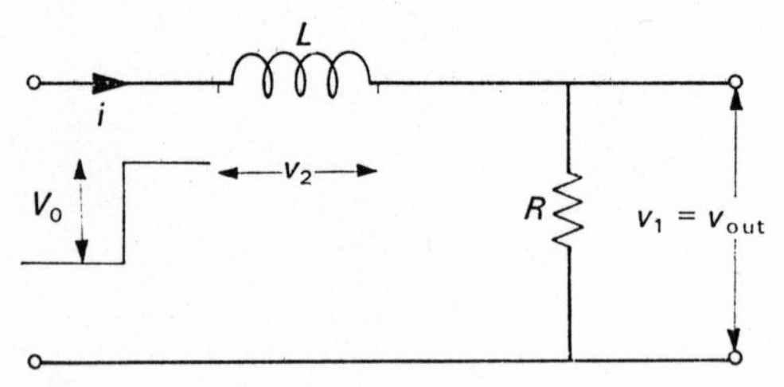

Figure 27. Series LR circuit.

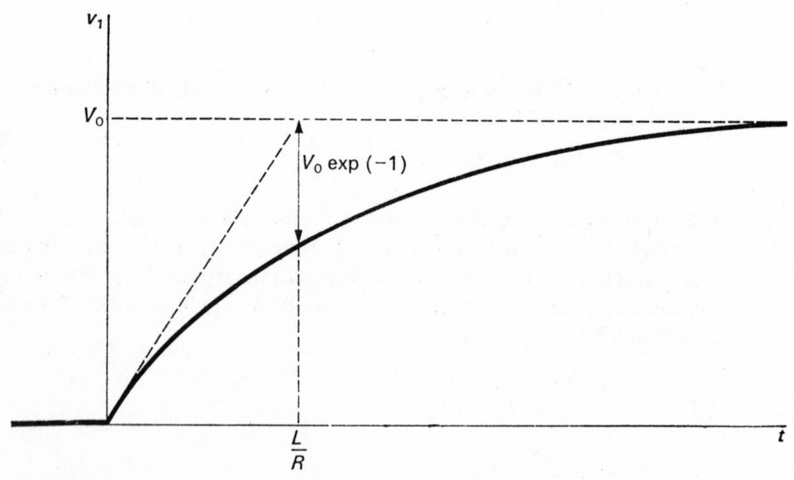

Figure 28. Output from series LR circuit.

with a solution

$$v_1 = v_{out} = iR = V_0\left\{1 - \exp\left(-\frac{R}{L}t\right)\right\} \qquad \textbf{3.11}$$

In this case our initial condition is that v_{out} ($=iR$) must be zero at $t=0$ since the current i through an inductance must start from zero. The output (Figure 28) is very similar to that of Figure 22 with the rise in this case governed by the time constant L/R.

3·2 Passage of a voltage step through an amplifier stage: the rise time

We can now consider the passage of a step of voltage through the amplifier stage of Figure 29 (which is merely Figure 9(c) redrawn), leaving out, as usual, components like R_s and C_s which are concerned only with setting the d.c. levels.

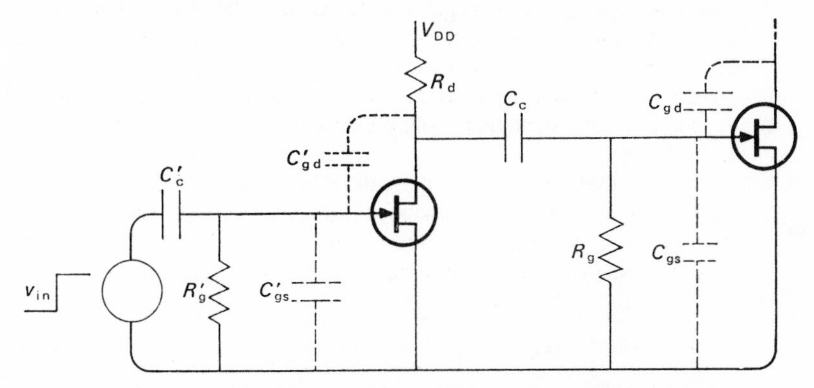

Figure 29. FET amplifier with voltage step input showing stray capacities.

We wish to investigate what happens when the step of voltage v_{in} coming from an ideal source and amplified by the left-hand FET, is applied through the coupling network to the input of the right-hand FET. When this happens the stray capacities and the coupling capacitor C_c begin to charge, but not instantaneously because they will each be charged through a resistance. In a small time after the application of the step (begging the question for the moment of what we mean by a 'small' time) the large capacitor C_c does not have time to charge appreciably. There can therefore be no appreciable voltage across it: that is, any voltage applied to one side of it will be transmitted faithfully onwards. Thus it plays no role in the circuit except the usual one of blocking off the d.c. voltage on the drain from the following gate, and we may consider it a short circuit in our equivalent circuit (Figure 30). Here we have used the current generator type of equivalent circuit (of Figure 11(e)) to represent the left-hand FET. The various stray capacities should be recognizable from our treatment of the sinusoidal case. They include C_{gd} augmented by the Miller factor $1 + A_0$, where A_0 is the

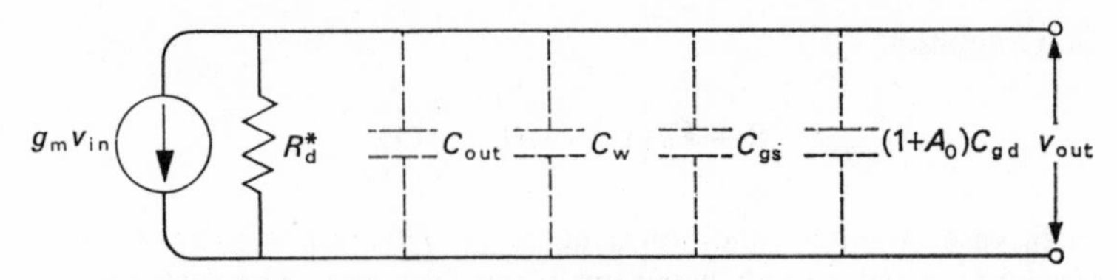

Figure 30. Equivalent circuit for the amplifier of Figure 29, a short time after application of the input.

gain of the following stage. It is appropriate to identify more closely what the capacity C_{out} is. In Figure 29 we see that a capacity C'_{gd} is connected from the output (that is, the drain) of the left-hand FET. The lower end of this capacity is effectively to ground, since it goes there via a large capacity C'_c (which can be considered a short circuit in the same way as C_c) and via a voltage generator of (ideally) zero impedance. C'_{gd} is thus the capacity between output and ground, and is therefore what we have called C_{out}. (Note that this applies only to a stage driven by an ideal voltage source.)

As before, we have neglected in our equivalent circuit, the large gate resistor R_g, but it could be included by slightly modifying the value of R_d^*, with which it is in parallel.

It is clear that our equivalent circuit of Figure 30 closely resembles the network of Figure 25 which we discussed previously, in that the assembly of stray capacities in parallel can be combined into a single capacity C_T. Therefore we expect to obtain an output from the left hand stage, v_{out} given by

$$v_{out} = -g_m \, v_{in} \, R_d^* \left\{1 - \exp\left(-\frac{t}{R_d^* \, C_T}\right)\right\} \tag{3.12}$$

The negative sign reflects the usual sign inversion produced by an amplifier stage, and is indicated in the equivalent circuit by the direction of flow from the current source.

Input and output signals are shown in Figure 31 although here for easier comparison we have shown the input and output with the same polarity. The rise time of the output signal is given as usual by $R_d^* \, C_T$. Since what we would ideally expect from an amplifier stage would be an accurate representation of the step merely increased in size, it is clear that we have considerable distortion here. If we make $R_d^* \, C_T$ smaller the output will rise more rapidly and approximate more closely to the ideal. Note how similar our difficulties are to those encountered with sinusoidal signals. There we had an upper half-power frequency containing the same constants as we now have in the rise time. (In comparing equation **2.18**, remenber that the r_s appearing there referred to the internal resistance of the source driving the FET on the right of Figure 15, and is therefore equivalent to the R_d^* of the present discussion.) The solution in this case must be similar: we shall try to reduce to a minimum all stray capacities like C_w, we must reduce R_d^* (by reducing R_d) to improve the rise time $R_d^* C_T$ and we must reduce

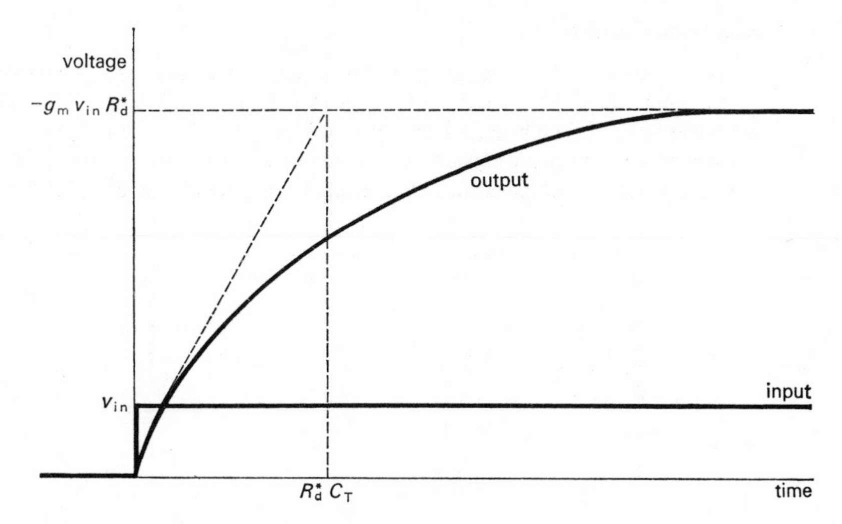

Figure 31. Input and output signals.

the gain A_0 of the following stage (by reducing its R_d) in order to reduce the Miller effect and thus the input capacity. As we reduce R_d, we reduce the gain of the stage, as can be seen from Figure 31, where the final level to which the output rises is $g_m v_{in} R_d^*$ (which is approximately $g_m v_{in} R_d$ for small values of R_d). So once again we must choose either good rise time or large amplification but not both. The figure of merit sometimes used to characterize the performance of the FET in the present circumstances is F = gain/rise time; a large value of F will imply reasonable gain with small rise time. For the ideal case of negligible wiring capacity

$$F = \frac{\text{gain}}{\text{rise time}} = \frac{g_m R_d^*}{R_d^* C_T} = \frac{g_m}{C_T} = \frac{g_m}{C_{out} + C_{gs} + (1 + A_0) C_{gd}}$$

$$= \frac{g_m}{C_{out} + C_{in}} \qquad \textbf{3.13}$$

The present figure of merit differs from the previous one of equation **2.21** only by the factor 2π. In general, also in this case, we should look for FETs with high values of g_m and low values of C_{gs} and particularly C_{gd}.

The figures used before ($C_{gd} = 2$ pF, $C_{gs} = 4$ pF, $C_{out} + C_w = 6$ pF) give a rise time of 0·7 μsec. for drain loads of both stages of 10 kΩ, and a rise time of 0·05 μsec. if these loads are both reduced to 2 kΩ. The corresponding values for the midband gains are as before, 30 and 6 respectively. Because the picture we have used for the FET is a simplified one, the figures quoted here are on the optimistic side.

3·3 Shunt compensation

It is possible to obtain some further improvement in rise time for a given value of the gain, by using an inductor in series with the load resistor R_d. The equivalent circuit then becomes as shown in Figure 32. Since the inductor is in parallel with the parasitic capacity C_T the method is referred to as 'shunt compensation' or 'shunt peaking'. (Series peaking circuits can also be used.)

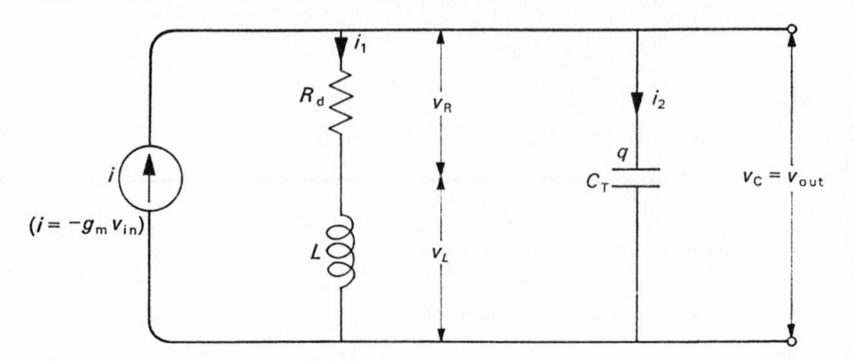

Figure 32. Circuit for shunt compensation.

Although we are discussing it in connexion with a f.e.t. stage, shunt compensation can also be used with bipolar transistors. In general it is an attempt to 'tune out' the unwanted stray capacity with an inductor L. We assume that we already have a small value of drain load, so we have R_d rather than R_d^* in Figure 32. Note also that to avoid a lot of negative signs in our equations the arrow in the current generator has been drawn in a different sense from usual. Later we shall restore the situation to normal by putting $i = -g_m v_{in}$. The equations describing the circuit are

$$i_1 + i_2 = i$$

$$v_R + v_L = v_C = v_{out}$$

$$v_R = i_1 R_d$$

$$v_L = L\frac{di_1}{dt}$$

$$v_C = \frac{1}{C_T}\int i_2\, dt$$

Eliminating all the variables except i_1

$$L\frac{d^2 i_1}{dt^2} + R_d\frac{di_1}{dt} + \frac{i_1}{C_T} = \frac{i}{C_T}$$

A particular solution of this is $i_1 = i$; the general solution is obtained by putting $i_1 = i + A\exp(-\mu t)$ giving

$$L\mu^2 - R_d\mu + \frac{1}{C_T} = 0$$

or

$$\mu = \frac{R_d \pm \left(R_d^2 - \frac{4L}{C_T}\right)^{\frac{1}{2}}}{2L} \qquad \textbf{3.14}$$

Hence

$$i_1 = i + A\exp(-\mu_1 t) + B\exp(-\mu_2 t)$$

where μ_1 and μ_2 are the values given by equation **3.14**.

Since $v_{out} = v_R + v_L = i_1 R_d + L\frac{di_1}{dt}$

$$v_{out} = iR_d + A(R_d - \mu_1 L)\exp(-\mu_1 t) + B(R_d - \mu_2 L)\exp(-\mu_2 t)$$

where $i = -g_m v_{in}$ (see Figure 32).

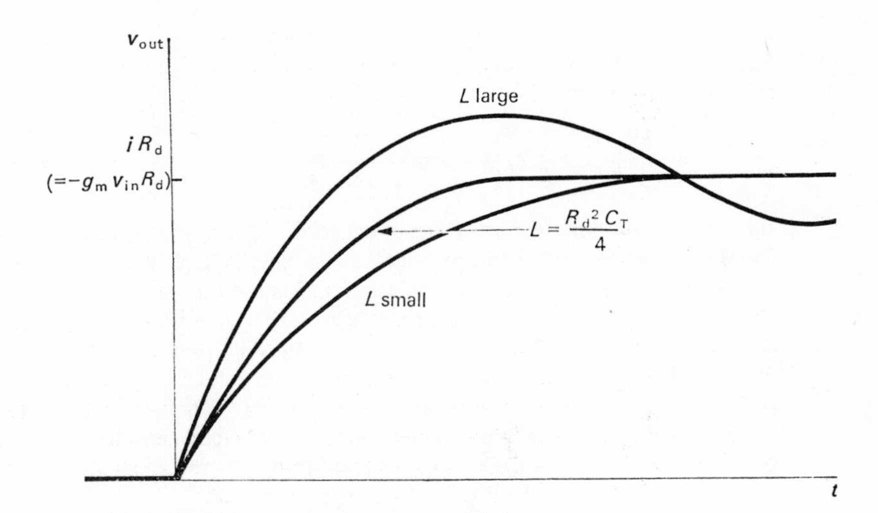

Figure 33. Output from shunt-compensated stage.

If L is small and $R_d^2 - 4L/C_T > 0$ then the solution is non-oscillatory; if L is large and $R_d^2 - 4L/C_T < 0$ we have damped oscillations (see Figure 33). The fastest rise we can obtain without any 'overshoot' is at the critical damping point where $R_d^2 = 4L/C_T$ (that is $\mu_1 = \mu_2$).

In this case of equal roots the solution is

$$i_1 = i + (A + Bt)\exp\left(-\frac{R_d t}{2L}\right)$$

or, since $R_d^2 = \dfrac{4L}{C_T}$

$$i_1 = i + (A + Bt)\exp\left(-\frac{2t}{R_d C_T}\right) \tag{3.15}$$

and

$$v_{out} = iR_d + \left\{\frac{(A + Bt)R_d}{2} + R_d^2\frac{C_T B}{4}\right\}\exp\left(-\frac{2t}{R_d C_T}\right) \tag{3.16}$$

To determine A and B we first note that i_1 must be zero at $t = 0$, since we cannot start a finite current through an inductor; this gives $A = -i$ (equation **3.15**). v_{out} must also be zero at $t = 0$ since the capacitor is assumed to be initially uncharged. From equation **3.16** we obtain $B = -2i/R_d C_T$ and the final solution

$$v_{out} = -g_m v_{in} R_d\left\{1 - \left(1 + \frac{t}{R_d C_T}\right)\exp\left(\frac{-2t}{R_d C_T}\right)\right\} \tag{3.17}$$

by putting $-g_m v_{in}$ for i. Compare this with the corresponding equation with no inductor present (that is, equation **3.12** with R_d^* replaced by R_d since this is the approximation we are now using)

$$v_{out} = -g_m v_{in} R_d\left\{1 - \exp\left(\frac{-t}{R_d C_T}\right)\right\} \tag{3.18}$$

It is easy to show that the curve given by equation **3.17** always lies above that of equation **3.18**, so we have improved the rise time. We shall calculate this improvement. Since the curve of equation **3.17** is not a pure exponential we shall use the '10 to 90 per cent' rise time for both curves. We have seen that for equation **3.18** it was $2{\cdot}2 R_d C_T$; similarly putting $v_{out} = 0{\cdot}1\ g_m v_{in} R_d$ and $0{\cdot}9\ g_m v_{in} R_d$ in turn in equation **3.17** we obtain a rise time of $1{\cdot}5\ R_d C_T$. For the circuit with this 'critical compensation', that is with $L = R_d^2 C_T/4$ we have thus obtained an appreciable improvement over the non-compensated case. In practice a little overshoot is usually accepted in the interest of slightly better rise time, and the final adjustment of L may be empirical.

There are further sophistications similar to 'shunt compensation' but involving more complicated networks. However we shall simply mention a more radical solution to the problem. This is the 'distributed amplifier' in which the stray capacities, which are the limiting factor in the situation, are made to play a useful role, by being incorporated as the capacities in a 'delay line' (see Chapter 8). The distributed amplifier has proved useful in considerably extending the frequency range of thermionic valves, but is no longer widely used because of the excellent high frequency performance available from bipolar transistors using more conventional methods.

3·4 The flat top of the step

Returning to Figure 29 let us ask how the stage reproduces the flat top of the step. We shall look at the output a considerable time after the step has been applied, in which case we can neglect different elements in our equivalent circuit. We now assume that the stray capacities C_{out}, and C_{in}, and C_w are fully charged and play no further active parts. The coupling capacitor C_c now, however, becomes of importance because, as previously mentioned, it will have had time to charge appreciably, and the voltage across it can no longer be neglected. As in the sinousidal case, the equivalent circuit reduces to that shown in Figure 34 but now v_{in} represents a voltage step. Since $R_d \ll R_g$ we can also

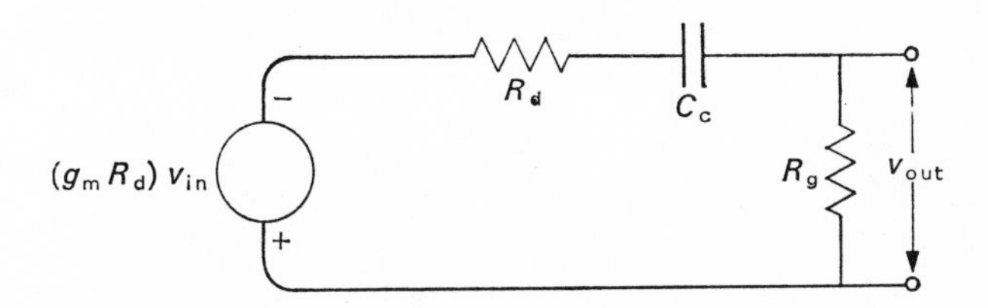

Figure 34. Equivalent circuit for FET stage a long time after application of the input.

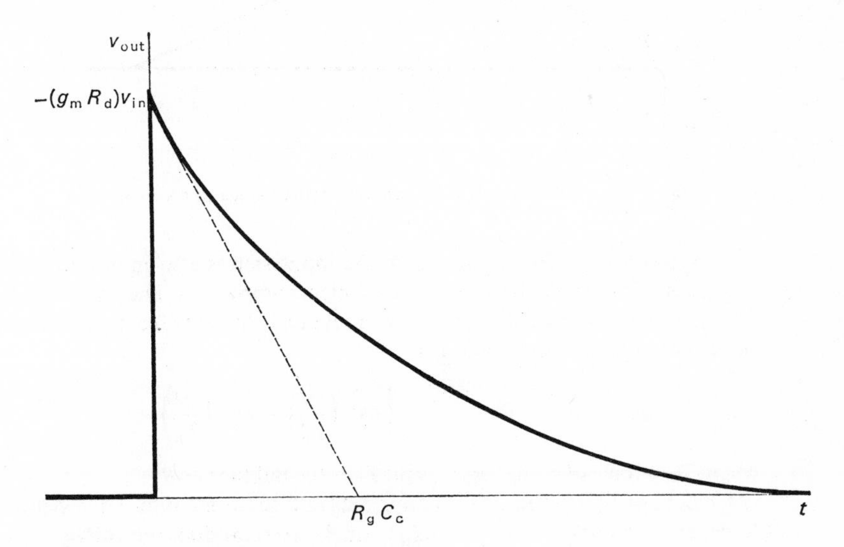

Figure 35. Output from circuit of Figure 34.

neglect R_d in the charging of C_c. Comparing Figure 34 with Figure 24 we can see from equation **3.9** that for the present case

$$v_{out} = -g_m v_{in} R_d \exp\left(\frac{-t}{R_g C_c}\right) \qquad \textbf{3.19}$$

that is, the signal decreases exponentially with a fall time of $R_g C_c$ (see Figure 35).

3·5 Overall response to a voltage step

A combination of the results of equations **3.12** and **3.19** (giving respectively the response to the rising edge and flat top of the step) gives a complete picture of what happens to a step of voltage on passing through an FET amplifier stage.

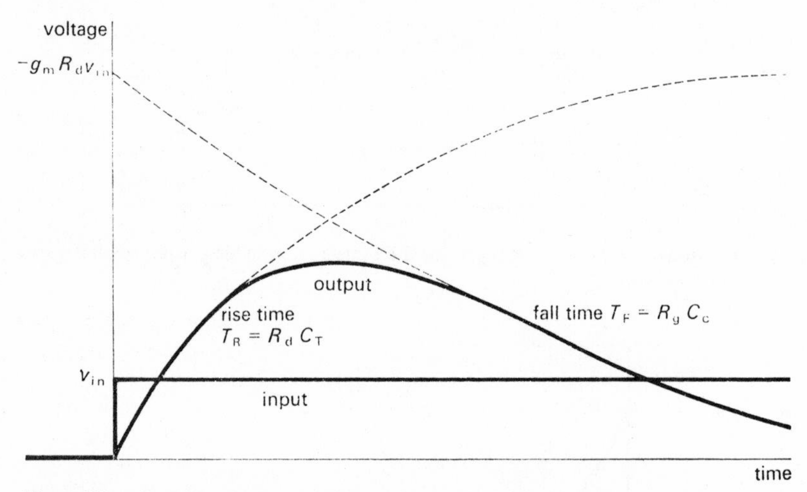

Figure 36. Overall response of FET common-source stage to a voltage step.

This is sketched in Figure 36 where the input step is also indicated – both polarities being shown the same for convenient comparison. The exact equation for the output curve can be deduced without much difficulty from an equivalent circuit containing both C_T and C_c

$$v_{out} = -g_m v_{in} R_d \frac{T_F}{T_F - T_R}\left\{\exp\left(\frac{-t}{T_F}\right) - \exp\left(\frac{-t}{T_R}\right)\right\} \qquad \textbf{3.20}$$

where T_R = the rise time = $R_d C_T$ and T_F = the fall time = $R_g C_c$.

It can be easily verified that equation **3.20** reduces to equation **3.12** when T_F is large, and to equation **3.19** when T_R is small. To reproduce the voltage step as accurately as possible, we need the shortest possible rise time and the longest possible fall time, but there are other more important considerations to be taken into account, which we shall deal with shortly.

3·6 Relationship between sinusoidal and step-voltage amplification

We digress here to emphasize what should be already obvious – the fact that the passage of step voltages through an amplifier stage is in fact controlled by the same parameters met with when dealing with sinusoidal signals. The upper half-power frequency is given by $1/(2\pi R_d C_T)$ and the rise time for a voltage step is $R_d C_T$. Similarly the lower half-power frequency is $1/(2\pi R_g C_c)$ and the fall time of a step is $R_g C_c$. It is not surprising that these similarities exist, as an amplifier stage capable of producing an output voltage step with a rise time of, say, 1 μsec should be able to deal with a sinusoidal signal of frequency $1/(2\pi \times 10^{-6})$, since such a signal rises from zero to its maximum in $(2\pi \times 10^{-6})/4$ seconds or 1·5 μsec. Alternatively we might consider the resolution of a perfect voltage step into a collection of sinusoidal waves by the Fourier transform method. The theory shows that there will be important components with frequencies extending from zero (that is, d.c.) to very large values. No amplifier stage can possibly reproduce faithfully this range of frequencies, and therefore the perfect step. But if we are content to accept an imperfect step output with a rise time of say 1 μsec and a fall time of say 1 msec, we can show by resolving into sinusoidal waves that the components much above 1 MHz and much below 1 kHz are not very important. Consequently an amplifier stage able to reproduce accurately a range of frequencies from 1 kHz to MHz should also faithfully amplify the imperfect step in question.

The problem of designing amplifiers for voltage steps is therefore closely related to the problem of designing wide band sinusoidal amplifiers. There are of course marginal differences – for example, the value of inductance used in shunt compensation to produce the fastest rising step with no overshoot is not quite the same as that which will produce a flat frequency response of the greatest bandwidth – but the similarities are far more important than the differences.

3·7 Choice of rise and fall times

Full consideration of the criteria governing the choice of rise and fall times can become quite complicated, and outputs from different kinds of detectors may need different treatments. The simple ideas presented here would be appropriate for signals coming from a scintillation counter and passing through an amplifier of moderate gain. In Chapter 1, we saw that in practice the signal from a detector is never a perfect step, but consists of a fast rising front edge followed by a gently sloping top. Clearly the amplifier rise time should be somewhat better than the rise time of the signal from the detector, to enable us to preserve a reasonably accurate representation of the leading edge of the signal, which may be necessary for timing purposes for example. There is no point in making the rise time of the amplifier very much better than that of the signal because not only does this introduce unnecessary complexity and cost into the amplifier, but it will increase the amount of spurious electrical disturbance ('noise') at the output as we shall see later.

Previous discussion about rise times has been concerned with a single stage and not a complete amplifier consisting of a number of stages in cascade. The problem of calculating the rise time in such a situation is extremely complicated. However, the rise time for a number of cascaded stages will certainly be worse than for any of the individual stages. In reference 1 it is shown that for n identical stages a good approximation for the overall rise time is $1{\cdot}1\sqrt{n}$ times the rise time of an individual stage.

The value of the fall time need not be large since all the basic information from the detector (that is, the charge Q produced by the incident particle or radiation) is conveyed by the height of the leading edge of the input signal, which equals Q/C, where C is the stray capacity shown in Figure 1. Indeed a large value can be positively harmful, as successive pulses may 'pile up' on one another's tails, and lead, among other things, to inaccuracies in the measurement of pulse heights. Exactly what happens at various points in the amplifier under 'pile-up' conditions is not at all obvious, and we shall deal with the matter in more detail later in this chapter. For the moment we are only concerned with avoiding this situation, and for this the effective fall time of the amplifier (which as we shall see shortly, can be concentrated in one stage) must be very much less than the average spacing between the arrival of signals. We say *very much less than* because the signals from the detector will not be arriving regularly, but at random time intervals: and if for example they are coming in at an average rate of 1000 per second (that is, separated on average by 1 millisecond) the percentage of pulses which follow previous pulses within a time interval of 10 μsec can be shown to be approximately 1 per cent.

A lower limitation on the size of the fall time is imposed by the fact that normally we do not want to reduce it to the same order as the rise time of our system. If we did we should have a loss of pulse height, because the step will not have time to rise to anything like its full height before it is clipped off (see for example Figure 36). Let us make a calculation on this point using equation **3.20** but replacing T_R by T_D to show it is the detector which is now controlling the rise time. The position of the maximum is found by differentiating the expression and setting the result equal to zero

$$\frac{1}{T_F}\exp\left(\frac{-t}{T_F}\right) = \frac{1}{T_D}\exp\left(\frac{-t}{T_D}\right)$$

If, for example, $T_F = 2T_D$, $\exp(-t/T_F) = 1/2$ and $\exp(-t/T_D) = 1/4$. Therefore the maximum value of the output voltage is $-g_m v_{in} R_d/2$ or only 50 per cent of the maximum obtainable. If $T_F = T_D$ the calculation is more difficult, since equation **3.20** becomes indeterminate. To resolve this we put $T_D = T_F + x$ and allow x to go to zero.

Equation **3.20** then becomes

$$v_{out} = -\frac{g_m v_{in} R_d t \left\{ \exp\left(\frac{-t}{T_F}\right)\right\}}{T_F} \qquad \textbf{3.21}$$

The maximum value for v_{out} now occurs at $t = T_F$ and its size is $-g_m v_{in} R_d \exp(-1)$, or only 37 per cent of the maximum pulse height obtainable. By comparison, for $T_F = 8T_D$ and $16T_D$ respectively, we obtain values of 85 per cent and 90 per cent of the maximum, using equation **3.20** in its original form for the calculation. As a numerical example, let us consider a detector with a rise time of 0·25 μsec (for example, the case of a scintillation counter with a sodium iodide crystal). Here we might choose an amplifier rise time of 0·1 μsec to reproduce fairly accurately the detector rise time of 0·25 μsec. The fall time could be ten times 0·25 μsec or 2·5 μsec, and we should be able to deal with signals coming at average separations of say 100 times 2·5 μsec with negligible pile up. This corresponds to a value of 4000 particles per second being detected by the counter.

For a practical amplifier with a number of stages, each with its own interstage coupling network and associated time constant $R_g C_c$, there may be further change in the pulse shape. We first investigate the result of putting a pulse with a fall time T_F (and for simplicity with a negligibly small rise time) through a second interstage coupling network with a time constant $R_g C_c = T^*$, say. It is simply necessary to repeat our calculations for the circuit of Figure 24 with V_0 replaced by $V_0 \exp(-t/T_F)$ and RC equal to T^*. The result is

$$v_{out} = V_0 \left(1 - \frac{T_F}{T^*}\right)^{-1} \left\{ \exp\left(\frac{-t}{T_F}\right) - \frac{T_F}{T^*} \exp\left(\frac{-t}{T^*}\right)\right\} \qquad \textbf{3.22}$$

If T^* is very large this reduces to

$$v_{out} = V_0 \exp\left(\frac{-t}{T_F}\right)$$

that is, just the signal we put in. Thus a signal with an exponential fall passing through an RC differentiating circuit of very long time constant comes out practically unaltered. (Note, however, for future reference that the exact equation **3.22** gives a curve which actually crosses the time axis and returns slowly to zero from underneath, as shown in Figure 37.) It is only necessary therefore (at least in this simple approach) to have one short coupling time constant in the amplifier, which will effectively control the fall time of the complete amplifier, and produce an output very similar in shape to that produced by the stage in question if operating alone. All the other coupling time constants will be made very much larger.

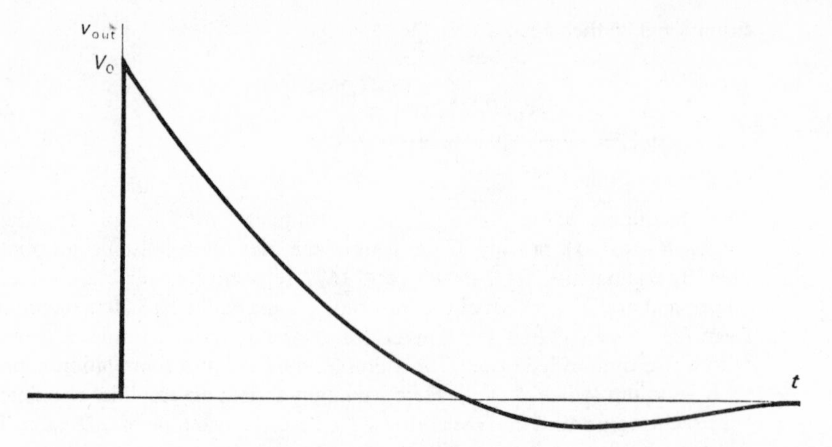

Figure 37. Exponential signal after passing through a long-time-constant, RC-coupling network.

The question arises as to where this short time constant is to be located. A first thought and a perfectly logical one would be to put it as near to the input of the amplifier as possible in order to eliminate pile up at an early stage. Indeed we might do this at the detector itself. In Figure 1 (which although representing an ionization chamber and not a scintillation counter, will nevertheless be appropriate for the remarks we wish to make) the incoming charge is deposited on the stray capacity C where it leaks off through the large resistance R. However if R is made small enough the charge will leak away quickly with the small time constant RC, and this can be made the basic fall time of the apparatus, all the amplifier interstage time constants being made much larger than RC. There is a fundamental objection (which we shall discuss in a later chapter on noise) to such a low value of R, particularly if the gain of the following amplifier is at all high. Apart from this, however, there is a reason for putting the short time constant towards the end of the amplifier. We saw earlier that a short fall time was equivalent to a poor low-frequency response. It is in fact a good thing to have such a poor low-frequency response, since this will attenuate unwanted sinusoidal signals at mains frequencies. These may be introduced via the d.c. power supply, which is usually derived from the mains by rectification and smoothing – the latter never being perfect – or by direct pick up. Even though these effects may be small they can ultimately be important if introduced at the first or an early stage of the amplifier, because they will then be multiplied by all or nearly all the gain of the amplifier. As the short-time-constant coupling will largely eliminate such spurious signals coming from all stages preceding it, one might put it as near to the output end of the amplifier as possible. This would mean that the pulses could be piled up all the way to the last stage, which is of itself, not necessarily harmful, since when they eventually pass through the short-time-constant coupling the long tails will be removed and the correct height relationships restored. But if the pile-up is considerable it tends to cause fluctuations in the *average* level at which the inputs of the various stages sit, in a manner which will be discussed later. Towards the output end of the amplifier where the signals are large, this may lead to shifts in d.c. levels large enough to cause malfunctioning of the stage. We compromise by putting the single short-time-constant coupling some way up from the input of the amplifier – often at the input of the main amplifier, if the complete amplifier system consists of a pre-amplifier plus a main amplifier. The pulses may be piled up before this coupling, but this is of little importance because in the early stages they are quite small. After the short-time-constant coupling there is little attenuation of unwanted low-frequency signals but this is not very important, since at the later stages there is not enough remaining amplification to make these small effects appreciable at the output.

The discussion on the choice of amplifier rise and fall times given above was a simple introduction only. We now mention two situations (to be discussed in more detail later in the book) in which the arrangement of rise and fall times is different from that given previously. In the first the detector signal is small and the gain of the amplifier has to be very large. Here the reduction of noise (spurious electrical disturbances of a basic nature) from the first amplifier stage is of vital importance. We shall show later that in these circumstances the best situation exists when the rise and fall times of the amplifier, T_R and T_F, are equal, and these will usually be larger than T_D, the rise time of the detector. The other case arises in dealing with signals arriving from a detector at an extremely fast rate. Then it is usual to have *two* short differentiating time constants of about the same value: the curve of Figure 37 then has a large dip below the axis, but returns to the zero line much more quickly. We shall mention this 'double-*RC* differentiation' again when dealing in some detail with the analogous 'double-delay-line differentiation' (or 'double-delay-line shaping').

Evidently a general purpose amplifier will need to have provision for varying both its rise and fall times to deal with various detectors and situations. The latter is effected by changing the value of a coupling capacitor, and the former by adding capacity to the existing stray capacity at the output of a stage.

3·8 Pulse shaping with delay lines

The converting of the basic detector signal with a long fall time into a more suitable pulse with a short fall time by means of a short-time-constant *RC* coupling is not the only method of pulse shaping available to us. Figure 38(a) shows a rectangular pulse which can be produced from the original voltage step by shaping with a single delay line in a manner to be discussed in Chapter 8. Figure 38(b) shows the pulse shape obtainable by the use of two delay lines, and is one which, as mentioned above, is particularly valuable when fast signal arrival rates are involved. In practice the ideally sharp rising and falling pulse edges shown in these diagrams will not be realized because of the finite rise times of the amplifier and the delay line itself. A more realistic picture of a step

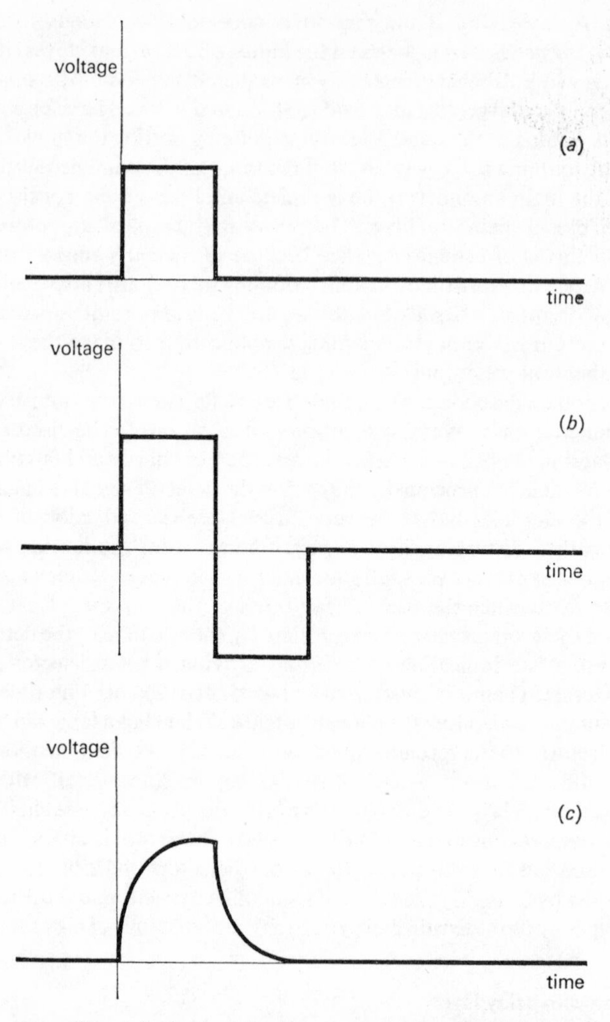

Figure 38. (a) Single delay-line-shaped pulse (idealized). (b) Double delay-line-shaped pulse (idealized). (c) Single delay-line-shaped pulse obtained in practice.

voltage shaped by a single delay line is shown in Figure 38(c), with the picture for the step with double delay line shaping being similarly modified.

We can examine some of the properties of these types of pulse before discussing the mechanism of their production. An important advantage of the single delay-line-shaped pulse is the flat top which helps in measuring its height, because the pulse height measuring device is allowed a finite time to make the determination. An *RC*-shaped pulse on the other hand, is only instantaneously at its maximum. Delay-line-shaped pulses share with RC-shaped pulses the advantage of a poor low-frequency response with its attendant attenuation of unwanted signals. They suffer somewhat compared with their exponential counterparts from the point of view of 'noise'. In Chapter 8 we shall see that a delay-line-shaped pulse is formed by subtracting from a step a delayed version of itself, and in this process random fluctuations in the signal are added and increased (see reference 3).

We shall now investigate what happens when a delay line shaped pulse passes through an interstage RC network. The result will lead to a more general discussion on 'pile-up' and 'base-line shift' both for these pulses and also for pulses with an exponential decay. A rectangular pulse can be considered as being made up of a positive step of height A followed after a time T (equal to the width of the pulse), by a negative step of similar magnitude. When the leading edge of such a pulse is applied to a coupling network like that of Figure 24 we know from our previous discussion that the output will be a step of the same size A which decays towards the base line with a time constant $RC(=T^*$ say) – see Figure 39. After a time T when the trailing edge of the pulse arrives the voltage has dropped to $A\exp(-T/T^*)$. The arrival of the trailing edge then forces the voltage negative by an amount A, from which point it decays back to the base line with the previous time constant T^*. The signal has an 'undershoot' below the base line of amount $A\{1-\exp(-T/T^*)\}$ and thus suffers from the same sort of trouble as an RC-shaped pulse passing through a further coupling network – and with the same possible consequences of mismeasurement of the height of following pulses sitting on the overshoot.

The first solution to this difficulty which might be considered, particularly for

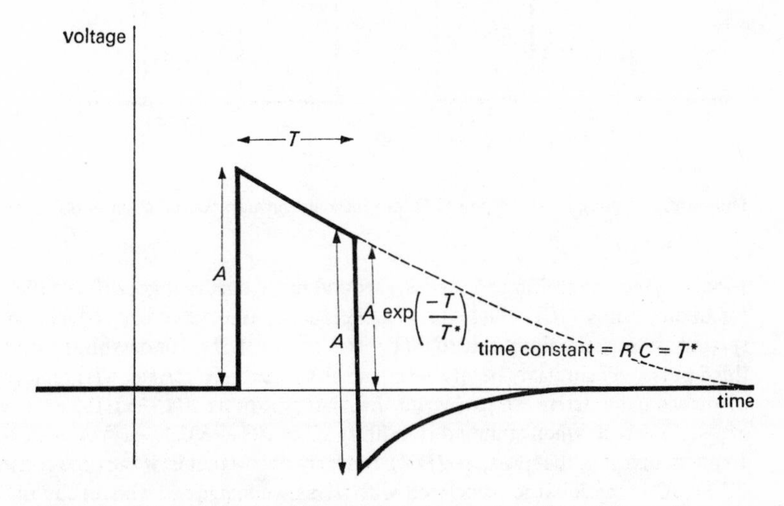

Figure 39. Delay-line-shaped pulse after passing through an RC-coupling network.

one isolated pulse, would be to make $T^*(=RC)$ very large compared with T, in order to reduce the 'droop' on the top of the pulse, and hence the undershoot. Although this certainly makes the magnitude of the undershoot smaller, it unfortunately makes it last much longer, since the final return of the signal from underneath the axis is governed by T^*. (Indeed a simple integration shows that the area contained under the part of the pulse above the axis is just equal to that in the undershoot – which incidentally is just equivalent to saying that the coupling capacitor will not transmit a d.c. voltage.) The danger of this situation for pulses closely following one another may be illustrated as follows. Let us take the particularly simple case of a train of pulses following one another at regular intervals equal to their own width T as in Figure 40. (In practice, we should never allow a situation to arise where signals from a nuclear radiation detector were arriving so close together, nor in any case would they be equally spaced apart. Nevertheless this simplification will give considerable insight into the sort of distortion that can occur in practice.) We have already seen that the undershoot on the first pulse of the train is $A(1-x)$ where for convenience we have written x for $\exp(-T/T^*)$. After a time T (when the next pulse arrives), this has decayed to $A(1-x)x$. The front edge of this next pulse brings this above the axis to a height $A - A(1-x)x$; it then decays for a time T during the period corresponding to the top of the second pulse, becoming $\{A - A(1-x)x\}x$.

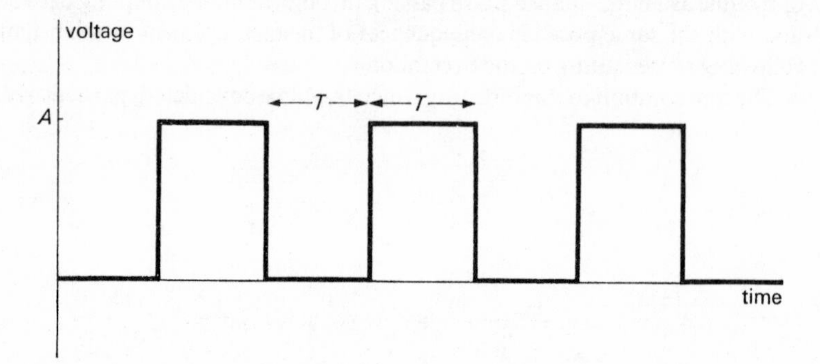

Figure 40. Regular train of pulses before passing through coupling network.

Finally when the trailing edge of the second pulse occurs it goes below the axis by an amount $A - \{A - A(1-x)x\}x$. This is the amount of undershoot on the second pulse, and simplifies to $A(1-x+x^2-x^3)$. The undershoot after the third pulse will similarly be $A(1-x+x^2-x^3+x^4-x^5)$. After a large number of pulses have arrive we can write the undershoot as $A(1-x)(1+x^2+x^4+x^6+\dots)$, which when summed to infinity gives $A(1-x)/(1-x^2)$ or $A/(1+x)$. In particular if x, that is $\exp(-T/T^*)$, is nearly unity (that is, if the time constant $T^* = RC$ is made large compared with T, as we suggested above) the undershoot eventually becomes approximately half the total height of the pulse.

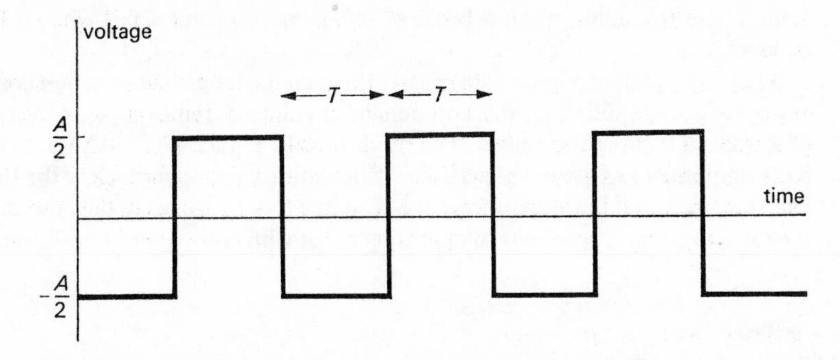

Figure 41. The previous train of pulses after passing through a long time-constant, RC-coupling network.

Figure 41 shows part of the output pulse train under these conditions, and after a large number of pulses have passed so that the voltages have settled down to their final levels. (Note that the tops of the pulses are almost flat because $x \approx 1$ and they thus decay very little in a time T.) Thus if we now make a measurement with an analyser of the pulse height from the zero line (ground) we shall have lost half of the pulse height, in other words, the base line of the pulses has been effectively shifted down through this amount. Figure 41 can again be regarded as an illustration of the inability of the coupling capacitior to transmit a d.c. voltage, and we may expect a similar, though smaller, base-line shift with pulses spaced considerably further apart, the area enclosed by the signals above the axis being equal to that below. Similar remarks clearly apply to pulses of unequal heights arriving in a random fashion, and to exponential RC-shaped pulses.

3·9 Pile-up and base-line shift

This section discusses in more detail the waveforms at various places in an a.c.-coupled amplifier system when connected to a nuclear radiation detector, such as that of Figure 1. When ionizing particles begin to arrive, the stray capacity C of the detector charges up to an equilibrium value, and the current through the resistor R becomes nQ, where n is the average number of particles arriving per second, and Q is the charge produced by each one (here assumed the same, for simplicity). The average voltage at the output from the detector is thus nQR, and the steps corresponding to the arrival of particles appear superimposed on this steady voltage (see Figure 42(a)). In Chapter 11 we shall show that a large value for R is required to reduce electrical 'noise' at the amplifier input. The time constant, RC, will thus be large and the pulses piled up as shown. In fact the ratio of the mean level nQR to the voltage step produced by a single isolated pulse Q/C, is nRC, which may be quite large. Of course, because of the random rate of arrival of the pulses, there are fluctuations of voltage about the mean level, a rise occurring when a burst of pulses arrives, and a fall when a lull occurs.

When the signal of Figure 42(a) passes through the long-time-constant coupling in the pre-amplifier, the d.c. component of voltage is removed, as in the case of a train of rectangular pulses. The result is as in Figure 42(b). Although the d.c. component has gone, the statistical fluctuations above and below the time axis remain, and if these are allowed to continue to later stages in the amplifier, they will produce large shifts in operating point with consequent non-linearity

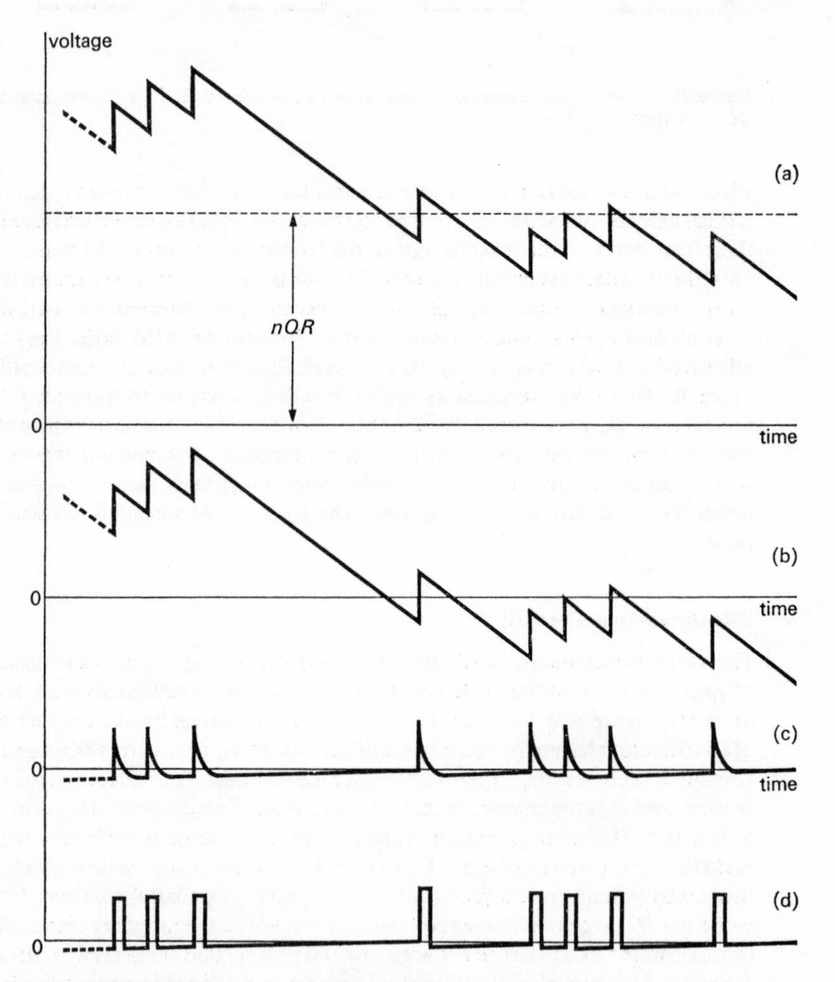

Figure 42. Output at high counting rate from
(a) nuclear radiation detector (b) pre-amplifier
(c) amplifier with RC shaping (d) amplifier with single-delay-line shaping.

and distortion of pulses. Before this happens we shall clip the pulses either with a short-time-constant *RC*-coupling network, or with a delay line, to produce the results shown in Figures 42(c) and 42(d), where the drastic fluctuations in level have been eliminated. However it is clear from previous discussion in connexion with Figure 39, that each pulse must still be followed by an undershoot as can be seen in the diagrams. A pulse arriving too soon after a previous one will have its height above the time axis reduced, and will thus be incorrectly measured by an analyser. At even higher count rates, this effect will be accentuated, and a result rather like that in Figure 41 will be obtained, although in this case the base-line shift will not be constant, because of the random arrival of the pulses. The obvious solution of reducing still further the short differentiating time constant for the exponential pulses, or the width of the delay-line-shaped pulses, will ultimately be frustrated in practice by loss of pulse height due to their finite rise time. This rise time will usually be a property of the detector itself, and thus is not under our control. So we must eventually expect trouble in pulse-height measurements at very high count rates, with the type of pulses we have been discussing. The base line can be restored to its correct position on the time axis by methods involving diodes ('base-line restoration') but these have their own peculiar difficulties. Instead we turn to methods using 'bipolar' pulses, that is, pulses with equal areas above and below the time axis, rather than the 'unipolar' pulses formed either by a single short-*RC*-time-constant network, or by a delay line, which, at least to start with, are on one side of the time axis only.

Bipolar pulses can be formed by two short-differentiating networks ('double-*RC* shaping'), as mentioned earlier, or by 'double-delay-line shaping', which produces a pulse of the type in Figure 38(b). The flat top of the latter makes it more suitable than the former for pulse height measurements, and as it is also rather easier to deal with mathematically, we shall confine our attention to it. As in the case of the single-delay-line-shaped pulse, we shall look at the passage of this double-delay-line-shaped pulse through an *RC*-differentiating network of time constant T^*. By a calculation like that at the end of section 3·8 the signal can be shown to be *above* the axis at the end of the pulse by an amount $A(1-2x+x^2)$ or $A(1-x)^2$, that is, it has an 'overshoot' rather than an 'undershoot'. The quantity x is, as before, $\exp(-T/T^*)$ and for a value of x of 0·9 there is only about 1 per cent overshoot. Furthermore by a similar but slightly more complicated calculation than before, we can show that for a train of pulses of this type succeeding one another immediately, with no time interval between them, the overshoot will settle down to a negligible value, for the usual case of $x \approx 1$ (T^* large). Even without a calculation it is easy to believe that this should be so, as such a train of double delay line shaped pulses would be identical to the waveform of Figure 41, which we have already seen is the kind of bipolar signal which a capacitor is capable of passing.

Again, however, the noise performance of an amplifier using double-delay-line shaping will be worse than when using single-delay-line shaping, for the sort of reason mentioned when discussing noise with the latter arrangement.

The double-delay-line amplifier will thus be preferred where high counting rates are unavoidable, but not where input signals are small and noise considerations predominant.

3·10 Pulse shaping by pole-zero cancellation

The term used to describe the technique known as 'pole-zero cancellation' will be more understandable when we have dealt with Laplace transforms in Chapter 12. A full discussion of this technique and its advantages in dealing with overload pulses in amplifiers is outside the scope of this book, but may be found in

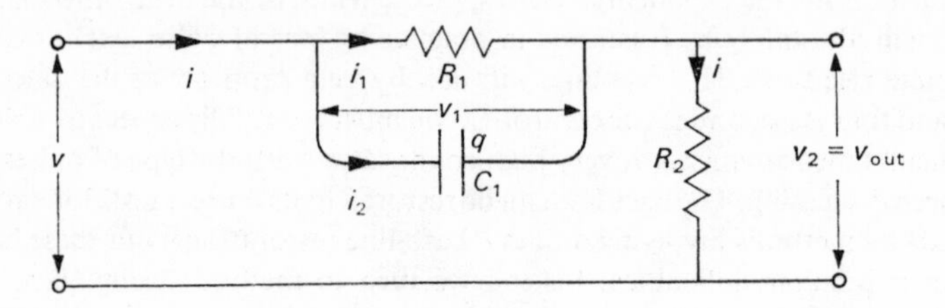

Figure 43. Circuit for pole-zero cancellation.

references 4 and 5. We shall examine in a simple way the circuit of this type shown in Figure 43. The equations describing the operation of this circuit are:

$$v = v_1 + v_2$$

$$v_1 = i_1 R_1 = \frac{1}{C_1} \int i_2 \, dt$$

$$i = i_1 + i_2$$

$$v_2(=v_{out}) = iR_2$$

Eliminating all the variables except v_{out} we obtain

$$\frac{dv_{out}}{dt} + \frac{v_{out}}{C_1} \frac{(R_1 + R_2)}{R_1 R_2} = \frac{v}{R_1 C_1} + \frac{dv}{dt} \qquad 3.23$$

If the input voltage v is an exponentially decaying signal given by $V_0 \exp(-t/T_F)$, and, for simplicity $R_1 C_1$ is written as T' and $R_2/(R_1 + R_2)$ as k $(k < 1)$, equation 3.23 becomes

$$\frac{dv_{out}}{dt} + \frac{v_{out}}{kT'} = \frac{V_0}{T'} \exp\left(\frac{-t}{T_F}\right) - \frac{V_0}{T_F} \exp\left(\frac{-t}{T_F}\right) \qquad 3.24$$

If the value of $T'(=R_1 C_1)$ is made equal to T_F (the fall time of the input pulse) the two terms on the right of equation **3.24** cancel (hence the 'cancellation' terminology). The solution for v_{out} then becomes

$$v_{out} = A \exp\left(\frac{-t}{kT_F}\right)$$

where the value of the constant A can be determined by the fact that at $t = 0$ the capacitor is uncharged, and therefore the whole of the initial voltage V_0 appears across the output. Thus $A = V_0$ and

$$v_{out} = V_0 \exp\left(\frac{-t}{kT_F}\right) \qquad \mathbf{3.25}$$

Equation **3.25** indicates that a signal with a long exponential fall time T_F can be shaped by this circuit into an exponential of much smaller fall time kT_F ($k < 1$); further it shows that the output is a true exponential with no crossing of the time axis as in Figure 37. However if this circuit is used as an interstage network it implies d.c. coupling between the stages, as it contains no blocking capacitor. We shall see as we proceed that such d.c. coupling is quite often possible.

References and further reading

1. J. MILLMAN and C. C. HALKIAS, *Integrated Electronics*, McGraw-Hill, 1972.
2. E. FAIRSTEIN and J. HAHN, 'Nuclear pulse amplifiers – fundamentals and design practice', Part 1, *Nucleonics*, vol. 23, no. 7, July 1965, p. 56; Part 2, *Nucleonics*, vol. 23, no. 9, September 1965, p. 81; Part 3, *Nucleonics*, vol. 23, no. 11, November 1965, p. 50; Part 4, *Nucleonics*, vol. 24, no. 1, January 1966, p. 54; Part 5, *Nucleonics*, vol. 24, no. 3, March 1966, p. 68.
3. R. L. CHASE, *Nuclear Pulse Spectrometry*, McGraw-Hill, 1961.
4. C. H. NOWLIN and J. L. BLANKENSHIP, 'Elimination of undesirable undershoot in the operation and testing of nuclear pulse amplifiers', *Review of Scientific Instruments*, vol. 36, 1965, p. 1830.
5. J. L. BLANKENSHIP and C. H. NOWLIN, 'New concepts in nuclear pulse amplifier design', *IEEE Transactions in Nuclear Science, NS*-13, no. 3, 1966, p. 495.

CHAPTER 4

The junction transistor

4·1 Principles of operation

The junction transistor may be manufactured in a variety of ways, and have many physical forms. Reference 1 in Chapter 2 and reference 1 of this chapter give a good account of some modern methods of production: however we shall describe a version which appeared fairly early in the transistor's history - the alloyed-junction type - chiefly because its construction is easy to visualize. As shown schematically in Figure 44, it consists of a thin slice of *n*-type material - the base - on either side of which are formed, by an alloying process with acceptor materials, two *p*-type regions, the 'emitter' and the 'collector', the latter being the larger in area. For obvious reasons this device is known as a *p-n-p* transistor. The width of *n*-type material remaining between the emitter and collector would typically be some tens of microns although with more sophisticated versions of this transistor, and with other types of transistor it might be as low as 1 μ. The diameter of the collector and emitter might typically be some hundreds of microns, so transistor construction clearly calls for extreme skill because of the small dimensions involved. External connexions to the three regions are also shown in Figure 44.

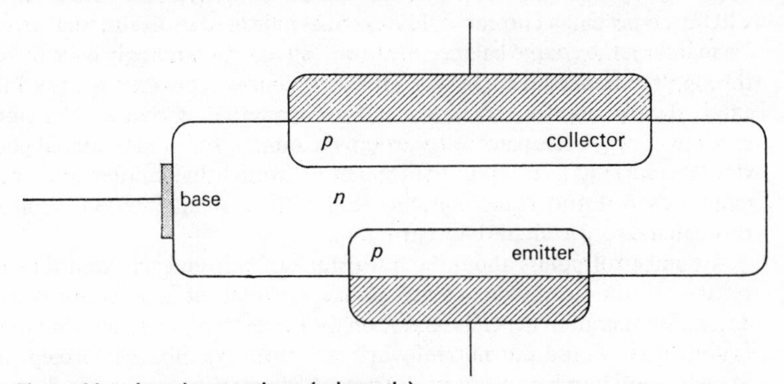

Figure 44. Junction transistor (schematic).

Our transistor then consists basically of two diode junctions, one between emitter and base, and the other between base and collector. In operation the emitter–base diode is biased on with a small negative voltage (a fraction of a volt, as we have seen, will suffice) between base and emitter: while the base–collector diode is biased off by a negative voltage of a few volts between

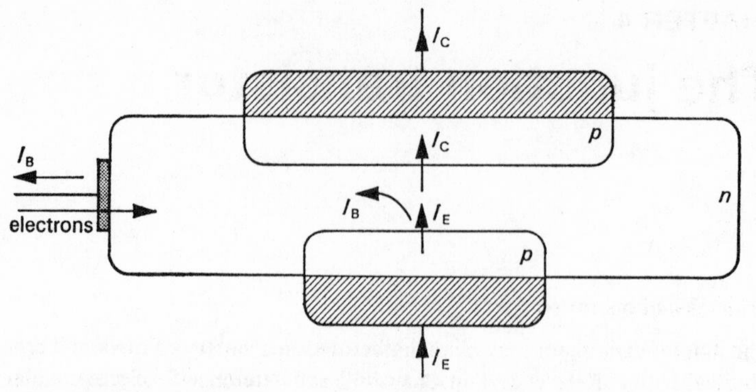

Figure 45. Current flow in *p-n-p* transistor.

collector and base. Figure 45 shows the current flow I_E across the forward-biased junction: because the base material is specifically made much less strongly *n*-type than the emitter is *p*-type, the junction is asymmetrical, and the current I_E consists almost entirely of holes flowing in the direction indicated by the arrow. The term 'emitter' is therefore appropriate for the region in question. Most of these holes will diffuse over and reach the base–collector junction. Although this junction is 'off' in the sense that it prevents collector–base current flow, it clearly represents a potential hill down which the holes coming from the emitter will be swept and 'collected'. This is indicated by the arrow I_C across the base–collector junction. Some of the holes from the emitter (typically 1 per cent) however will recombine with electrons in the base and will thus represent a current to this region as indicated by the internal arrow I_B. To maintain the charge balance, electrons will correspondingly have to flow in through the base lead as shown, which of course represents a conventional external current in the base lead I_B. (The corresponding change from holes to electrons for I_C takes place at the external contact.) From the external point of view the current I_E entering the transistor through the emitter lead emerges mainly as I_C through the collector lead, with a small fraction coming out through the base lead, the base current I_B.

A number of points about the transistor can be seen even from this simple picture of this operation. Firstly, unlike the FET, it is a 'minority-carrier' device. Its operation depends largely on the presence of holes in the *n*-type base region: these would not normally appear in a *n*-type material except in extremely small numbers as thermally generated minority carriers. In the present instance they are of course injected from the emitter. However the electrons do play a role – in flowing into the base region from the base contact to compensate for those negative charges neutralized by the recombining holes, so we can refer to these devices as 'bipolar transistors'. This description of their operation has been in terms of currents, not voltages, so we assume, and shall see a little more clearly later, that they are 'current-controlled' devices.

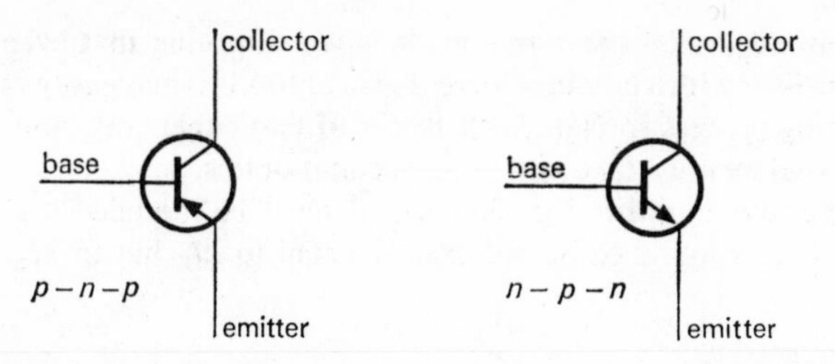

Figure 46. Transistor symbols.

Transistors of the *n-p-n* type can obviously also be made starting with a *p*-type base. Now we shall be dealing internally with electron flow, so the directions of the external currents I_E, I_B, and I_C will be reversed for such a device, when compared with Figure 45, as will the polarities of the batteries for biasing the junctions correctly, but the basic mechanism will be quite analogous. Figure 46 shows the symbols used in circuitry for the two types of transistor, the identification being by means of the arrow on the emitter, which shows in both cases the direction of flow of the conventional current.

4·2 Basic current relationships

The basic statement about transistor action is that, of the total hole (or electron) current leaving the emitter, a constant fraction, α, reaches the collector. Thus

$$I_C = \alpha I_E \qquad \textbf{4.1}$$

where α is the current gain between collector and emitter, also known as the common-base current gain. (The word 'gain' is used in its general sense, since we know in fact that α is less than unity, typically 0·99.) The third current involved I_B may be found straightforwardly by subtraction, as the total current flowing into the transistor must equal that flowing out. Thus

$$I_E = I_C + I_B$$

or

$$I_B = I_E - I_C \qquad \textbf{4.2}$$

Eliminating I_E between equations **4.1** and **4.2**

$$I_C = \frac{\alpha}{(1-\alpha)} I_B$$

or

$$I_C = \beta I_B \qquad \textbf{4.3}$$

where we have written β for $\alpha/(1-\alpha)$. β is the current gain between collector and base (also known as the common-emitter current gain) and from the previous value for α is typically 100, although transistors for various purposes with gains from 30 to 300 are common. It is worth noting that even if we purchase a transistor with a nominal current gain of 100, this may easily vary due to manufacturing process spreads, by a factor of two either way, and in our later circuit design we must take particular account of this.

The picture above is rather too simple and must be modified slightly as follows. In fact I_C is found to be not exactly equal to αI_E but to αI_E plus a constant

$$I_C = \alpha I_E + I_{CO} \tag{4.4}$$

where I_{CO} (also written I_{CBO}) represents the reverse current in the back-biased base–collector diode. Even if the emitter were open circuited ($I_E = 0$), we should therefore still have a collector current of I_{CO}, which might be of the order of 0·1μA for a silicon transistor, and very much larger for a germanium type. I_{CO} is a thermally generated reverse current and increases rapidly with temperature; for a silicon device a change in ambient temperature from 25° to 85°C would typically increase the reverse collector–base current by more than an order of magnitude. However the full implications of the presence of this current are not apparent until we determine the new relationship between I_C and I_B corresponding to the previous equation **4.3**. This is obtained from equations **4.4** and **4.2**, and is

$$I_C = \frac{\alpha}{1-\alpha} I_B + \frac{I_{CO}}{1-\alpha}$$

or

$$I_C = \beta I_B + (\beta + 1) I_{CO} \tag{4.5}$$

The arrangement in which I_B rather than I_E is the important parameter is one in which we shall be particularly interested; but equation **4.5** tells us that under these conditions, the reverse current has been effectively increased by the factor $\beta + 1$, which we know to be typically of the order of 100. Changes in I_{CO}

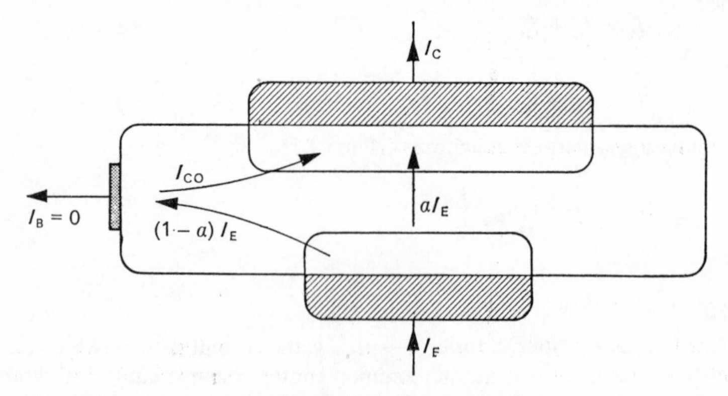

Figure 47. Current relationships with base open-circuited.

with temperature, particularly with germanium transistors, may thus appreciably alter the operating conditions, and must be allowed for in the circuit design.

It is instructive to check directly how this magnification of I_{CO} arises. Equation **4.5** shows that even with I_B zero we are still left with the relatively large current $(\beta+1)I_{CO}$. In Figure 47 we see that to make I_B zero we must balance I_{CO} with a current $(1-\alpha)I_E$. This implies the existence of the much larger current αI_E (plus of course I_{CO}) going to the collector.

4·3 Setting the operating point

To discuss the setting of the operating point of a transistor amplifier, we must first look at the characteristic curves of Figures 48(a) and 48(b), which are typical of a low power silicon *p-n-p* transistor. As there are four variables involved here, two families of curves are needed to describe the situation, unlike Figure 8 for the FET. The characteristics are given for the 'common-emitter' or 'grounded-emitter' arrangement in which all voltages are referred to the emitter. The curves shown are not the only possible ones by which to display the necessary information for this particular arrangement, but they are the ones we shall find most suitable for our purpose. Figure 48(a), which, as we shall later see represents the output characteristics of the device, gives the relation between the collector current I_C, and the collector to emitter voltage V_{CE} for various values of the base current I_B. (In line with our previous ideas about the action of a transistor, a current I_B is taken as the controlling parameter, rather than a voltage as for the FET.) The curves themselves, apart from an initial region where the collector voltage is close to zero, are a series of equally spaced,

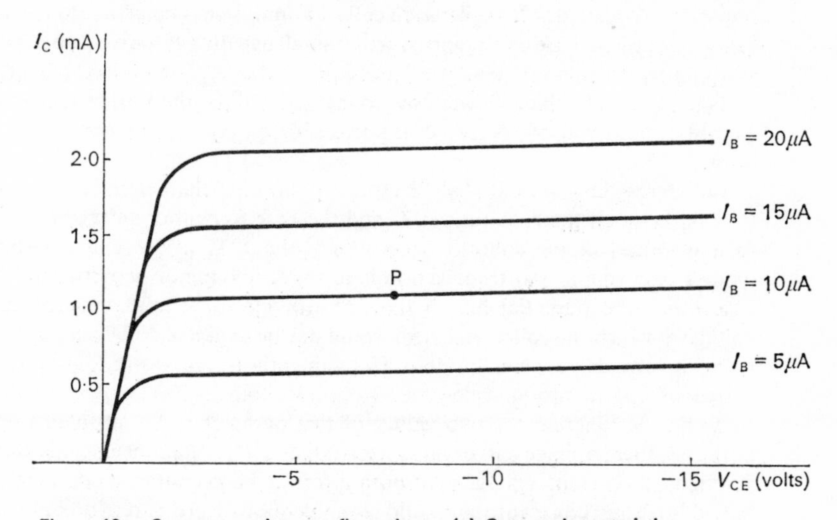

Figure 48. Common-emitter configuration. (a) Output characteristics.

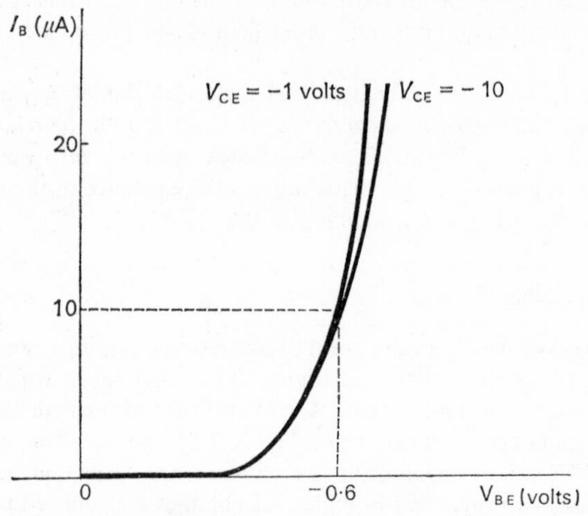

Figure 48. Common-emitter configuration. (b) Input characteristics.

almost horizontal lines. Clearly this reflects the basic transistor equation $I_C = \beta I_B$, where for simplicity we have assumed that the reverse current component $(\beta + 1) I_{CO}$ can be neglected (this is true for a silicon transistor at room temperature). By looking across to the collector current opposite any one of these horizontal lines we can see that the value of β is 100 for the transistor in question. For example, for the second line from the bottom we see that $\beta = I_C/I_B = 1\ \text{mA}/10\ \mu\text{A} = 10^{-3}/10^{-5} = 100$, as stated. Note also that the polarity of V_{CE} is negative for the *p-n-p* transistor in question. We have already seen that a fairly large negative voltage between collector and base is necessary to reverse bias the collector–base diode, and a further small negative voltage between base and emitter to forward bias the base–emitter diode. The collector–emitter voltage V_{CE}, with which we are now concerned, will be the sum of these two quantities, and will of course be negative. For an *n-p-n* transistor it will be positive.

The curves of Figure 48(b) (which represent the input characteristics) give the relation between the base current I_B and the base to emitter voltage V_{BE} for various values of the collector to emitter voltage V_{CE}. The collector–base diode's only role in transistor action is to sweep the minority carriers in the base down the potential hill to the collector electrode, and therefore the voltage at which the collector is held would not be expected to be very critical unless it is made very small indeed. Consequently the curves in Figure 48(b) corresponding to various values of V_{CE} almost coincide, as we have tried to show on the diagram for two values of this parameter. To all intents and purposes, then, a single curve will represent the I_B/V_{BE} relationship; this curve describes the current–voltage relationship for the base–emitter diode which is biased 'on', and consequently should resemble the forward part of an ordinary diode characteristic – which indeed it does.

Let us now suppose that we have decided to work at some point in the central part of the characteristics such as P in Figure 48(a), where $I_C = 1$ mA, $V_{CE} = -7{\cdot}5$ volts, and $I_B = 10\ \mu$A. Figure 48(b) shows that such a value for I_B requires a value for V_{BE} of approximately $-0{\cdot}6$ volts (for the silicon transistor in question); fortunately, however, the exact determination of the value of this voltage is unnecessary. To set the voltages just specified we should in general need two supplies, one for V_{CE} and the other for V_{BE}, but as in the case of the f.e.t., it is easy to obtain the necessary voltages from a single supply as shown in Figure 49. This in fact represents a simple but complete common-emitter amplifier stage, with a collector resistor R_c and a base resistor R_b, the latter being used to set the base voltage. Input and output coupling capacitors are also shown. The arrangement bears some resemblance to the common-source FET, it is easy to obtain the necessary voltages from a single supply as shown in counterpart the gate, is held at a voltage of the same polarity as that of the collector, but of a very much smaller magnitude.

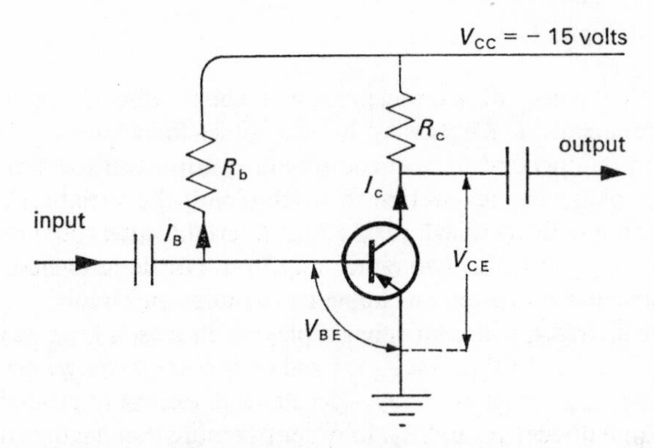

Figure 49. Simple biasing circuit.

What then should be the size of R_b? The current through it we decided should be 10 μA; the voltage across it is almost exactly V_{CC} (which we have arbitrarily chosen as -15 volts) since V_{BE} is so small. Thus R_b is $15/(10 \times 10^{-6}) = 1{\cdot}5$ megohms, or $(15 - 0{\cdot}6)/(10 \times 10^{-6}) = 1{\cdot}44$ megohms, to be exact. Using the 'wrong' value of 1·5 megohms which we obtained by ignoring V_{BE}, instead of the correct one of 1·44 megohms will result merely in V_{BE} shifting very slightly from $-0{\cdot}6$ volts to accommodate to the altered conditions. The important point is that the exact value of V_{BE} need not be known, merely that it is very small compared with V_{CE}.

The value of the collector resistor R_c can be determined similarly, as we know that the current flowing through it, I_C, is to be 1 mA, and the voltage across it is the supply voltage, 15 volts, less the voltage across the transistor, which we decided should be 7·5 volts. Thus $R_c = (15 - 7{\cdot}5)/(1 \times 10^{-3})$ or 7500 ohms. This completes the design of the circuit, although from a very simple point of view, since for example the value of R_c which we have obtained might not be acceptable on grounds say of frequency response, and further adjustments, including the selection of a different voltage supply might be necessary before a suitable arrangement could be arrived at. A load line can be drawn on the curves of Figure 48(a) in the usual way, intersecting the voltage axis at -15 volts, and the current axis at $15/7500 = 2$ mA. This will allow us to obtain the limiting values of output voltage and current, and the values of I_B corresponding to them. In this case the output can swing through very nearly $\pm 7{\cdot}5$ volts, although we shall have more to say about this point later.

Although the circuit of Figure 49 is a simple one with few components, it has two important disadvantages which stem from the first step in its design – that of fixing I_B by means of R_b. Referring to equation **4.5** we see that if I_{CO} varies, as it will rapidly with temperature, then,

$$\frac{dI_C}{dI_{CO}} = \beta + 1 \text{ (since } I_B \text{ is fixed)}$$

$$\approx 100 \text{ typically.}$$

We thus have a rapid change of operating point with temperature. Even in the case where such temperature effects may be negligible, for example with a silicon transistor to be operated only at room temperature, and the simpler relation $I_C = \beta I_B$ applies, we may still be in trouble with the variation of β among transistors nominally identical. Fixing I_B at a certain value could result in a difference in I_C by a factor of two either way from our design value, depending on the particular transistor we happened to put in the circuit.

Figure 50 shows a circuit, still relatively simple, which goes a long way to overcome these difficulties. In this case I_E is fixed as follows. First we set the base of the transistor at a potential of, say, -5 volts with respect to ground by means of the potential divider R_1 and R_2. In order to ensure that changes in I_B with temperature do not affect the value of this chosen voltage we must ensure that the current through the potential divider is very much larger than the design point value of $I_B = 10\ \mu$A. Values of R_1 and R_2 of 80,000 ohms and 40,000 ohms respectively will fulfil this condition. Since the potential of the base is now fixed at -5 volts, and the base to emitter potential is always small, the emitter is also fixed at -5 volts, approximately. From the previously selected value of 1 mA for the collector current we can now deduce the value for the emitter resistor R_e. Since $I_C = \alpha I_E + I_{CO}$ it is evident that I_E is practically identical with I_C for the usual values of α and I_{CO} and is thus 1 mA in the present case, giving a value for R_e of $5/(1 \times 10^{-3})$ or 5000 ohms. Before discussing the effect of this arrangement, let us fix R_c. If we stick to our previous value of $7{\cdot}5$ volts for V_{CE} we shall be left with $2{\cdot}5$ volts across R_c, which at a current of 1 mA implies a value of 2500 ohms for this resistor. It might be better if we changed our operating conditions slightly to allow a little more voltage across R_c, because with a value of $2{\cdot}5$ volts, it is clear that we cannot have an output swing of more than this value in one direction before the transistor collector

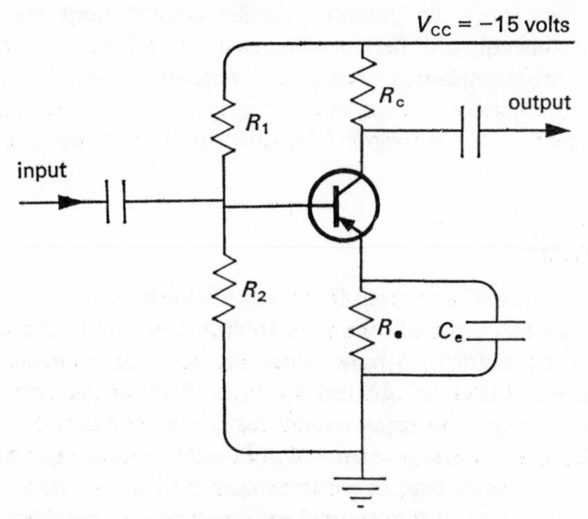

Figure 50. Biasing circuit for greater stability.

reaches the supply voltage. If we arrange to have only 5 volts across the transistor – still a perfectly acceptable working point – we shall have also 5 volts across R_c, implying a value of 5000 ohms for the resistor. The design is now complete. I_E is now essentially fixed, because it is determined by the –5 volt potential initially defined by the potential divider, and by the fixed emitter resistor R_e. (The only role of the large capacitor C_e is the usual one of ensuring that the emitter is still grounded from the point of view of signals, a matter which does not concern us at the moment.)

The value of this arrangement can be seen by reference to equation **4.4** (rather than to equation **4.5** as previously). Since $I_C = \alpha I_E + I_{CO}$ and I_E is now fixed, $dI_C/dI_{CO} = 1$, which implies that the collector current is less dependent on changes in I_{CO} with temperature by the large factor $\beta + 1$, compared with the previous case. Furthermore the variation from the designed value of I_C with different transistors is similarly reduced. Suppose we have two transistors with values of α differing by one per cent – say $\alpha = 0{\cdot}99$ and $\alpha = 0{\cdot}98$ respectively. Since $\beta = \alpha/(1 - \alpha)$, the corresponding values of β will be 100 and 50. With the present arrangement (and assuming for simplicity that I_{CO} is negligible) $I_C = \alpha I_E$, and for a fixed value of I_E, as we have, I_C differs also by only 1 per cent for the two transistors. Comparing this with the previous arrangement, where the relevant equation is $I_C = \beta I_B$ we note that for a fixed value of I_B, as the circuit provides, the value of I_C for one transistor will be half that of the other, and a circuit with quite different characteristics may be obtained.

Two final remarks should be made about the circuit of Figure 50. Firstly the choice of –5 volts for the potential of the base was fairly arbitrary, but it represents a reasonable compromise between too small a value, which would have been unable to 'swallow up' possible variations in V_{BE} with temperature, and too large a value which would have left insufficient voltages across the transistor and R_c. Secondly, the potential divider cannot hold the base potential absolutely constant, and therefore the result of unity for dI_C/dI_{CO} is clearly optimistic. A more sophisticated analysis would show that in fact this factor would be approximately four for the component values used, which is still a big improvement on the corresponding value of 100 for the circuit of Figure 49.

4·4 The *T*-equivalent circuit

Now that the static operating conditions for the transistor have been fixed we must look at its performance when a signal is applied to it and in particular devise an equivalent circuit for it. A basic statement to be incorporated into any model we construct is that of equation **4.4** or, in its alternate form, that of equation **4.5**. The latter is a statement about steady d.c. currents, but if the base current I_B undergoes a change of size δI_B produced by some input signal, clearly, this is related to the change in collector current δI_C by the equation $\delta I_C = \beta \delta I_B$, since I_{CO} is constant. In our previous notation this can be written simply as $i_c = \beta i_b$. But the equivalent circuit must also include a description of the input side of the transistor, and the actual circuit we shall employ is shown in Figure 51. This representation is referred to as the *T*-equivalent circuit, because of its obvious resemblance to the letter *T* (turned on its side, as we have drawn it here). It contains, first of all, a current generator of size βi_b in the collector lead, to satisfy the relation $i_c = \beta i_b$. The resistor r_e represents the impedance of the 'on' emitter–base diode, while r_b represents the ohmic resistance of the base of the transistor – which is manufactured from quite feebly doped material – between the base lead and the actual 'works' of the transistor. A representative value for r_b might be 500 ohms. According to simple diode theory the value of r_e should be $25/I_E$ or 25 ohms for a current of 1 mA, and this order of magnitude is worth remembering. Since r_e is strongly dependent on I_E, r_e cannot be assumed to be even approximately constant unless we restrict current changes, and therefore the magnitude of the signals,

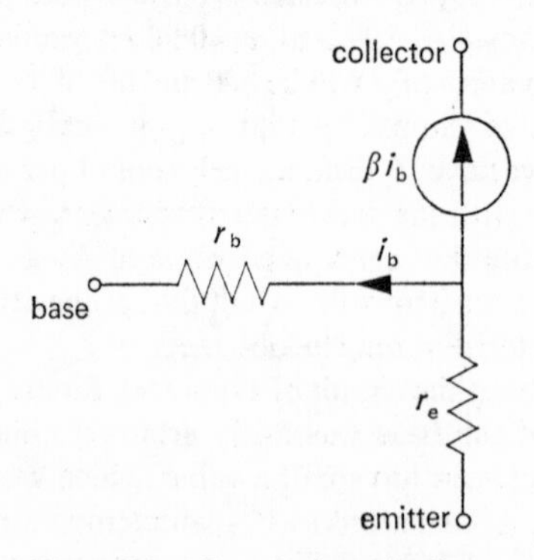

Figure 51. *T*-equivalent circuit.

to a very small value – a point to which we shall return later. We shall take as before the value of β to be typically 100. We have not, as usual, in our equivalent circuits, included any of the components of Figure 50 such as R_1, R_2, R_e and C_e which are concerned with setting the d.c. conditions, although we shall refer to R_1 and R_2 later as also having some part in determining the input impedance.

The model for the transistor we have just discussed is an extremely simple one, and differs appreciably, even in notation, from more complicated *T*-equivalent circuits. It will however be quite adequate for our purposes, although we shall now mention briefly how this model could be made more sophisticated if necessary. Firstly, we have neglected the effect of the collector voltage on the collector current: allowance could be made for this small effect by placing a large resistance in parallel with the current generator, thus acknowledging its imperfect character. This resistance would typically be around 50,000 ohms, so it is not an unreasonable approximation to neglect it in comparison with a collector resistor of some thousands of ohms effectively in parallel with it. A more sophisticated analysis of transistor action shows that the collector voltage also has a small influence on the emitter–base diode; this can be allowed for by including in the equivalent circuit a small voltage source – or, what turns out to be the same thing, by altering the values of r_b and r_e from those derived from simple ideas. The general order of magnitude of these quantities will not be changed however, which is the most important point.

4·5 The common-emitter amplifier

The *T*-equivalent circuit can be used to calculate the characteristics of the common-emitter, or grounded-emitter amplifier, where the emitter is to ground or common, the input is between base and emitter, and the output taken between collector and emitter – that is, the circuit of Figure 50 (or 49). The *T*-equivalent circuit of this is shown in Figure 52 with the collector resistor R_c in place; again we are not interested in the collector d.c. supply, so the bottom end of R_c goes directly to ground. The directions of the arrows indicating i_b and βi_b have been reversed compared with Figure 51. In Figure 51, the arrow for i_b was in the same direction as that for I_B for the *p-n-p* transistor with which we were concerned. As i_b represents the change in I_B, this can be taken to imply that such a direction for the i_b arrow corresponds to an increase in I_B. An increase in I_B must have meant that an additional negative voltage was applied to the base since, for a *p-n-p* transistor, the base is held at a negative d.c. potential with respect to the emitter (to facilitate the flow of holes to the base contact). The direction of the i_b arrow in Figure 51 thus corresponds to a negative signal voltage on the base. The application of a positive signal at the input, as in Figure 52, is generally taken as the standard case, so in this case I_B will be reduced, as shown by the arrow i_b now pointing inwards. I_B itself does not change direction, but merely suffers a small reduction in its magnitude. Note that the relation between i_b and the signal voltage is the correct one, with i_b

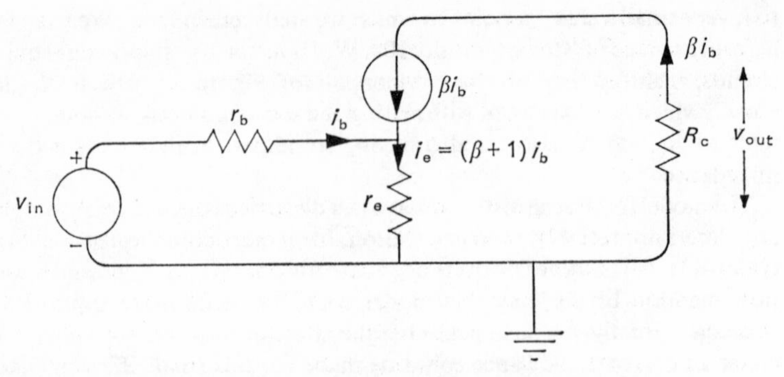

Figure 52. Equivalent circuit for common-emitter amplifier stage.

flowing away from the positive terminal. The current generator βi_b must now also be reversed in direction, as must i_e, the emitter current which must be equal to the sum of the other two currents. The circuit of Figure 52 and the direction of the arrows, although derived for the *p-n-p* case, are equally valid for an *n-p-n* transistor. In that case the base current I_B would have been due to a flow of electrons out from the transistor, that is to a flow of conventional current into the transistor: therefore I_B would point inwards. We have seen that a positive d.c. voltage on the base is required to maintain I_B, so a small added positive signal voltage would further increase I_B. Such an increase is indicated by an arrow i_b in the same direction as I_B, that is, inwards, which is in the same sense as for the *p-n-p* case. So although the direction of the steady d.c. currents is different for the two types of transistor, the same equivalent circuit will represent both from the point of view of signals.

The relation which determines the behaviour of this circuit is obtained by considering the voltages around the loop formed by the input signal source, r_b and r_e. Hence

$$v_{in} = i_b r_b + (\beta + 1)\, i_b r_e \qquad \textbf{4.6}$$

To obtain the voltage gain of the stage this is combined with the relation

$$v_{out} = -\beta i_b R_c \qquad \textbf{4.7}$$

This latter follows simply from the fact that the current from the current generator βi_b flows through R_c, the minus sign occurring because the flow is from the grounded end of the resistor. Eliminating i_b between equations **4.6** and **4.7** and solving for the voltage gain v_{out}/v_{in} we obtain

$$\frac{v_{out}}{v_{in}} = -\frac{\beta R_c}{r_b + (\beta + 1)\, r_e} \qquad \textbf{4.8}$$

$$= -A_v \text{ say,}$$

the minus sign denoting, as in the FET case, polarity reversal at the output. To bring this formally into line with the result for voltage-controlled devices, we may write

$$A_v = g_m R_c \tag{4.9}$$

where g_m, the transconductance for the transistor is given by

$$g_m = \frac{\beta}{r_b + (\beta + 1) r_e} \tag{4.10}$$

To obtain an even simpler formula we note, from the values previously given as typical for r_b, r_e and β, that it is not too bad an approximation to neglect the r_b term compared with $(\beta + 1) r_e$, and then to take $\beta + 1 \approx \beta$ to obtain finally

$$g_m = \frac{1}{r_e} \tag{4.11}$$

Not only is this a simple relation which will often be good enough for our calculations, but also as from our simple ideas $r_e = kT/eI$ (equation **1.3**) and k, T, and e do not depend on the transistor in question, we can deduce that all transistors, at least to the fairly rough approximation we are using, have the same value of g_m for a given value of the current. In addition, as r_e is numerically 25 ohms at 1 mA, the corresponding value of g_m is 40×10^{-3} ohms^{-1}, or 40 millimhos which is a value very much larger than the corresponding one for a typical FET, or even for a typical thermionic valve. In principle therefore, very high voltage gains can be obtained with the common-emitter stage, typically 200 for $R_c = 5000$ ohms.

We shall also be interested in the current gain of the arrangement, that is the ratio i_{out}/i_{in}. The current through R_c, i_{out}, is βi_b, and i_{in} the current drawn from the signal source is i_b, so the current gain A_i for the common-emitter stage is β, the same as that of the transistor itself. (This rather obvious result will not be true for cases where the large resistance, which more sophisticated equivalent circuits place across the current generator, cannot be neglected. But as in the case of voltage-controlled devices good high-frequency response requires low values of R_c, and consequently all but a negligible fraction of the current βi_b will in fact flow through R_c.)

We next calculate the input and output impedances. The input impedance, or the impedance felt by the source of signals is just v_{in}/i_{in} or v_{in}/i_b. From equation **4.6** this is $r_b + (\beta + 1) r_e$ (or approximately βr_e), which for values of the quantities previously indicated works out at a few thousand ohms. This relatively low input impedance contrasts markedly with the very large values obtained for voltage-controlled devices; indeed the transistor could be thought of as a device with a near-zero input impedance, compared with the near-infinite value for the FET. (We shall return to this problem after calculating the output impedance.) In determining the input impedance of this stage we cannot neglect the resistors R_1 and R_2 of Figure 50, which for reasons explained earlier are also relatively low. R_2 is connected between base and ground, as is R_1 effectively since the top end of this resistor goes to the supply voltage, and thus to ground from the point of view of signals. The load thrown on the input source by the potential divider is thus equal to R_1 and R_2 in parallel, which for the values of 80,000 and 40,000 ohms previously decided means an overall resistance of 27,000 ohms. This is not entirely negligible compared with the value of a few thousand ohms for the input impedance of the transistor itself, and must thus be included in parallel with it.

The output impedance of the stage is, as discussed in section 2·6, the open-circuit voltage divided by the short-circuit current. Equation **4.8** gives the expression for the open-circuit voltage, but the form in equation **4.7** will be more appropriate for our present purpose. This is, apart from the minus sign, $\beta i_b R_c$. The short-circuit current is found by putting a short-circuit across the output, which is equivalent to making R_c zero. However, because of the perfect nature of the current source in our simple picture, the output current βi_b remains unchanged, and this is the short-circuit current. The output impedance is thus simply R_c, the value of the load resistor.

The results for the input impedance, current and voltage gain enable the action of the common-emitter amplifier stage to be simply visualized. A voltage v_{in} is applied to the device, and the current produced in the input impedance $r_b + r_e(\beta + 1)$ is $v_{in}/\{r_b + r_e(\beta + 1)\}$. The transistor action multiplies this current by β to give an output current of $\beta v_{in}/\{r_b + r_e(\beta + 1)\}$ which, flowing through the resistor R_c, gives an output voltage of $-R_c \beta v_{in}/\{r_b + r_e(\beta + 1)\}$ in accordance with equation **4.8**.

We now consider some implications of the low input impedance of the transistor in the common-emitter mode. The first is that the impedance of the source of signals must be taken into account, as even if this is small, it may still be comparable with, or even larger than, the low value of the transistor input impedance. We can see how this would affect our results by imagining that in Figure 52 the voltage generator v_{in} had in series with it an internal resistance of size r_s. This is also directly in series with r_b, so the formulae we previously derived will still be valid if r_b is replaced by $r_s + r_b$. The voltage gain, for example, will now be $-\beta R_c/\{r_s + r_b + (\beta + 1)r_e\}$ which is less than the previous value. But there are compensations. r_e is not a constant, but depends strongly on the current through the transistor; therefore the voltage amplification is a function of the size of the input signal – a form of distortion that restricts us to very small output signals. The larger r_s is the more important it becomes in determining the voltage gain, and the less the dependence on r_e. Indeed pushing this argument to the limit, if r_s is so large (say 10,000 ohms) that it is much greater than $(\beta + 1)r_e$ and of course than r_b, the expression for the voltage gain is then $-\beta R_c/r_s$ approximately, which is independent of r_e and thus of the sort of distortion we previously mentioned, while at the same time it is not unreasonably small. For $R_c = 5000$ ohms, the voltage gain would be approximately 50. Thus there are advantages in distortion free amplification to be gained from feeding the transistor from a source with an internal resistance of size much larger than its own input impedance (that is, from something which is a constant current source from the transistor's point of view).

Even if the source has a negligibly low internal resistance, there is another way to get over the variation of r_e with current, and that is by adding in series with r_e a comparatively large physical resistor – say $10r_e$ or 250 ohms. This resistor will 'swamp' the variations in the much smaller resistance r_e. The term $(\beta+1)R_e'$, where R_e' is this added resistor, will now be dominant in the denominator of the expression for the voltage gain, and the value for the gain will be approximately $-R_c/R_e'$, which would still leave the reasonable value of 20, for $R_c = 5000$ ohms. The device of adding a resistor in series with r_e is an example of 'current feedback'. It can be achieved very simply in practice for the circuit of Figure 50 by leaving say 250 ohms of the total of R_e undecoupled by the capacitor C_e, that is by having in series 250 ohms and $R_e - 250$ ohms, with the capacitor placed across the latter resistor. The signal then 'sees' only $R_e' = 250$ ohms as required.

A final example of the effect of the low input impedance of the common-emitter stage is shown by taking two (or more) stages in cascade (Figure 53).

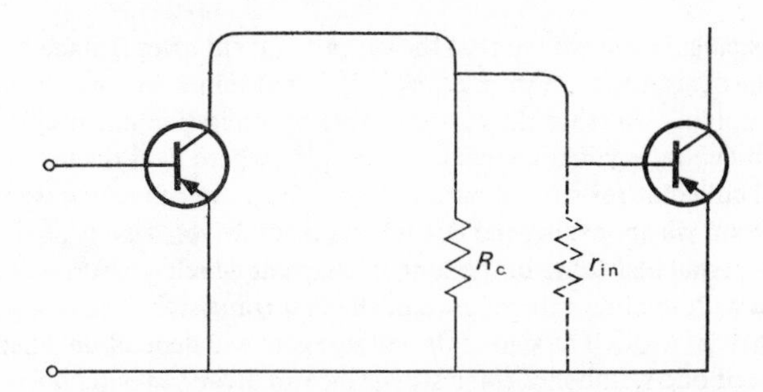

Figure 53. Common-emitter stages in cascade.

Here as usual we have not shown components associated solely with fixing d.c. levels. In parallel with the collector resistor R_c of the left-hand transistor, there is, as expected, the input impedance r_{in} of the second transistor. This is $r_b + (\beta+1)r_e$ or a few thousand ohms. (Here we assume that we do not add any external resistance R_e' in series with r_e.) Thus the input impedance is comparable with a typical value for R_c; a sizeable fraction of the signal current formerly passing through R_c is diverted through r_{in} and the value of R_c may be effectively halved. (Remember that we are dealing with *signal* currents; the d.c. current in the left-hand transistor is blocked off from the input of the other transistor by the interstage coupling capacitor, not shown in this diagram.) It is not inconceivable, with a higher value of R_c, and a lower value of r_{in} (obtained for example if r_e were made much smaller by increasing the d.c. current through the transistor in question), that most of the signal current in the left-hand transistor would by-pass R_c and flow through r_{in}. If in such circumstances we have a series of cascaded common-emitter stages (see Figure 54), and the input current to the first is i_{in}, its output current βi_{in} will flow almost entirely in the input circuit of the second stage. The output current of this stage will be $\beta(\beta i_{in})$ or $\beta^2 i_{in}$, and will flow almost entirely in the input circuit of the third stage.

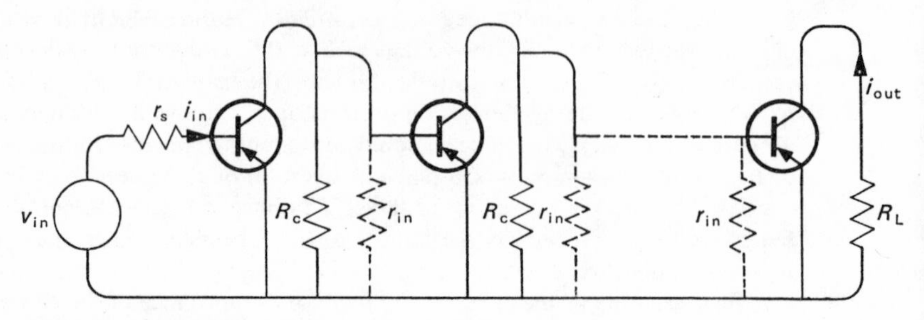

Figure 54. Multistage common-emitter amplifier.

Continuing this calculation we see that the output current after n stages, i_{out}, will be $\beta^n i_{in}$. The overall current gain, $i_{out}/i_{in} = \beta^n$, is obtained in these circumstances by multiplying together the current gains of the individual stages. To obtain the overall voltage gain note that $v_{out} = i_{out} R_L$, where R_L is the final load (that is the final collector resistor, in parallel, from the point of view of signals, with whatever load we are asking the last transistor to drive); also $v_{in} = i_{in} r_s$, where r_s is the internal resistance of the source of signals which is assumed to be large compared with the input impedance of the first transistor. Thus v_{out}/v_{in}, the voltage gain, is $\beta^n R_L/r_s$. The sign of the voltage gain will depend on whether we have an even or odd number of transistors, since an inversion will take place at each stage.

4·6 High-frequency characteristics of the common-emitter amplifier

Some modifications must be made to the previous treatment of the common-emitter stage for very high and very low frequencies; we shall start with the former. Clearly, as with voltage-controlled devices the capacities of the base–collector and base–emitter diodes, which we have labelled C_{bc} and C_{bem} in Figure 55, must be taken into account. They would be typically ten picofarads (although we should expect the collector capacity to be the smaller, since with a back-biased diode, the capacitor 'plates' would be farther apart). In parallel with C_{bem} there is another capacity labelled C_d, the emitter-diffusion capacity, which arises because of the nature of the transfer of charge across the base of the transistor. This takes place by diffusion, not, as in a thermionic valve, under the action of an electric field. For diffusion to take place there must be a non-uniform distribution of carriers, with the current flowing from the more dense

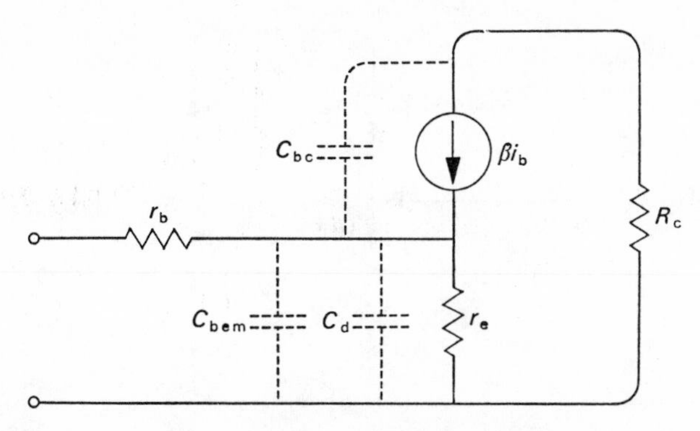

Figure 55. Stray capacities in the junction transistor.

to the less dense region. It can be shown that the density of minority carriers in the base of a transistor falls steadily from its initial value at the emitter junction (which depends on the current flowing) to zero at the collector, where the carriers are swept away. Before changing the emitter current of a transistor by a change in applied voltage, we must change the charge distribution in the base, and this cannot be done instantaneously because of the finite transit time of the carriers across this region. The effect is analogous to the relation between voltage and charge on a capacitor, and hence can be represented by a 'diffusion capacity' C_d. It may typically be of the order of 100 picofarads, so the total emitter to base capacity, C_{be}, will be made up largely of diffusion capacity. Since the emitter capacity C_{be} is so much larger than C_{bc}, the former might be expected to be the controlling factor in the high-frequency performance of the transistor. This is not entirely so, since the capacity C_{be} though large, is shunted by the small resistance r_e, whereas the capacity C_{bc} is shunted by the larger resistance R_c. This is equivalent to saying that the capacity C_{bc} is effectively multiplied by the factor R_c/r_e – which turns out to be none other than the Miller effect, met in voltage-controlled devices.

Initially, C_{bc} will be ignored; we shall see later the alterations necessary when it is re-introduced. Figure 56 shows the situation with a current i_1 flowing through the impedance Z_e of the capacity C_{be}. This current enters the transistor through the base lead, together with the conventional base current i_b. The current gain i_{out}/i_{in} is given by

$$A_i = \frac{i_{out}}{i_{in}} = \frac{\beta i_b}{i_b + i_1} \qquad 4.12$$

Further, since r_e and Z_e are in parallel, the voltage across them must be the same

$$(\beta + 1)\, i_b r_e = i_1 Z_e \qquad 4.13$$

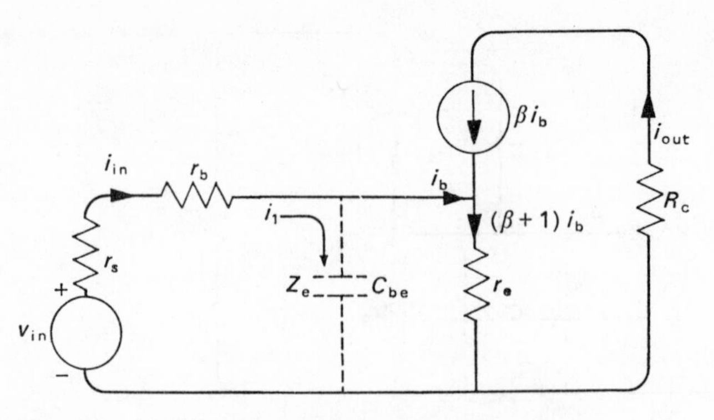

Figure 56. High-frequency equivalent circuit, ignoring C_{bc}

Eliminating i_b between equations **4.12** and **4.13** and solving for i_{out}/i_{in} we obtain

$$A_i = \frac{\beta}{1 + \dfrac{(\beta+1)\,r_e}{Z_e}} \qquad \textbf{4.14}$$

or putting in the value for $Z_e = \dfrac{1}{j\omega C_{be}}$

$$A_i = \frac{\beta}{1 + j\omega(\beta+1)\,r_e\,C_{be}} \qquad \textbf{4.14}$$

Thus, for moderate frequencies where the second term in the denominator is small, the expression for the current gain reduces to β, the value previously found: as ω is increased the magnitude of A_i drops, until, when $\omega = 1/[(\beta+1)r_e C_{be}]$, it has dropped to $1/\sqrt{2}$ of its midband value. We refer to this value of ω as ω_β: thus $\omega_\beta = 1/[(\beta+1)r_e C_{be}] \approx 1/(\beta r_e C_{be})$. We shall see that although we have met ω_β in a rather restricted context (where we neglected C_{bc}) it is nonetheless an important parameter in specifying the performance of the transistor at high frequencies. At frequencies very much greater than that corresponding to ω_β, the second term in the denominator of the right-hand side of equation **4.14** becomes dominant and we have

$$A_i \approx \frac{\beta}{j\omega(\beta+1)\,r_e\,C_{be}} \approx \frac{1}{j\omega r_e\,C_{be}}$$

or, taking the magnitude of A_i and using the frequency f instead of ω,

$$|A_i|\,f = \frac{1}{2\pi r_e\,C_{be}} \qquad \textbf{4.15}$$

Equation **4.15** gives the 'gain–bandwidth product' and in particular it tells us that when the frequency is $1/(2\pi r_e C_{be})$, the magnitude of the gain has dropped to unity, which clearly represents the end of the useful performance of the transistor. This value of f is usually written as f_T and although it has been introduced here in restricted circumstances, it is an important (and commonly used) parameter for specifying the ultimate performance of the transistor at high frequencies.

The effect of the collector to base capacity C_{bc} must now be taken into account. Figure 57 shows the situation, with an additional current i_2 flowing in the impedance Z_c of the collector to base capacity, and entering as did i_1, through the base lead. Now the value of the current gain, A_i is given by

$$A_i = \frac{i_{out}}{i_{in}} = \frac{\beta i_b - i_2}{i_b + i_1 + i_2} \qquad \textbf{4.16}$$

where the numerator of the expression is obtained by considering the currents arriving at the point E. As before, since r_e and Z_e are in parallel we have

$$(\beta + 1)\, i_b\, r_e = i_1\, Z_e \qquad \textbf{4.17}$$

There is no simple equation for Z_c corresponding to equation **4.17**. Instead we must consider voltages round the loop ABDEFGA giving

$$-(\beta + 1)\, i_b\, r_e + i_2\, Z_c - i_{out}\, R_c = 0$$

or substituting the value for i_{out} of $\beta i_b - i_2$

$$-(\beta + 1)\, i_b\, r_e + i_2\, Z_c - (\beta i_b - i_2)\, R_c = 0 \qquad \textbf{4.18}$$

From equation **4.18**, i_2 can be expressed in terms of i_b, and from equation **4.17**, i_1 can be expressed in terms of i_b: substituting in equation **4.16** gives, after multiplying above and below by $R_c + Z_c$,

$$A_i = \frac{\beta Z_c - (\beta + 1)\, r_e}{Z_c + (\beta + 1)\left\{\dfrac{(Z_c + R_c)\, r_e}{Z_e} + R_c + r_e\right\}}$$

or dividing above and below by Z_c

$$A_i = \frac{\beta - \dfrac{(\beta + 1)\, r_e}{Z_c}}{1 + (\beta + 1)\left\{\left(1 + \dfrac{R_c}{Z_c}\right)\dfrac{r_e}{Z_e} + \dfrac{R_c}{Z_c} + \dfrac{r_e}{Z_c}\right\}} \qquad \textbf{4.19}$$

This looks complicated, but unless the frequencies involved are extremely high, Z_c is a very large quantity, so the terms in r_e/Z_c can be entirely neglected. The term R_c/Z_c is a small quantity which can be neglected with respect to unity in the term $1 + R_c/Z_c$, but not where it stands by itself further on. Thus the expression simplifies to

$$A_i \approx \frac{\beta}{1 + (\beta + 1)\left(\frac{r_e}{Z_e} + \frac{R_c}{Z_c}\right)}$$

or putting in the values for Z_e and Z_c we have finally

$$A_i = \frac{\beta}{1 + j\omega(\beta + 1)(r_e C_{be} + R_c C_{bc})} \qquad \textbf{4.20}$$

Thus the product $r_e C_{be}$ again appears with a new term $R_c C_{be}$. In fact it can now be seen that our previous treatment leading to equation **4.14**, and our related definitions of ω_β and f_T can be derived from equation **4.20** by putting $R_c = 0$. So ω_β and f_T are controlling parameters for the frequency dependence of current gain when the collector load is zero, or at least negligibly small. The value corresponding to ω_β in the present case where the term $R_c C_{bc}$ is not negligible is obviously

$$\omega = \frac{1}{(\beta + 1)(r_e C_{be} + R_c C_{bc})} \qquad \textbf{4.21}$$

which is smaller than ω_β, and means that the presence of an appreciable load R_c has reduced the high-frequency performance.

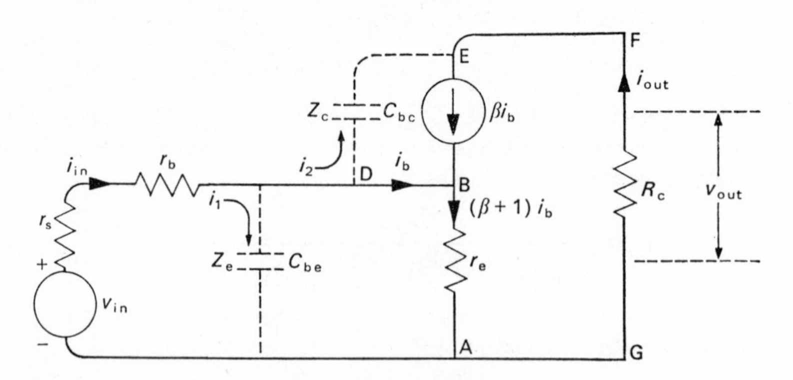

Figure 57. Complete high-frequency equivalent circuit.

What about the voltage gain? To find this the input voltage v_{in} (see Figure 57) must first be related to the input current i_{in} by the relation

$$v_{in} = i_{in}(r_s + r_b) + (\beta + 1)\, i_b\, r_e \qquad \textbf{4.22}$$

where r_s is the resistance of the source of signals. Using equation **4.22** with the previous equations **4.17** and **4.18**, we may, in principle, obtain the voltage gain

$v_{out}/v_{in} = -i_{out} R_c/v_{in}$. The answer is obviously simple if r_s is large, because then the other terms on the right hand side of equation **4.22** can be neglected and v_{in} is simply $i_{in} r_s$. The voltage gain in this case is R_c/r_s times the current gain and thus depends on frequency in the same way.

If the signals come from a voltage source, that is with $r_s = 0$, the full form of equation **4.22** must be used. In this case the working is straightforward but tedious, so we shall just quote the result, obtained after making the sort of approximations used in connexion with equation **4.19**.

$$\text{voltage gain} = \frac{-\beta R_c}{r_b + (\beta+1) r_e + j\omega(\beta+1) r_b(r_e C_{be} + R_c C_{bc})}$$

or, dividing above and below by $r_b + (\beta+1) r_e$

$$\text{voltage gain} = -\frac{\dfrac{\beta R_c}{r_b + (\beta+1) r_e}}{1 + \dfrac{j\omega(\beta+1) r_b(r_e C_{be} + R_c C_{bc})}{r_b + (\beta+1) r_e}} \qquad 4.23$$

The numerator of equation **4.23** is the midband voltage gain obtained in equation **4.8** and which can here be called A_{v0}. Thus

$$\text{voltage gain} = \frac{-A_{v0}}{1 + \dfrac{j\omega(\beta+1) r_b(r_e C_{be} + R_c C_{bc})}{r_b + (\beta+1) r_e}}$$

This reduces, as we expect, to $-A_{v0}$ for small values of ω. The magnitude of the voltage gain drops to $1/\sqrt{2}$ of its midband value for ω given by

$$\omega = \frac{r_b + (\beta+1) r_e}{(\beta+1) r_b(r_e C_{be} + R_c C_{bc})} \qquad 4.24$$

It is not difficult to see that this represents an improvement in frequency response over the case when r_s is large (equation **4.21**). Equation **4.24** also shows that the requirements for good high-frequency response for a common-emitter stage driven from a voltage source are a low value of R_c and a transistor with low values of r_b, C_{be} and C_{bc}. Unfortunately the requirement of low r_b conflicts to some extent with the requirement of a thin base region to minimize transit time effects. In practice transistors with values of f_T in the kilomegahertz region are commercially available.

To see most clearly how the Miller effect appears for the transistor common-emitter amplifier, it will be convenient to use in equation **4.23** the approximation previously mentioned of neglecting r_b in the expression $r_b + (\beta+1) r_e$.

With the additional simplification of assuming $\beta/(\beta+1) \approx 1$ this equation becomes

$$\text{voltage gain} = \frac{-\dfrac{R_c}{r_e}}{1 + j\omega r_b \left\{ C_{be} + \dfrac{R_c}{r_e} C_{bc} \right\}} \qquad \mathbf{4.25}$$

R_c/r_e is the midband voltage gain, with the approximation we are using, and thus equation **4.25** can be written as

$$\text{voltage gain} = -\frac{A_{v0}}{1 + j\omega r_b (C_{be} + A_{v0}\, C_{bc})} \qquad \mathbf{4.26}$$

Equation **4.26** illustrates the Miller effect multiplication of the base–collector capacity C_{bc} by the factor A_{v0}, and underlines the need to keep this capacity small for good performance. The role of r_b as a series resistance even in the absence of any signal source resistance r_s is also evident.

So far we have been talking in terms of frequency response but we expect from our previous discussion in connexion with voltage-controlled devices, that the factors discussed should also control the response of the transistor to voltage steps and pulses. While this is generally true, the approach to the problem of rapidly switching the transistor on or off is often rather different. For a discussion of this see, for example, reference 1. One final remark should be made. In our simple model of a transistor, minority carriers diffused across a uniform, field free base region, and were swept away when they reached the base-collector junction by the field there. If however we make the base non-uniform, that is with a drop in the doping as we go from emitter to collector, then we can show (see reference 2) that such an impurity gradient creates a built-in electric field which helps the carriers on their way. (We could naively think of this concentration change as a series of mini junctions right across the base.) Modern transistors using such graded base techniques will clearly have shorter carrier transit times across the base, and thus better step and high frequency responses.

4·7 Low-frequency characteristics of the common-emitter amplifier

The same general conditions obtain for the low-frequency response of the common-emitter amplifier stage (and what is equivalent, the reproduction of the flat top of a pulse) as for voltage-controlled devices. There the controlling factor was the product $R_g C_c$, where R_g was the gate resistor, and C_c the coupling capacitor. In the case of a transistor stage R_g will be replaced by R_1 and R_2 of Figure 50, which are effectively in parallel, together with r_{in} which is also in parallel with R_1 and R_2 (see Figure 58(a)). Since r_{in} $(=r_b + (\beta+1)r_e)$ is normally of the order of a few thousand ohms and much less than the parallel combination of R_1 and R_2, it is the product $(r_s + r_{in})\,C_c$ which will control the low-frequency response in this case, as can be seen from the equivalent circuit of

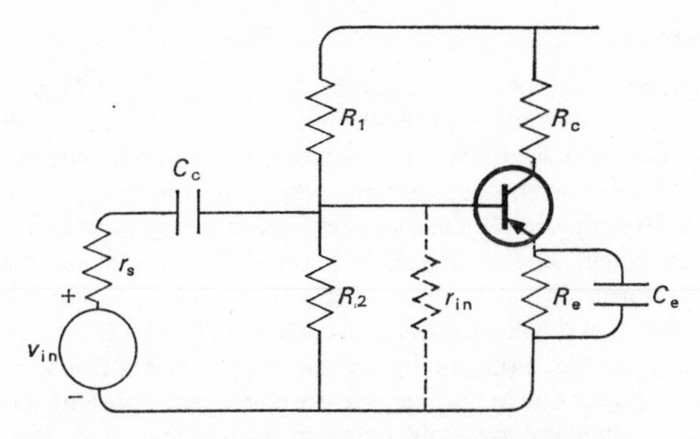

Figure 58. (a) Circuit for low-frequency calculations.

Figure 58(b). For the case of a voltage source with $r_s = 0$ we are left with the quantity $r_{in} C_c$. Because of the very small value of r_{in} compared with R_g for the f.e.t. case, the value of C_c must be correspondingly larger to provide the same response. (In all the above it has been assumed that the capacitor C_e, in Figure 58(a), is sufficiently large to effectively short circuit R_e from the point of view of signals; otherwise the time constant $R_e C_e$ must also be taken into account.)

It is often possible to connect transistors together without a coupling capacitor at all. This arises for a number of reasons, for example, because the base of a transistor is biased with the same polarity of voltage as the collector, and because there exist both *p-n-p* and *n-p-n* transistors with complementary voltage requirements. Examples of this type of direct coupling will be found in circuits later in this book. While such direct coupling may be advantageous in removing the low-frequency drop due to the coupling capacitor, and also extending the response right down to d.c., it has some disadvantages. For example, residual drifts in d.c. level due say to temperature changes, will be transmitted right through a direct-coupled multistage amplifier with consequent amplification and a large effect on the operating conditions of later stages. Some additional form of corrective 'feedback' network must thus be added.

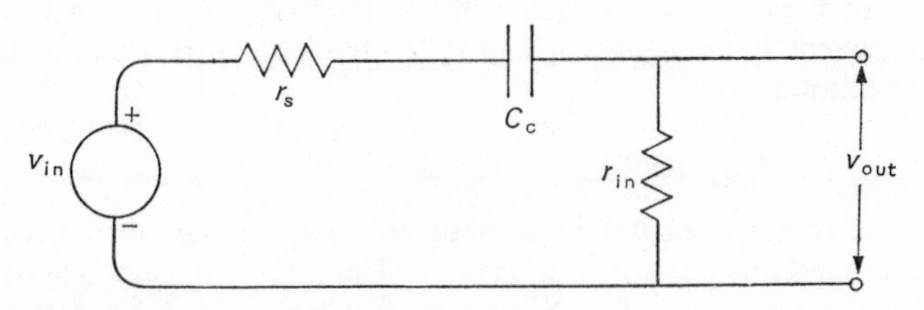

Figure 58. (b) Low-frequency equivalent circuit.

4·8 The *h* parameters

The representation of the transistor used so far involving r_b, r_e and the current generator βi_b, has the advantage, particularly for an introductory treatment, that it allows an easy visualization of transistor action. However the quantities r_b and r_e are not easy to determine experimentally as it is impossible to 'get at' the central point (B in Figure 57). On the other hand it is easy to measure input and output impedances, and it is from this type of measurement that the behaviour of the transistor may be alternatively characterized. These relations between input current and voltage, or output current and voltage, can be represented graphically, by equations, or by an equivalent circuit. This approach differs from the previous one in that here we are concerned with how the transistor reacts to applied voltages and currents, rather than with the basic processes occurring inside it. For this reason it is often known as the 'black-box' approach – we do not care what exactly is happening inside the 'box'. We have in fact used this sort of approach to the field effect transistor. In the present case our description will be rather more complicated as there are four basic quantities with which we must concern ourselves – the input voltage and input current, and the output voltage and output current. In the case of the FET, one of these quantities, the input current can be neglected. Here we need two equations, or two families of curves such as those we have already met in Figures 48(a) and 48(b).

We must first decide which to choose as independent and which as dependent variables. This choice is of course, to a large extent, arbitrary. In the case of voltage-controlled devices the input voltage (that is, the gate–emitter voltage of an FET) and the output voltage (that is, the drain-emitter voltage) were taken as the independent variables, and we may do this again. However it is more usual to replace the input voltage with the input current (that is, the base current for the common-emitter stage we were discussing); this is a reasonable change as the basic transistor operation has already been shown to be that of current amplification. We shall retain the output voltage as the other parameter; in the common-emitter configuration this is the collector voltage. This choice of independent variables is the one used in Figures 48(a) and 48(b), to which we now refer, although in the latter I_B, the independent variable, was plotted along the vertical axis to show the relationship between the curves there and that for a simple diode.

Looking at Figure 48(a) first, we see that, in an analogous manner to our treatment of Figure 8, we can relate the small change i_c occurring in the collector current I_C to changes i_b and v_{ce} in the base current and collector voltage as follows

$$i_c = h_{21e}\, i_b + h_{22e}\, v_{ce} \qquad \textbf{4.27}$$

where the *h*s represent partial differentials, tolerably constant over a reasonable area around the chosen operating point. The suffix $_e$ indicates that we are discussing the common-emitter arrangement. Clearly there are two other arrangements that have not concerned us so far: the common-collector and common-base configurations, and corresponding hs associated with them. The reasons for the other suffices on the hs will be clear in a moment. As in the treatment of Figure 48(a) we may write down from Figure 48(b) the relation

$$v_{be} = h_{11e} i_b + h_{12e} v_{ce} \qquad \textbf{4.28}$$

although, because of the strong curvature of the characteristics in this case, we expect these hs to be approximately constant only over a very restricted area near the operating point.

It is standard practice when using the h parameters to use the suffix $_1$ to indicate input voltage and current, and the suffix $_2$ to indicate output voltage and current. Rewriting equations **4.27** and **4.28** with this modification (and placing equation **4.28** first) we have

$$v_1 = h_{11e} i_1 + h_{12e} v_2 \qquad \textbf{4.29}$$

and

$$i_2 = h_{21e} i_1 + h_{22e} v_2 \qquad \textbf{4.30}$$

The matrix form of the suffices on the hs is now obvious.

Let us next look at the physical meaning and typical values for the common emitter hs. From equation **4.29** we see that if v_2 were zero then h_{11e} would equal v_1/i_1. As v_1, i_1 and v_2 all refer to changes in the related quantities, h_{11e} is evidently the change in input voltage divided by the change in input current, under conditions where there is no change in output voltage ($v_2 = 0$): that is h_{11e} is the input impedance when the output is shorted from the point of view of signals – for example, by placing a large capacitor across it. Because it is an input impedance, h_{11e} is often written as h_{ie}. Its value may be deduced from the inverse of the slope of the characteristic curve of Figure 48(b) at the operating point, and will have a value of a few thousand ohms typically. In a similar way h_{12e} equals v_1/v_2 with $i_1 = 0$, that is, the change in input voltage which must be made after a change in output voltage to keep the input current constant. h_{12e} is given by the spacing of the curves of Figure 48(b) at the operating point. It is extremely small, and will typically be in the 10^{-3} to 10^{-4} range. This h parameter is known as the 'reverse-voltage transfer ratio' ('reverse' because it is input over output) and is therefore alternatively written h_{re} (a dimensionless quantity). The parameter h_{21e} from equation **4.30** is i_2/i_1 with $v_2 = 0$, that is the ratio of output current to input current change while the output voltage is held constant. It is therefore referred to as the 'forward-current transfer ratio' and alternatively written h_{fe}, again a dimensionless quantity. We recognize it as the β of the T-equivalent circuit, and as such it typically has a value of 100. Its value in any particular case could be deduced from the spacing of the curves of Figure 48(a). Lastly we come to h_{22e}, which is i_2/v_2 with $i_1 = 0$, that is, the change in output current with output voltage, the input current being held constant. It is the slope at the operating point of the particular curve of Figure 48(a) in which we are interested, and clearly has dimensions of $(\text{resistance})^{-1}$. It is known as the output admittance – written alternatively h_{oe} – and from the slope of the curves is obviously a small quantity, typically in the range 10^{-4} to 10^{-5} ohms^{-1}. The use of the symbol h for these parameters is now clear. They are 'hybrid' parameters – one is a resistance, one the inverse of a resistance, and two dimensionless. It is worth emphasizing that the h parameters vary with operating point, some of them quite markedly, and the manufacturer will usually supply not only the values for the hs at a particular set of operating conditions, but also curves giving the variation of the hs with change in these conditions.

The h parameters, although the most popular, are not the only set of parameters which can be used to describe transistors and similar devices. There exist also the z parameters, which use i_1 and i_2 as the independent variables, and have the dimensions of resistance, and the y parameters which use v_1 and v_2 as the independent variables, and which have the dimensions of $(\text{resistance})^{-1}$. We have in fact met y parameters when discussing the field effect transistor: g_m for example was also written as y_{fs}. The advantage of h (and other parameters) is that we can formally work out properties of a circuit using these parameters, and it will be equally applicable to the common-emitter, common-collector or common-base configurations. Of course we must put in the parameter values appropriate to the configuration we are working with at the time. Indeed we can describe the properties, not only of single transistors, but of complete amplifiers, by giving the values of one of these sets of parameters. However we shall be most concerned with the h parameter system for the common-emitter configuration.

Since these various systems of parameters are simply different ways of describing the same device, relationships must exist between them, and also between them and the parameters of the T-equivalent circuit. We have already come across one of these, when we saw that h_{fe} is just β, and we quote one further case to show the sort of relationship that exists: $r_e = h_{12e}/h_{22e}$.

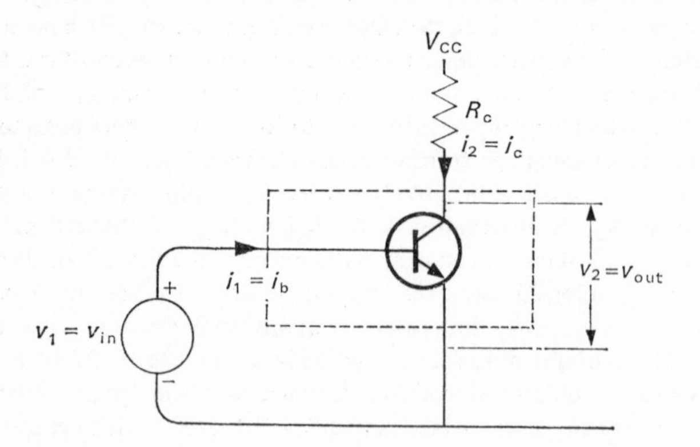

Figure 59. Common-emitter stage for h-parameter calculations.

4·9 The common-emitter amplifier using h parameters

We now attempt to obtain the characteristics – voltage and current gain, input and output impedance – for a common-emitter amplifier stage with a collector resistor R_c as shown in Figure 59, using the h parameters. We have shown the transistor enclosed in a box since we need not concern ourselves with the physical processes inside it. The equations for dealing with the circuit are firstly **4.29** and **4.30** which we repeat here for convenience in the alternative notation

$$v_1 = h_{ie}\, i_1 + h_{re}\, v_2 \quad \textbf{4.31}$$

and

$$i_2 = h_{fe}\, i_1 + h_{oe}\, v_2 \quad \textbf{4.32}$$

The third equation we require is that involving the resistor R_c

$$v_2 = -i_2\, R_c \quad \textbf{4.33}$$

This follows from the relation $V_2 = V_{CC} - I_2 R_c$ where V_2 and I_2 are the steady values of these quantities. As the supply voltage V_{CC} is constant we have for changes in V_2 and I_2, $dV_2 = -(dI_2) R_c$ which is equation **4.33** stated in another notation. Equations **4.31, 4.32,** and **4.33** can now be solved for v_2, i_2 and i_1 in terms of the input voltage v_1, and the various gains and impedances thus deduced. Because we shall soon apply the approximation of neglecting the small quantities h_{re} and h_{oe} the precise results will not be quoted in full, but one of the simpler ones will be noted

$$\text{input impedance} = h_{ie} - \frac{h_{re}\, h_{fe}\, R_c}{1 + h_{oe}\, R_c} \quad \textbf{4.34}$$

We now proceed to the approximation mentioned above, making the reasonable qualification that R_c should not be too large – say in the thousands of ohms range – as otherwise terms like $h_{oe} R_c$, which will occur in the exact results will not be negligible. This restriction on the value of R_c will nearly always be obeyed for other reasons. With the approximation in question, the input impedance, as given by equation **4.34** becomes simply h_{ie}, but in order to calculate the other quantities we must return to equations **4.31** and **4.32**, which now become

$$v_1 = h_{ie}\, i_1 \quad \textbf{4.35}$$

and

$$i_2 = h_{fe}\, i_1 \quad \textbf{4.36}$$

These together with equation **4.33** give the following relations

$$\text{current gain} = \frac{i_2}{i_1} = h_{fe}$$

$$\text{voltage gain} = \frac{v_2}{v_1} = -\frac{h_{fe}}{h_{ie}} R_c$$

$$\text{input impedance} = \frac{v_1}{i_1} = h_{ie} \quad \text{(compare equation 4.34)}$$

and

$$\text{output impedance} = \frac{\text{open-circuit voltage}}{\text{short-circuit current}}$$

$$= \frac{h_{fe}}{h_{ie}} \frac{R_c v_1}{i_2}$$

$$= R_c$$

the last step being obtained by expressing i_2 in terms of i_1 and thence in terms of v_1.

These results are very similar to those obtained with the T-equivalent circuit. The output impedance is naturally R_c again. The current gain is h_{fe} (or h_{21e}) which we already know to be the same as the β of our previous method. Comparing the expression for the input impedance here with the value obtained previously, another relation can be deduced between the h and T parameters (which is true within the general approximations to which we are working)

$$h_{ie} = r_b + (\beta + 1) r_e$$

4·10 The h-equivalent circuit

In the last section we have been deducing facts about the common-emitter amplifier stage from the h equations and the relation $v_2 = -i_2 R_c$, but just as with voltage-controlled devices, the same results can be obtained from an equivalent circuit reflecting these equations. Such a circuit is shown in Figure 60, where the resistor R_c is also shown (dashed to indicate that it is not an integral part of the equivalent circuit). On the input side there is a resistance of size h_{ie} in series with a voltage source $h_{re} v_2$, while on the output side there is a current source $h_{fe} i_1$ in parallel with a resistance of size $1/h_{oe}$. It is not hard to see that this circuit is in agreement with equations **4.31** and **4.32**. The total voltage in the input circuit is $v_1 - h_{re} v_2$. This drives a current i_1 through a resistance h_{ie} so we have

$$v_1 - h_{re} v_2 = i_1 h_{ie}$$

which rearranges to give equation **4.31**. On the output side, the current in the resistance of size $1/h_{oe}$ (which flows from emitter to collector) is the difference between the current produced by the generator and that flowing in the external circuit, or $h_{fe} i_1 - i_2$. Since, however, there is a voltage v_2 across this resistance the current through it is also given by $-v_2/(1/h_{oe}) = -v_2 h_{oe}$, the minus sign taking account of the direction of current flow as mentioned above. Equating these two values for the current we obtain

$$h_{fe} i_1 - i_2 = -v_2 h_{oe}$$

which becomes identical with equation **4.32** on rearrangement. Note also that equation **4.33** may be deduced from the equivalent circuit by looking at the voltage and current relations for R_c. The equivalent circuit thus fully represents the equations defining the action of the circuit, and can be used in their place. The equivalence of this circuit to our previous calculations could also have been shown by looking in turn at the definitions of the various h parameters themselves. For example h_{ie} ($=h_{11e}$) is the input impedance with $v_2 = 0$. This is obviously what the circuit of Figure 60 represents it to be, since with $v_2 = 0$ the voltage generator in the input circuit disappears. The other parameters may be dealt with similarly.

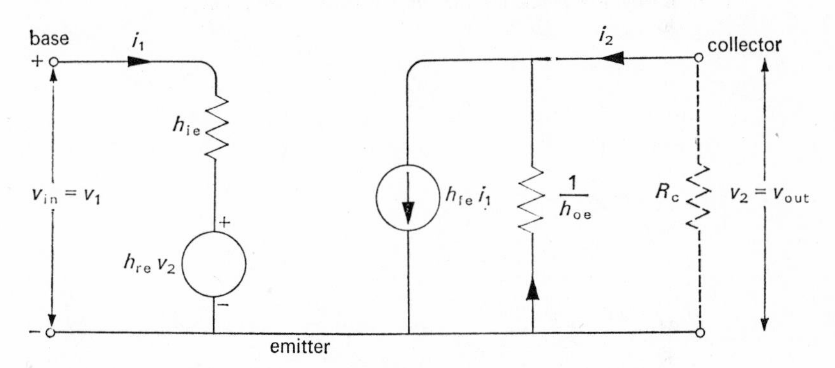

Figure 60. Common-emitter, h-equivalent circuit.

The approximations made in the previous section in which h_{re} and h_{oe} were neglected, are reflected in the very simple equivalent circuit of Figure 61, again

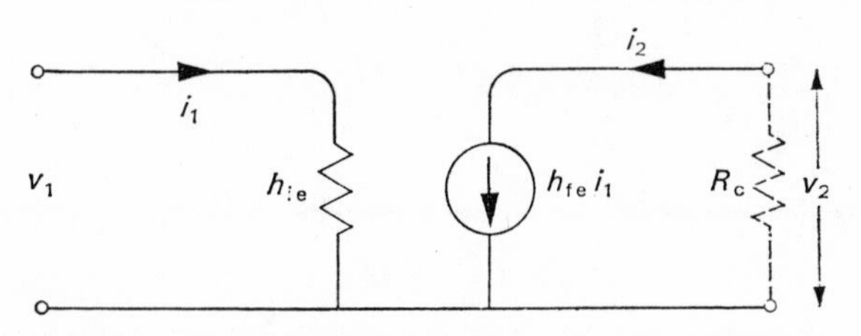

Figure 61. Simplified common-emitter, h-equivalent circuit.

shown with the collector resistor in place. The action is correspondingly simple to understand. The input voltage v_1 causes a current i_1 of size v_1/h_{ie} to flow in the input circuit. The current generator in the output thus produces a current of size $h_{fe} i_1 = h_{fe}(v_1/h_{ie})$. This current, also labelled i_2, flows through the resistor R_c to produce an output voltage $-h_{fe}(v_1/h_{ie}) R_c$. We met this simple picture of transistor action in the common-emitter mode in another form in section 4·5, where on that occasion h_{ie} appeared as $r_b + (\beta + 1) r_e$ and h_{fe} as β.

The full circuit of Figure 60 underlines again the appropriateness of the term 'hybrid' parameters. Appearing there we have a voltage generator, a current generator, a resistance, and a conductance (the inverse of a resistance). With the z parameters, the equivalent circuit contains voltage generators and resistances only, while in the y parameter equivalent circuit only current generators and conductances appear.

References and further reading

1. J. MILLMAN and C. C. HALKIAS, *Integrated Electronics,* McGraw Hill, 1972.
2. A. BAR-LEV, *Semiconductor and Electronic Devices,* Prentice Hall, 1979.
3. R. KING, *Electronic Circuits and Systems,* Nelson, 1975.

CHAPTER 5

Some important circuits

5·1 Introduction

In this chapter we discuss a number of important elements which appear so frequently in circuits of interest to the physicist that they may be considered as 'building blocks'. Indeed they are building blocks in a very real sense since when they contain two or more transistors, they can be obtained in integrated circuit form – that is with the transistors and associated circuit resistors manufactured on the same silicon 'chip', and contained in a single compact package. Examples of these circuits in both current and voltage-controlled device form will be discussed and although here, as previously, these must often be dealt with separately, we shall try to emphasize the properties they have in common. This will be particularly applicable for the output side of the devices. For a voltage-controlled device like an FET in the common-source arrangement we have (see equation **2.2**)

$$i_d = g_m v_{gs} + g_d v_{ds} \qquad \textbf{5.1}$$

where the symbols have the meaning explained in that chapter, and in particular $g_d = 1/r_d$, r_d being the drain resistance. With an obvious change in notation equation **5.1** becomes

$$i_2 = g_m v_1 + g_d v_2 \qquad \textbf{5.2}$$

For a transistor in the common-emitter mode we have from the previous chapter

$$v_1 = h_{ie} i_1 + h_{re} v_2 \qquad \textbf{5.3}$$

and

$$i_2 = h_{fe} i_1 + h_{oe} v_2 \qquad \textbf{5.4}$$

If, as before h_{re} is neglected, then combining equations **5.3** and **5.4** we have

$$i_2 = \frac{h_{fe}}{h_{ie}} v_1 + h_{oe} v_2 \qquad \textbf{5.5}$$

which closely resembles equation **5.2** for a voltage-controlled device. Similar results for the two types of device would be expected if g_m, the transconductance were replaced by h_{fe}/h_{ie} $(=\beta/\{r_b + (\beta+1)r_e\} \approx 1/r_e)$, and g_d by h_{oe}. Indeed it was shown in the last chapter that it is also quite a good approximation to neglect h_{oe}, and the same applies to g_d; the only proviso is that the resistors we have in the circuit in question should all be small compared with $1/h_{oe}$, or $1/g_d$ $(=r_d)$, as the case may be.

On the input side we expect some differences, since for an FET the input impedance is very large, while for a transistor in the common-emitter mode it is approximately h_{ie}, a relatively small quantity.

5·2 Follower circuits: the source follower

The first example we shall discuss is the common-drain configuration (for an FET) and the corresponding common-collector configuration for the bipolar transistor. These are also known as the grounded-drain and grounded-collector configuration, the names implying that the drain (or collector) is to ground – or, what is the same thing from the point of view of signals, directly connected to the voltage supply. The input is between gate (or base) and ground, and the output is taken between source (or emitter) and ground. More usual names are source and emitter follower, since the output electrode (source or emitter, as the case may be) 'follows' the input voltage closely, as we shall see. (In addition to the common-source and common-emitter arrangements discussed in Chapters 3 and 4 respectively and the common-drain and common collector arrangements now to be discussed, there is another FET configuration, the common gate, and its junction-transistor analogue, the common base. While these have their own peculiar advantages and disadvantages, they will not be discussed in detail as they do not figure often in the circuits in which we are interested.) The source follower is shown in Figure 62. It is drawn as a *n*-channel device, but by considering the physical processes involved, it can be shown that the action as regards *signal* voltages and currents would be identical for the *p*-channel type. R_g is the usual gate resistor in the hundreds of thousands to megohm range. The value of the resistor R_s between source and ground across which the output is taken, is not critical,

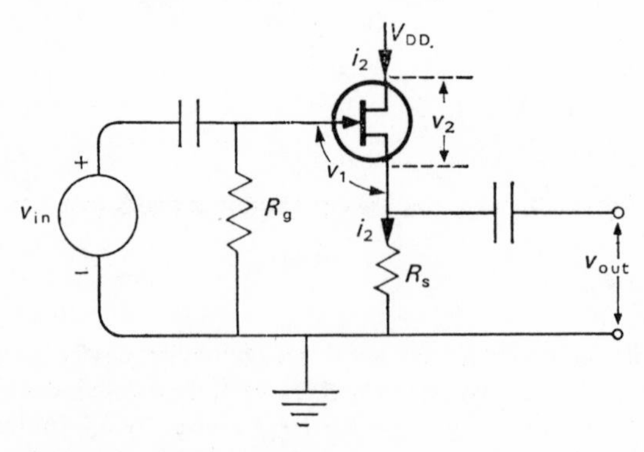

Figure 62. The source follower.

and is usually between one thousand and ten thousand ohms. The voltages v_1 and v_2 of Figure 62 are those referred to in equation **5.2**, and from the forms v_{gs} and v_{ds} in which they appeared in equation **5.1**, they evidently are voltages with respect to the source, not with respect to ground. When dealing with the common source stage in Chapter 2, 'source' and 'ground' were identical, but in this case they are not, as we have interposed an unbypassed resistor R_s between source and ground. The position of the lower reference point of v_1 and v_2 in Figure 62 is thus understandable.

The problem may be dealt with adequately by equation **5.2** in its approximate form

$$i_2 = g_m v_1 \qquad \mathbf{5.6}$$

When the signal voltage v_{in} is applied a current i_2 flows through R_s, and the source voltage rises by an amount $i_2 R_s$. Thus

$$v_1 = v_{in} - i_2 R_s \qquad \mathbf{5.7}$$

(If we were using the exact equation 5.2 and required v_2, this would clearly be $-i_2 R_s$, the fall in voltage across the FET corresponding to the rise across R_s.) From equations **5.6** and **5.7**, on eliminating v_1, and solving for i_2, we have

$$i_2 = \frac{g_m v_{in}}{1 + g_m R_s} \qquad \mathbf{5.8}$$

and since $v_{out} = i_2 R_s$, the voltage gain is given by

$$\text{voltage gain} = \frac{v_{out}}{v_{in}} = \frac{g_m R_s}{1 + g_m R_s} \qquad \mathbf{5.9}$$

For a value of g_m of 3 millimhos, and with R_s of the size indicated above – say 5000 ohms – we have $g_m R_s \gg 1$ and the voltage gain is just smaller than unity. This, together with the non-inversion of the output signal (indicated by the positive value for the voltage gain), justifies the use of the 'source-follower' terminology. With such a value for the voltage gain, the source follower must clearly have other virtues to explain its usefulness, and these will be seen when we calculate the input and output impedances.

The output impedance is the open-circuit voltage divided by the short-circuit current. The open-circuit voltage is equal to the input voltage v_{in}, since the voltage gain is close to one. The short-circuit current can be obtained by putting a short across the output – through a large coupling capacitor, of course, to prevent the d.c. levels being disturbed. Since the short is now, from the point of view of signals, directly in parallel with R_s, the short-circuit current can be obtained by putting $R_s = 0$ in equation **5.8**, to give a value of $g_m v_{in}$ for this quantity. The output impedance is thus $v_{in}/g_m v_{in}$ or $1/g_m$.

Negligible currents flow in the input of the FET itself, so the input impedance of the source follower is R_g, the gate resistor, which as we have already noted, may be of the order of a megohm. The source follower can thus feed a low-

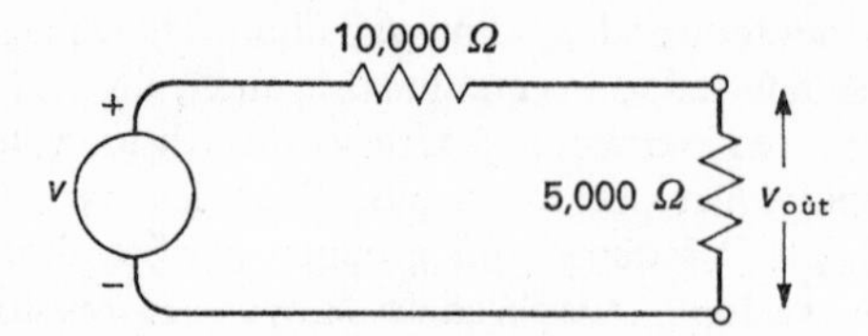

Figure 63. Common-source-stage driving load.

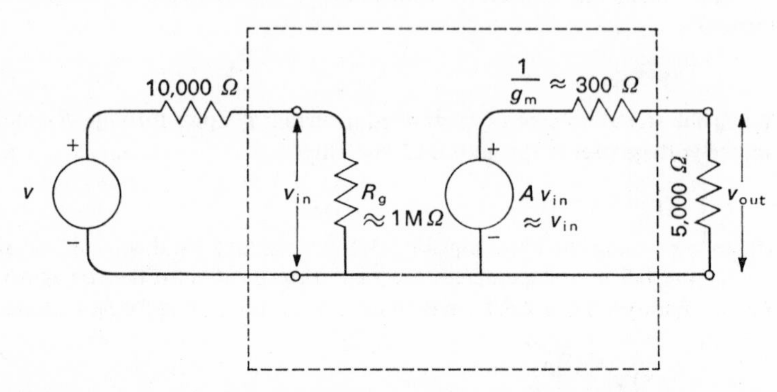

Figure 64. Source follower as buffer between common-source amplifier and load.

impedance load because of its low value of output impedance and in addition, because of its high value of input impedance, it throws a negligible load on the circuit preceding it. For this reason it is used as an 'impedance transformer' or 'buffer' to be inserted between a source with high or moderately high output impedance and a low impedance load. If, for example, we have an FET common-source amplifier stage with a drain resistor of 10,000 ohms, we have seen that its output impedance is approximately this 10,000 ohms. If this amplifier were to drive directly a load of 5000 ohms, clearly, as seen from Figure 63, the output will be only $v \times 5000/15{,}000 = v/3$. But if the source follower whose equivalent circuit is given in the dashed box in Figure 64, were interposed between the amplifier stage and load there is negligible loss of signal at the input, and from the output side the final output is $(5000/5300)v \approx v$. The substantial improvement in output obtained in such cases by the inclusion of a source follower can be looked at from another point of view: even though the source follower provides no voltage amplification it nonetheless provides power amplification, as can easily be verified by calculating the power abstracted from the source in Figure 64, and comparing it with that delivered to the load. The concept of a buffer stage will, however, be more useful in the applications we shall meet later.

At higher frequencies the source follower continues to be useful. On the output side it is capable of charging stray capacities quickly from its low output impedance and thus producing pulses with fast rise times, although we have a reservation to make about this later. On the input side its mode of operation also minimizes the effect of stray capacities, as can be seen from Figure 65. As there is no drain resistor, the drain stays at a constant potential V_{DD}, consequently there is no Miller effect multiplication of the capacity C_{gd} between gate and drain. Furthermore by a sort of inverse Miller effect, the effective value of C_{gs}, the gate to source capacity, may be reduced by a large factor. From Figure 65 it can be seen that since the source follows the signal on the gate almost perfectly, the voltage across the capacity C_{gs} is very small, typically less than 10 per cent of v_{in}. From the point of view of the input signal this is equivalent to charging a capacity of less than 10 per cent of C_{gs} to a voltage v_{in}, which is a measure of the effective reduction of C_{gs}. The small capacitative load thrown on the previous stage as a result of these properties means a corresponding improvement in the rise time of a step at the gate of the source follower.

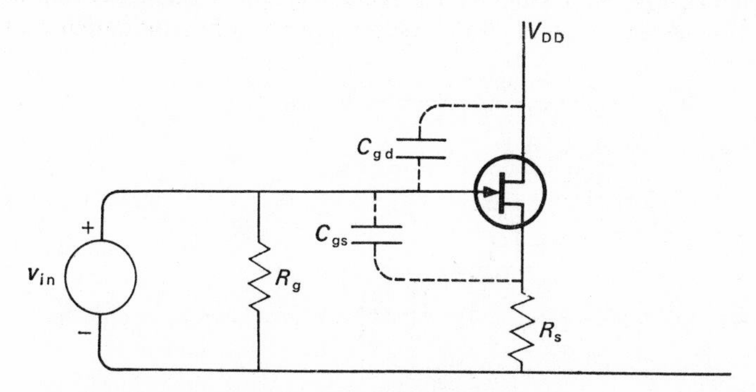

Figure 65. Stray capacities in the source follower.

The simple circuit of Figure 62, is not really a practical one, as can be seen by considering the d.c. operating point. Because of the large source resistor R_s, required for other reasons, the source follower will set itself so as to pass a very small d.c. current. Although Figure 8 is not on a large enough scale for a precise determination, it is clear that with a 5000 ohm value for R_s, a d.c. operating current of about 0·4 mA, giving a source bias of $5000 \times 0{\cdot}4 \times 10^{-3} = 2$ volts, would fall on the characteristics and represent the approximate operating point. Such a low operating value of the current has two disadvantages. Firstly we can produce reasonably large swings in one direction only, and secondly, because of the narrow spacing of successive characteristic curves in that region we can deduce that g_m is small there. This in turn would be reflected in an increased value for the output impedance. The solution is to return the lower end of the gate resistor R_g, not to ground, as in Figure 62, but to a positive potential (for an *n*-channel device) of, say, about one half of V_{DD} – which may be obtained from a potential divider between V_{DD} and ground.

This may seem a strange thing to do, as the gate is normally held negative with respect to the source. In fact the source follower regulates its own operating point. When the power is switched on, gate and source both rise, the former to the potential set by the divider (with a negligible current flowing in R_g), and the latter to a volt or so higher: the required negative voltage of the gate with respect to the source is thus assured. For example suppose that $V_{DD} = 36$ volts, and that we have chosen the potential divider, as suggested, to make the gate voltage $V_{DD}/2$ or 18 volts. The source sets itself also at approximately 18 volts, or to be precise at a rather higher value which we need not attempt to discover. The current through R_s and the FET is thus a little more than $18/5000 = 3{\cdot}6$ mA, a much more appropriate current. The voltage across the f.e.t. is clearly a little less than 18 volts, which is again quite suitable.

5·3 The emitter follower

The emitter follower, the bipolar transistor analogue of the source follower is shown in Figure 66, with an emitter resistor R_e across which the output is taken.

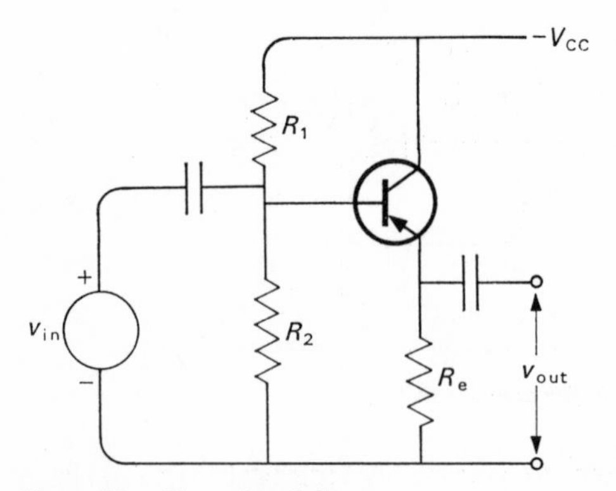

Figure 66. The emitter follower.

Resistors R_1 and R_2 have the same stabilizing role as in Figure 50. R_e will have approximately the same size as the corresponding source resistor R_s of the last section. Following the analogy mentioned at the beginning of this chapter, the gain of the emitter follower would be expected to be close to unity, and its output impedance would be $1/g_m = r_e$. This is very much smaller than the corresponding value for the source follower, as we know that even for a current of 1 mA, $r_e \approx 25$ ohms; we can thus drive very much smaller loads without appreciable loss. We shall see in a later chapter that connecting cables are often required to be terminated in a particular fashion, and under these conditions they act as quite a small resistive load on whatever is sending signals through them. Values in the range 50 to 100 ohms are common. The superiority of the emitter follower over the source follower for driving such cables is apparent.

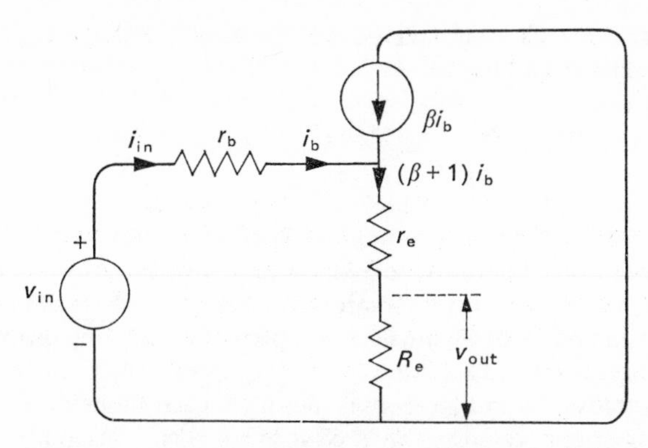

Figure 67. The emitter follower in *T*-equivalent circuit form.

To discuss the input impedance of the emitter follower we must write down some equations, as the analogy with the source follower fails on the input side of the *T*-equivalent circuit (Figure 67). The equation governing operation is

$$v_{in} = i_b r_b + (\beta + 1)\, i_b(r_e + R_e) \qquad \textbf{5.10}$$

giving

$$i_b = \frac{v_{in}}{r_b + (\beta + 1)(r_e + R_e)} \qquad \textbf{5.11}$$

A current $(\beta + 1)i_b$ flows through R_e so the output voltage is $(\beta + 1)i_b R_e$ and the voltage gain $= v_{out}/v_{in} = (\beta + 1)i_b R_e/v_{in}$. Therefore from equation **5.11** we have

$$\text{voltage gain} = \frac{v_{out}}{v_{in}} = \frac{(\beta + 1)\, R_e}{r_b + (\beta + 1)(r_e + R_e)}$$

Remembering the relative sizes of the various quantities involved, we see that the term $(\beta + 1)\, R_e$ is dominant in the denominator, and the gain is thus very little short of unity, as anticipated.

The output impedance (open-circuit voltage over short-circuit current) is $v_{out}\,(\approx v_{in})$, divided by $(\beta + 1)\, i_b$, where the value of i_b is obtained from equation **5.11** with $R_e = 0$. Thus

$$\text{output impedance} = \frac{r_b + (\beta + 1)\, r_e}{(\beta + 1)} = r_e + \frac{r_b}{\beta + 1} \qquad \textbf{5.12}$$

Since β is large this reduces to approximately r_e, as envisaged in the previous discussion. But if the internal resistance r_s of the source of signals driving the emitter follower is appreciable, it must be included with r_b in equation **5.12** as it is in series with it. The output impedance will thus be somewhat larger than r_e.

The input impedance (the input voltage over the input current) $= v_{in}/i_{in} = v_{in}/i_b$, is given by equation **5.11** as

$$\text{input impedance} = r_b + (\beta + 1)(r_e + R_e)$$
$$\approx \beta R_e$$

with the usual approximations. For example if R_e is 5000 ohms and β is 100, the input impedance would appear to be 0·5 megohms, a vast improvement on the common-emitter case. But with the improved performance of the transistor itself, the resistors R_1 and R_2 of Figure 66 now represent the limiting factor on the input impedance. If high input impedance is at a premium, a return to a single large base resistor, as in Figure 49 is indicated, even though this may mean reduced temperature stability. Figure 68 gives a numerical example.

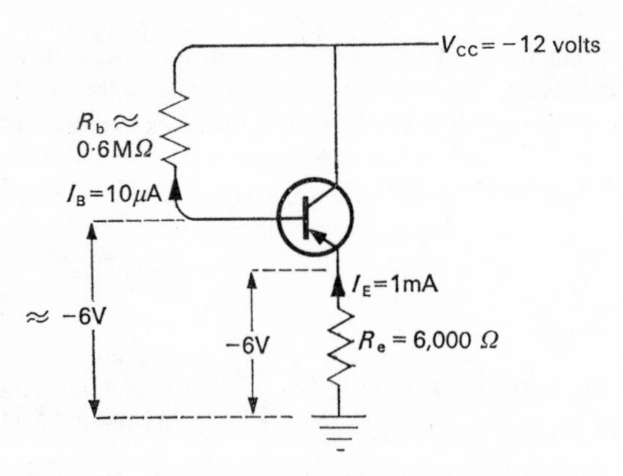

Figure 68. Emitter follower with one large base resistor.

A more serious difficulty arises when the emitter follower is driving a load of say 50 ohms, which it can easily do because of its low output impedance. The 50 ohms resistor and the emitter resistor R_e, though separated by a coupling capacitor, are in parallel from the point of view of signals, the effective resistance between source and ground being thus somewhat less than 50 ohms. The input impedance is now only $\beta \times 50$ or 5000 ohms which is low enough to affect a preceding common-emitter stage. To remedy this two emitter followers are often connected in cascade, as shown in Figure 69. The input impedance of the right-hand emitter follower is $\beta \times R_L = 100 \times 50 = 5000$ ohms. (R_{b_2} is large enough to be neglected.) This 5000 ohms is now in parallel with the emitter resistor R_{e_1} of the left-hand transistor, which if R_{e_2} is also 5000 ohms gives an effective value of 2500 ohms. The input impedance at the base of the left-hand transistor is thus $\beta \times 2500$ or 0·25 megohm. Although this may be reduced somewhat by the resistor R_{b_1} (about 1 megohm) which is in parallel with it, the overall input impedance is still extremely good. For an even higher input

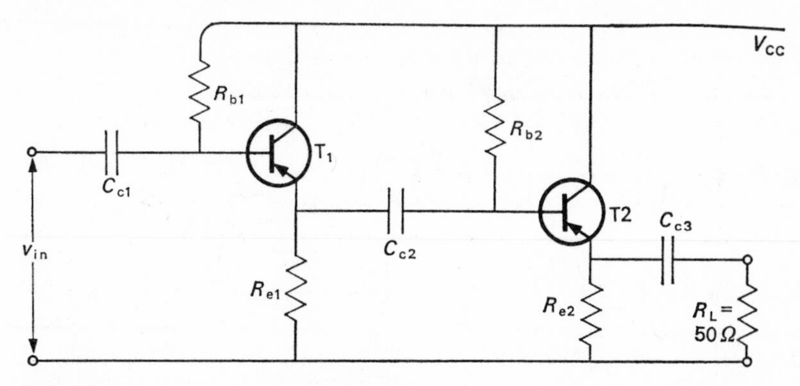

Figure 69. Emitter followers in cascade.

impedance, with low output impedance, the first emitter follower could be replaced by a source follower.

This discussion underlies what might be called the 'transparency' of bipolar transistors. We have seen that the load on the output of an emitter follower affects the input impedance and also that the internal resistance of the source of signals affects the output impedance. The FET does not suffer from such interactions.

Some considerable simplifications can be made in the circuit of Figure 69. The capacitor C_{c_2} and the base resistor R_{b_2} can be eliminated and a direct connexion made between the emitter of the left-hand transistor T_1, and the base of T_2. This is possible because the d.c. emitter voltage of T_1 is at a level suitable for connexion to the following base (Figure 68). The base current of T_2 is small so we need not worry that its presence will upset the voltage level of the emitter of T_1. (Indeed, it might be possible to eliminate C_{c_1} and R_{b_1} by an analogous d.c. connexion to the collector of a preceding common-emitter stage.) Figure 69 may be further simplified by eliminating R_{e_1} and using the base current of T_2 as the emitter current of T_1. This will be particularly appropriate when T_2 is passing a big d.c. current, and its base current is correspondingly large, thus providing a reasonable emitter current for T_1. Such an arrangement of transistors (Figure 70) is known as a Darlington pair, or compound amplifier. It is available in integrated circuit form, and in this case, as can be seen from the dashed box in Figure 70 from which one base, one emitter and one collector connexion project, it can be thought of as a single transistor of superior properties. The input impedance, for example, will be $\beta^2 R_L$ (for $R_L \ll R_{e_2}$); in this and other respects it acts as a single transistor with a current gain of β^2, and an equivalent value of α extremely close to unity. For this reason it is also known as the 'super-alpha' arrangement.

Although different results or further information would not be expected by using the h parameters rather than the T-equivalent circuit for dealing with the emitter follower, it might be helpful in understanding their application to make

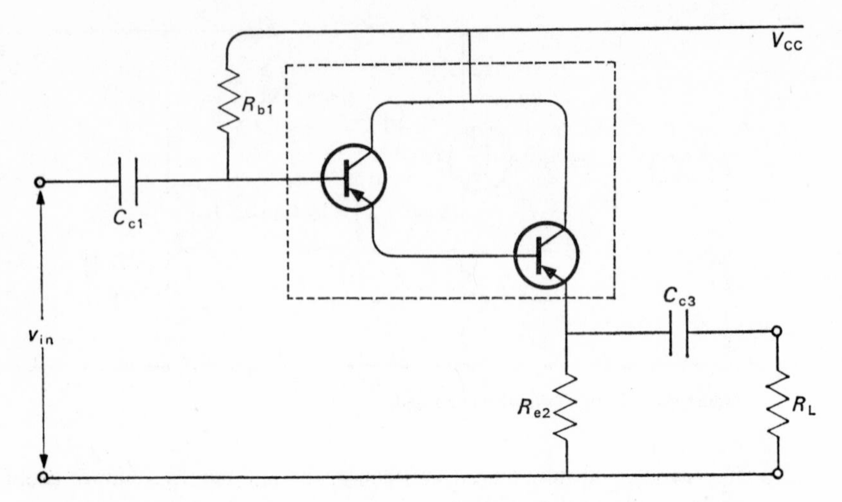

Figure 70. The Darlington pair.

one or two calculations with them. For example, the input impedance for the common-emitter stage is given by equation **4.34** as

$$\text{input impedance} = h_{ie} - \frac{h_{re}\, h_{fe}\, R_c}{1 + h_{oe}\, R_c}$$

and because of the formal nature of the definition of the *h*s we expect the input impedance of the common-collector (that is, emitter-follower) arrangement to be given by

$$\text{input impedance} = h_{ic} - \frac{h_{rc}\, h_{fc}\, R_e}{1 + h_{oc}\, R_e}$$

where the suffix $_c$ on the *h*s indicates that we are now operating in the common-collector mode. So if someone has already measured or calculated the values of the h_cs for us, we may obtain this and similar results without further calculation. In fact h_{oc} has the same small value as h_{oe} and therefore $h_{oc}\, R_e$ is negligible with respect to unity. h_{ic} has the same value as h_{ie}, a few thousand ohms. h_{rc} is however unity, and $h_{fc} = -\beta = -100$, say. Thus the input impedance is approximately $h_{ic} + \beta R_e \approx \beta R_e$ as above. Calculations on the voltage gain and output impedance will be similar.

To avoid carrying in our heads values for the common-collector *h* parameters as well as those for the common-emitter case, we add here a treatment of the emitter follower using the equivalent circuit with the common-emitter *h*

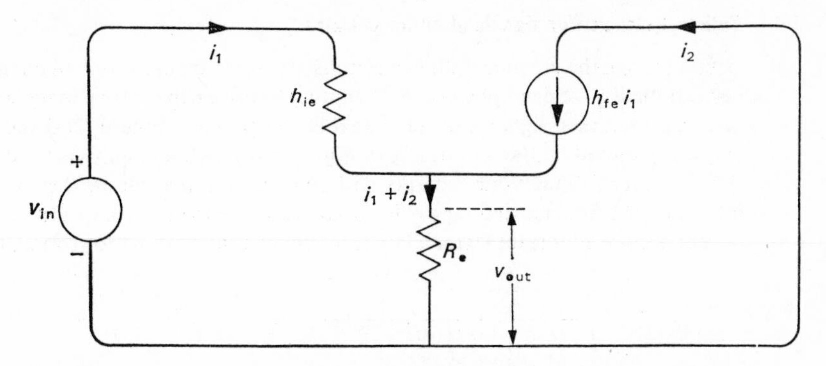

Figure 71. Equivalent circuit for the emitter follower using common-emitter h parameters.

parameters we already know. Figure 71 shows the situation where the simplified equivalent circuit of Figure 61 is used. The equations describing its operation are

$$v_{in} = i_1\, h_{ie} + (i_1 + i_2)\, R_e$$

and

$$i_2 = h_{fe}\, i_1$$

Solving these for i_1 we obtain

$$i_1 = \frac{v_{in}}{h_{ie} + (1 + h_{fe})\, R_e}$$

and hence, using the usual methods

$$\text{voltage gain} = \frac{(1 + h_{fe})\, R_e}{h_{ie} + (1 + h_{fe})\, R_e}$$

$$\text{input impedance} = h_{ie} + (1 + h_{fe})\, R_e$$

and

$$\text{output impedance} = \frac{h_{ie} R_e}{h_{ie} + (1 + h_{fe}) R_e}$$

Remembering the relative values of the quantities involved, the expression for the voltage gain reduces approximately to unity, that for the input impedance to $h_{fe} R_e = \beta R_e$ as before and that for the output impedance to h_{ie}/h_{fe}, which from our remarks at the beginning of this chapter is $1/g_m$ ($\approx r_e$).

5·4 Follower circuits for signals of either polarity

Useful though the previous follower circuits are, there are occasions when they perform badly, for example, when attempting to drive capacitative loads with certain polarities of signal. Consider the follower circuit of Figure 72(a) and its equivalent circuit on the output side of Figure 72(b). (Although this is the case of an emitter follower our remarks will be equally applicable to the source follower.) We assume that, apart from any resistive load on the output, we have as a more important factor a large amount of stray capacity to ground, C_s.

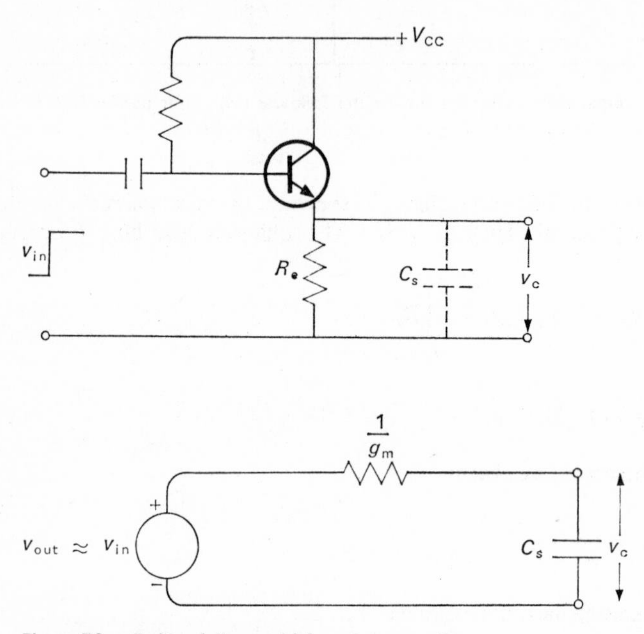

Figure 72. Emitter follower driving stray capacity
(a) circuit (b) equivalent circuit.

Such a case could arise for example when driving a long length of coaxial cable with a large resistor at the far end, the capacity between the central conductor and the case being the stray capacity in question. If we put a positive step of voltage v_{in} on the base of the transistor, we see from Figure 72(b) that the output v_C across the capacity will rise exponentially with a time constant $(1/g_m) C_s$ which, because of the low value of $1/g_m$, can be quite small. For example if C_s were 1000 pF and $1/g_m$ were 25 ohms the rise time would be 0·025 μsec. If the rise time of the input step is faster than this, the voltage on the base gets ahead of that of the emitter. For the *n-p-n* transistor as drawn, this is of no importance – indeed, it is actually beneficial, as the increased base to emitter potential tends to increase the current in the transistor, and thus reduce

$1/g_m$ the output impedance. This further facilitates the charging of C_s. It is otherwise with a negative step. There, if the base runs ahead of the emitter by the order of a volt, the transistor is cut off. To see what happens here imagine the transistor itself removed from Figure 72(a). The capacity C_s is charged to some positive voltage (the ordinary d.c. voltage on the emitter, less whatever small negative change has already taken place) and it can now discharge itself only through the emitter resistor R_e in its attempt to catch up with the negative-going signal on the base. The time constant in this case however is $R_e C_s$, which is a couple of orders of magnitude larger than the value for the positive step of $(1/g_m)\,C_s$. The negative step is thus very poorly reproduced, as will be the trailing edge of a positive rectangular pulse. There will always be some stray capacity on the output of a follower circuit so the problem is always with us to a greater or lesser extent. It should also be clear that the presence of a coupling capacitor between the output and C_s does not invalidate in any way the arguments already used.

From similar arguments, *p-n-p* transistors (and *p*-channel f.e.t.s) should be able to reproduce accurately large, fast, negative going signals, while behaving poorly with positive signals. Thus we can reproduce well either polarity of

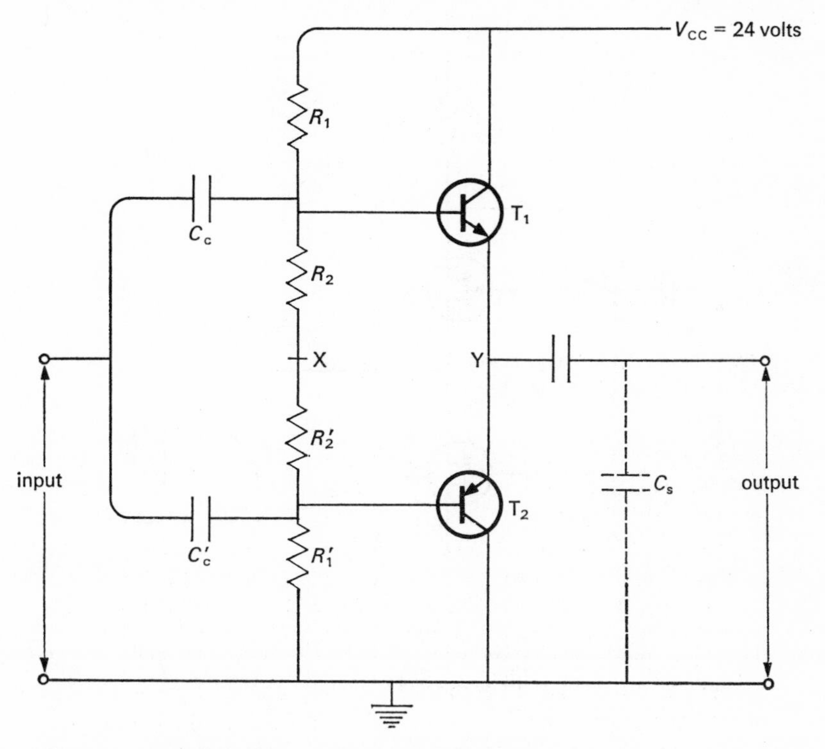

Figure 73. Complementary symmetry emitter follower.

signal due to the existence of devices with both positive and negative charge carriers. But the problem still remains of devising a circuit which will be able to deal with both polarities, and hence, for example, faithfully reproduce both the leading and trailing edges of a rectangular pulse. The obvious solution of combining together a *p-n-p* and a *n-p-n* transistor in a 'complementary-symmetry' emitter follower is shown in Figure 73. Here, from symmetry, the points X and Y are at half the potential V_{CC}, twelve volts in this case, the resistors R_1 and R_2 ($R_2 \ll R_1$) bias T_1 slightly on, while R_1' and R_2' perform a similar function for T_2. The input signal goes to the bases of the transistors by means of the two capacitors C_c and C_c', although because of the low values of R_2 and R_2', a single capacitor to the point X would be quite in order. The action should be fairly obvious: when a positive step is applied at the input, T_1 is driven on (the large but not infinite resistance of T_2, now cut off, acting as its emitter load) making possible the rapid charging of any stray capacity C_s present on the output. With a negative step, the process is reversed, the charging of C_s being accomplished by T_2 in this case. As each transistor is required to respond to only one polarity of signal, this circuit has the additional advantage that the transistors can be biased to a low value of standing d.c. current, with resulting economy in

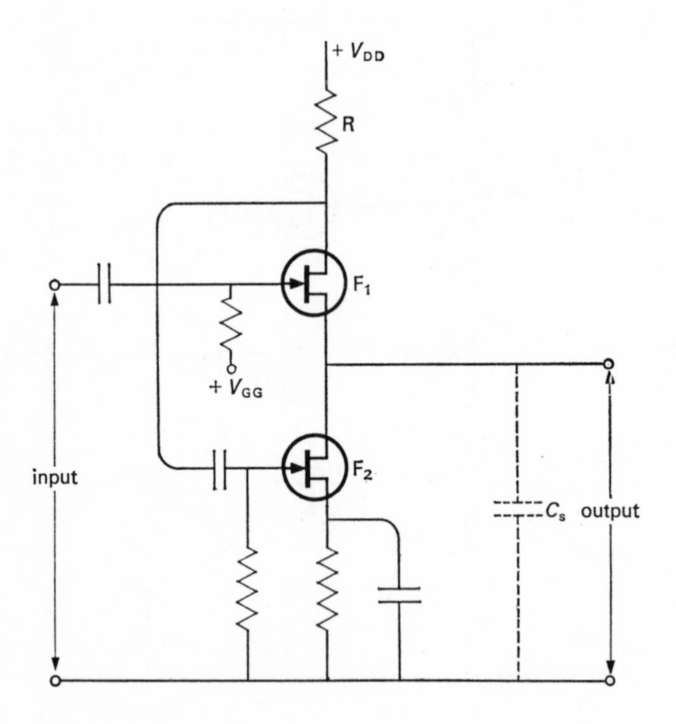

Figure 74. White source follower.

consumption, which may be important for a battery-powered instrument, or for a circuit with a large number of components.

Although the availability of complementary transistors makes the design of followers for either polarity of signal particularly easy, it is not essential to make use of such an arrangement, nor indeed in the days of thermionic valves was it possible. All we really require is a circuit with two valves, transistors or FETs coupled in such a way that when one goes off the other will come on, so that the stray capacity will always have a low-impedance source from which to charge. Figure 74 shows an FET version of the 'White cathode follower' which was widely used with valves, and in a modified version is often found with bipolar transistors. It can be thought of as a source follower with the normal source resistor of F_1 replaced by F_2. The positive voltage V_{GG} on the gate of F_1 is necessary because its source, being connected to the drain of F_2, is already at a fairly high potential. The signal developed across the small resistor R (≈ 1000 ohms) is coupled to the input of F_2, thus ensuring that when F_1 is driven off by a negative input signal F_2 is driven on, as required to charge C_s. With a positive input, F_1 behaves very like a simple source follower. A more detailed analysis of the action of this circuit shows that it has a gain closer to unity, and an output impedance considerably lower than for the source follower with a single FET.

5·5 The cascode amplifier

We mentioned the cascode circuit in section 2·8 as an arrangement in which the unfortunate results of the Miller effect were largely eliminated. It is shown in Figure 75 in FET form, although it could equally well have been drawn with bipolar transistors. A hybrid version, with a high input impedance FET in the lower position, and a bipolar transistor in the upper, less critical position is also common. Figure 75 can be thought of as consisting of an FET in the common source configuration (F_2), having another FET (F_1) as a load. For the reason discussed in connexion with Figure 74, the gate of F_1 must be held at a substantial positive voltage V_{GG} with respect to ground; in this case, however, the gate is connected directly to this voltage supply, and is therefore, from the point of view of signals, at ground potential. On the application of an input signal v_{in}, a common signal current i flows in both FETs and the voltage at the drain of F_1 drops by an amount v, say. The equations describing the action are then (using the approximate form of equation **5.2** which neglects g_d), for F_2

$$i = g_m v_{in} \qquad \textbf{5.13}$$

and for F_1, rembering that its gate is at a fixed potential, but that the drop of voltage v at its source will nevertheless produce an increase in gate-source potential

$$i = g_m v \qquad \textbf{5.14}$$

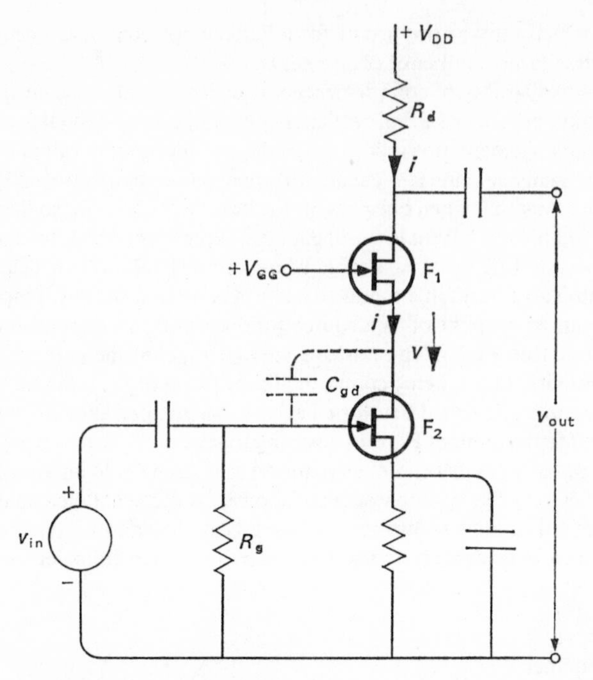

Figure 75. Cascode amplifier.

(The FETs are assumed to be identical and therefore the values of g_m appearing in equations **5.13** and **5.14** are the same.) The voltage gain $= v_{out}/v_{in} = -iR_d/v_{in}$ = (from equation **5.13**) $-g_m R_d$. The output impedance (open-circuit voltage over short-circuit current) is easily seen to be R_d, while the input impedance, at moderate frequencies, is obviously R_g. The use of a second FET in the circuit seems, so far, to have produced results differing in no way from those with the simple common-source stage. The difference becomes apparent when we consider the size of the voltage change v, in the drain potential of F_2. From equations **5.13** and **5.14** $v = v_{in}$, so when the gate moves up by an amount v_{in} the drain drops by a similar amount, and the total voltage across the stray capacity C_{gd} in Figure 75 is $2v_{in}$. The size of C_{gd} has been effectively increased by a factor of only two, rather than by the large factor $1 + A_0$ of the conventional common-source stage (see for example equation **2.16**). We thus possess an amplifier giving the same voltage gain as the simpler configuration, but with practically none of the Miller effect associated with that gain.

5·6 The 'long-tailed pair'

The circuit with this rather curious name shown in its bipolar transistor form in

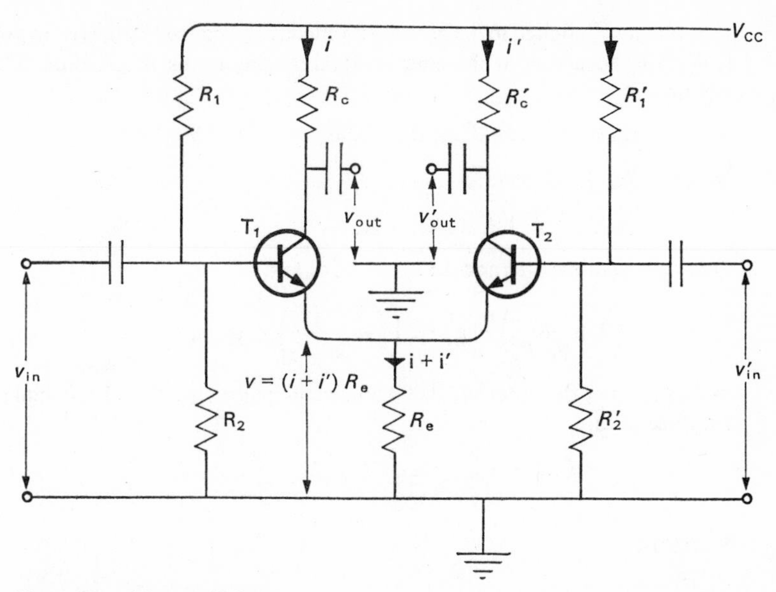

Figure 76. A long-tailed pair.

Figure 76, can be used in a variety of ways, some of which will be discussed after its general properties have been investigated. It is, at least in the form we shall deal with initially, completely symmetrical, with two inputs and two outputs. The two equal resistors R_c and R_c' are typically of the order of 1000 ohms, while the large resistor R_e, which is unbypassed, would be an order of magnitude greater. R_1 and R_2 (and their counterparts R_1' and R_2') are chosen to hold the bases of the transistors, and hence their emitters, not too far from the supply voltage V_{CC}. Most of V_{CC} thus appears across the large 'tail' resistor R_e, establishing in it an almost constant current $\approx V_{CC}/R_e$. In the quiescent state this current is shared equally between the two transistors – or because transistors and resistors never in practice have characteristics exactly in accord with their nominal values, one of the biasing resistors is adjusted initially until such a condition prevails. When signals are applied to the inputs, this condition of symmetrical sharing is disturbed and this action is responsible for the characteristic behaviour of the circuit.

The application of voltage signals v_{in} and v'_{in} (not necessarily equal) to the inputs produces signal currents i and i' in the transistors, and a signal current of size $i + i'$ in R_e. The voltage of the emitters rises by an amount $v = (i + i')\,R_e$. Equation **5.5** is now applied in its approximate form to the transistors in turn, although in this case the approximation may demand a closer scrutiny because R_e is large, and terms like $h_{oe} R_e$ might not be negligible with respect to unity. As usual h_{fe}/h_{ie} will be written as g_m. Thus for T_1

$$i = g_m\{v_{in} - (i + i')\,R_e\} \qquad \textbf{5.15}$$

the voltage v_{in} applied to the base, being effectively reduced by the rise in voltage $(i + i')R_e$ occurring at the emitter. Rearranging terms in equation **5.15** we obtain

$$i(1 + g_m R_e) + i' g_m R_e = g_m v_{in} \quad \textbf{5.16}$$

Similarly for T_2 we have

$$i g_m R_e + i'(1 + g_m R_e) = g_m v'_{in} \quad \textbf{5.17}$$

Solving equations **5.16** and **5.17** for i we obtain

$$i\left(2 + \frac{1}{g_m R_e}\right) = g_m v_{in}\left(1 + \frac{1}{g_m R_e}\right) - g_m v'_{in} \quad \textbf{5.18}$$

and as R_e is of the order of 10,000 ohms and hence $1/g_m R_e \ll 1$ the final result for i becomes

$$i \approx \frac{g_m(v_{in} - v'_{in})}{2} \quad \textbf{5.19}$$

Similarly,

$$i' \approx \frac{g_m(v'_{in} - v_{in})}{2} \quad \textbf{5.20}$$

$$= -i$$

It need not surprise us that $i' = -i$ since this means that the current change $i + i'$ in R_e is zero (that is, that the current through R_e is, to our approximation, constant, since we saw previously that this was a basic characteristic of the circuit).

We are now in a position to discuss ways in which the long-tailed pair may be used. The current change in either transistor is proportional to the difference in the voltages applied to the bases. Taking the output from, say, the collector of T_2 we obtain an output voltage of $-i'R'_c = g_m(v_{in} - v'_{in})/2$ which is again proportional to the difference between the input voltages. If $v_{in} = v'_{in}$, we obtain no output signal – at least to the approximation to which we are working. We thus have a 'difference amplifier', one application of which we shall find later in the design of stabilized power supplies.

On the other hand we may make use of only one input, shorting the other to ground from the point of view of signals with a large capacitor. If it is the base of T_2 which is shorted to ground, $v'_{in} = 0$, and the output from the two collectors, $-iR_c$ and $-i'R'_c$ respectively, will be $(-g_m v_{in} R_c)/2$ and $(g_m v_{in} R'_c)/2$. Thus, as $R_c = R'_c$ we obtain from a single input v_{in} two outputs of equal size but of opposite polarity (and each equal to half the output $g_m v_{in} R_c$, which one would obtain from a simple common-emitter stage with a collector resistor R_c). The availability of alternative outputs of opposite polarity could be of advantage, for example, for a general purpose pre-amplifier required to drive, on different occasions, main amplifiers, some requiring a positive, and some a negative input. By simply taking the output from one or other of the collectors of the long-tailed pair, we could obtain the appropriate polarity; the magnitude of the signal being identical in both cases. There are also occasions when a balanced or 'push–pull' output, making use simultaneously of the signals from both collectors, could be of value.

Although the facility of alternative outputs is lost, there are certain advantages in operating a long-tailed pair with the collector load R_c of T_1 equal to zero, and with the base of T_2 still shorted to ground for signals as before. In the first place, as there is now no load in the collector circuit of the transistor T_1, to whose base we apply the input, the collector potential does not change and there is no Miller effect multiplication of the base to collector stray capacity. But we should check that, as we now have a resistor in the emitter circuit, there is no analogous effect occurring with the base to emitter capacity. Since the voltage change at the emitters is $(i+i')R_e$, which to our approximation is zero, and is at worst small, there is clearly no problem here. The long-tailed pair, in the form we are now discussing, is thus free from the Miller effect. The output signal taken from the collector of T_2 has the same polarity as the input, so we could construct an amplifier consisting of a number of long-tailed pair stages, in which the polarity of the signal was the same all the way through and not alternating in polarity at each stage, as with a conventional arrangement. Such an amplifier can be shown to have advantages in the case of overload signals.

We close with some general remarks about the long-tailed pair. The assumption made in deriving the approximate equations **5.19** and **5.20** on which the subsequent discussion was based, really amounts to considering R_e to be infinitely large. Practical values of R_e must fall below this ideal, so the performance of long-tailed pairs must be somewhat worse than that indicated. For example, in a difference amplifier, the application of equal signals to both inputs, will in practice result in a small output. (A measure of such imperfection in real difference amplifiers is known as the 'common-mode rejection factor'. This factor is defined as $(v_{in})_2/(v_{in})_1$ where $(v_{in})_2$ is the voltage which when applied to both input terminals will produce the same output as a voltage $(v_{in})_1$ applied to one terminal only, the other being to ground for signals. For an ideal difference amplifier the common-mode rejection factor is infinite. For a simple long-tailed pair difference amplifier the factor might be of the order of 100, though with more sophisticated arrangements, values several orders of magnitude higher are obtainable.) In view of this, one might be tempted for best results to make R_e very large, with almost all of V_{CC} across it and only a small fraction of this voltage across the transistors and R_c and R'_c. But small voltages across the collector resistors will mean limited output swings in certain directions, and very small voltages across the transistors may impair their operation. An arrangement which provides a constant current without taking too large a share of V_{CC} is to replace the resistor R_e with a transistor, T_3, of the same type (*p-n-p* or *n-p-n*) as T_1 and T_2, with its collector connected directly to the emitters of these transistors. The operating point of T_3 will be set with a resistor network as in Figure 50. We cannot discuss in detail the action in the present case as it is one in which the base is not an active electrode, unlike our normal practice. But, generally, the current in a transistor is very insensitive to the voltage applied to the collector, so any changes in the voltage at the emitters of T_1 and T_2, that is at the collector of T_3, will result in very little change in the current in T_3. Thus T_3 is acting as a very large resistance for signals, as required for good long-tailed pair action. On the other hand the actual d.c. voltage across the transistor may be only a few volts, and thus a reasonable fraction of V_{CC}; this, coupled with a transistor current of a few milliamps, represents quite a low resistance for d.c. We thus have the best of both worlds – a high resistance to signals for good long-tailed pair action, while at the same time a low d.c. resistance allowing an adequate voltage drop across the other components.

References and further reading

1. J. MILLMAN and C. C. HALKIAS, *Integrated Electronics,* McGraw-Hill, 1972.
2. A. BAR-LEV, *Semiconductor and Electronic Devices,* Prentice Hall, 1979.
3. R. KING, *Electronic Circuits and Systems,* Nelson, 1975.

CHAPTER 6

Negative feedback

6·1 Introduction

When discussing the gains and input and output impedances of the common-emitter amplifier stage, we remarked on the unfortunate fact that these depended on quantities like β and r_e, which vary not only from transistor to transistor, but even with the operating point of a particular transistor – and as a corollary of this with the size of the signal. Similar remarks apply to the FET. We found a happy exception in the case of a common-emitter stage with an unbypassed emitter resistor R_e' where the voltage gain, $-R_c/R_e'$, although reduced in size from its normal value, was independent of transistor parameters. This was an example of 'feedback' where interaction between output and input of an amplifier was deliberately introduced. In the case in question the output current, which also flows through R_e', interacts with the input signal on the base. The properties of such 'negative current feedback' will be considered later in this chapter.

6·2 Negative voltage feedback

The terminology here implies that a fraction of the output *voltage* of an amplifier is fed back to the input in such a way ('negative') as to oppose the input

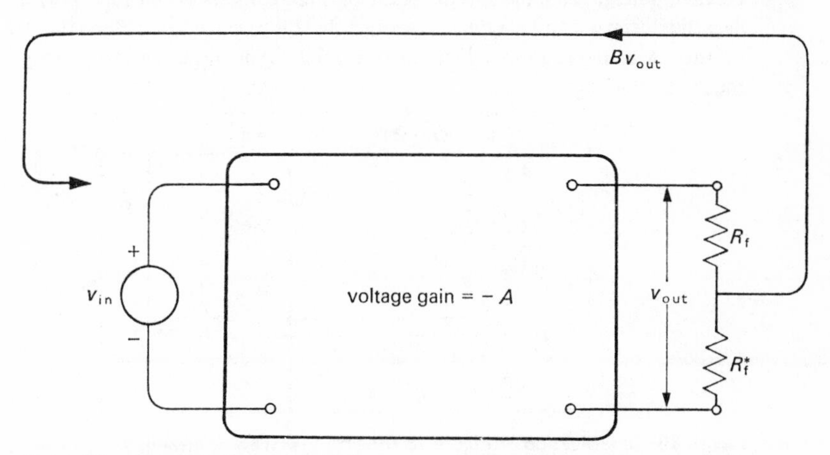

Figure 77. Simple (but unworkable) arrangement for negative voltage feedback.

signal there. First thoughts on how to do this might perhaps be as in Figure 77, where a fraction B of the output, defined by the potential divider R_f and R_f^* is fed back. (The letter B is used for this fraction, rather than the more common β, to avoid confusion with the transistor current gain.) But the scheme of Figure 77 is unworkable because the low, ideally zero, internal resistance of the voltage source v_{in} is shorted across R_f^*. Figure 78 shows a practical arrangement, which also includes the input and output impedances of the amplifier,

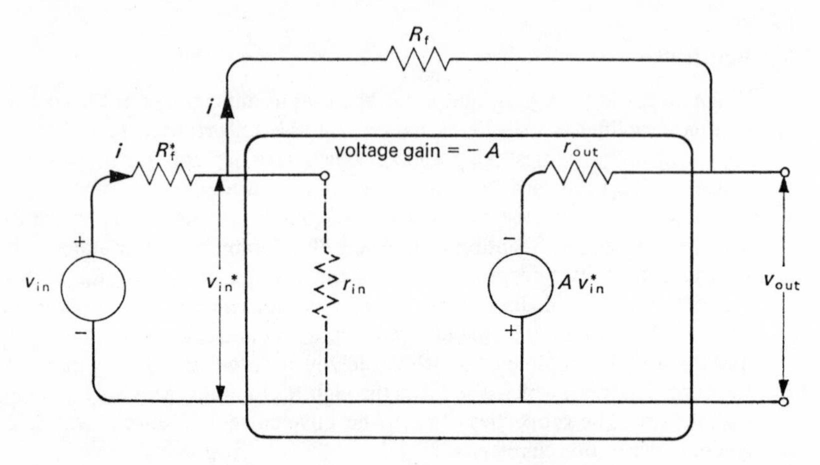

Figure 78. Practical arrangement for negative voltage feedback.

r_{in} and r_{out}. The gain of the amplifier is negative, thus ensuring that the voltage fed back does in fact oppose that at the input. That is, the voltage v_{in}^* actually on the input will be much less than the voltage v_{in} applied. If the amplifier has a positive voltage gain the modification of Figure 78 shown in Figure 79 allows negative feedback to be applied here also. In this case feedback raises the voltage of the lower input terminal of the amplifier, again resulting in v_{in}^* being very much less than v_{in}.

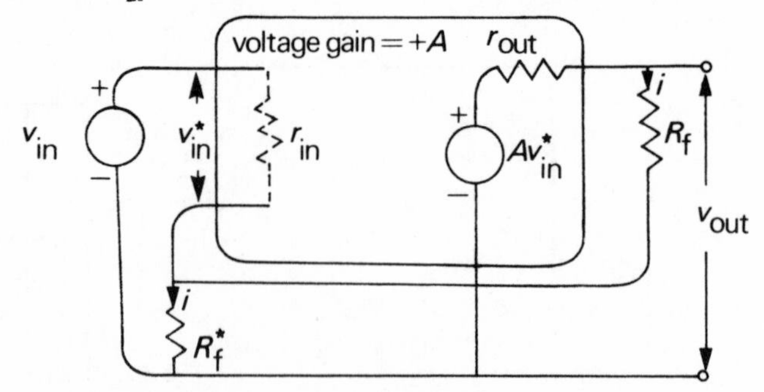

Figure 79. Modification of Figure 78 for amplifier with positive gain.

Since there is a slight complication, which we will deal with later, in the circuit of Figure 78, we start with an analysis of that of Figure 79. To simplify the discussion of this circuit we make in addition two assumptions which are always well fulfilled by any reasonable voltage amplifier - that r_{in} is large (with respect to R_f^*, say) and that r_{out} is small (with respect to R_f). Looking at the input section of the amplifier we can then write

$$v_{in}^* = v_{in} - iR_f^* \qquad \textbf{6.1}$$

(where i is the current flowing through R_f and R_f^*) and looking at the output section, we can write down for the loop containing the voltage generator Av_{in}^* and the resistors r_{out}, R_f, and R_f^*

$$Av_{in}^* = i(r_{out} + R_f + R_f^*) \simeq i(R_f + R_f^*) \qquad \textbf{6.2}$$

Eliminating i between equations **6.1** and **6.2** one obtains

$$v_{in}^* = v_{in}/(1 + AB) \qquad \textbf{6.3}$$

where we have written B for the feedback fraction $R_f^*/(R_f + R_f^*)$. Now the output voltage v_{out} is given by Av_{in}^* (since any voltage drop in r_{out} is negligible due to the negligible size of this resistance) so the voltage gain for the complete system is given by:

$$\text{voltage gain} = v_{out}/v_{in} = A/(1 + AB) \qquad \textbf{6.4}$$

We now apply the procedure which is at the heart of feedback theory. Suppose A is large enough for $AB \gg 1$. Then the voltage gain reduces simply to $1/B$. This is independent of A and therefore also of the characteristics of the transistors or FETs composing the amplifier, and our purpose is achieved. Note of course that the resistors R_f and R_f^* which define B and thus the amplifier gain, must now be precise, high-stability devices, but it is very much easier to obtain such qualities in resistors than in transistors or FETs.

A numerical example will illustrate our results so far. If A is 1000, a value for B of 0·1 will ensure that the condition $AB \gg 1$ is fulfilled. The gain of the system is thus $1/B = 10$. If through some change in characteristics or replacement of a transistor the unfedback gain A of the amplifier drops by one half to 500, as the quantity AB is still very much greater than unity, it is clear that the overall gain is still $1/B = 10$. Using the more precise formula **6.4** for the gain we see that it has in fact changed by approximately 1 per cent. This is still extremely small compared with the percentage change in A, but we have apparently paid dearly for this insensitivity to component changes in that the gain has dropped from its unfedback value of 1000 to a mere 10. This loss may be compensated for in other ways as will be seen when the output impedance of our new system is calculated.

For this we need the open-circuit voltage and the short-circuit current. The former is that already obtained in equation **6.4**. With the output shorted there can be no voltage to feed back and thus no feedback action. In particular, the current i which previously flowed through R_f and R_f^* is now zero. There is no voltage drop across the small resistor R_f^*, and hence $v_{in}^* = v_{in}$. The voltage Av_{in}^* in the output becomes Av_{in}, and as the output is shorted, the short circuit current through r_{out} is Av_{in}/r_{out}. Thus the output impedance is given by

$$\text{output impedance} = r_{out}/(1 + AB)$$

$$\simeq r_{out}/AB \qquad \textbf{6.5}$$

Thus, although the gain in going from A to $1/B$ has dropped by a factor AB, the output impedance has fallen by the same factor, which affords considerable compensation for the gain loss when low impedance loads are driven. For the extreme case where the load is very much smaller than r_{out}, the fedback amplifier can be shown to be at no disadvantage at all: we shall, however, content ourselves with a numerical example. Taking the values previously used of $A = 1000$ and $B = 0{\cdot}1$, and assuming the output impedance without feedback to be 5000 ohms, it is clear that the output impedance with feedback will be only 50 ohms. Thus if a load of 100 ohms must be driven, the voltage appearing across it with the amplifier in its original form would be $(v_{in} \times 1000) \times 100/(5000 + 100)$ or approximately $20v_{in}$. With feedback it would be $(v_{in} \times 10) \times 100/(50 + 100)$ or approximately $7v_{in}$. So now we are only paying with a factor of three in gain for the very important advantages of feedback.

We now return to equation **6.3** to discuss the input impedance with feedback. The voltage v_{in}^* which appears directly across the terminals of the amplifier itself, is smaller, by the large factor of approximately AB, than the input voltage v_{in}. The input current taken by the input impedance r_{in} (Figure 79) is thus reduced by a factor AB compared with the case where v_{in} is placed across it directly - that is for an amplifier without feedback. This reduction in current is equivalent to saying that feedback of this type effectively increases the input impedance by the factor AB. These two effects of negative voltage feedback, the raising of the input impedance and the lowering of the output impedance combine to make the fedback amplifier an even better voltage amplifying device. The higher input impedance means it will throw little load on the signal source preceeding it, and the input signal will not be much reduced even if the internal resistance of the source of signals is high: the lower output impedance means that the amplifier can, as we have seen, drive a low value load, without much loss.

The treatment of the case of an inverting amplifier with feedback (Figure 78) parallels fairly closely what we have already done. The relevant equations are:

$$v_{in}^* = v_{in} - iR_f^* \text{ and}$$

$$v_{in} + Av_{in}^* = i(R_f^* + R_f + r_{out}) \simeq i(R_f^* + R_f)$$

the second equation being obtained, fairly obviously, by considering the loop containing v_{in}, R_f^*, R_f, r_{out}, and Av_{in}^*. Hence eliminating i we obtain

$$v_{in}^* = \frac{v_{in}}{1 + (1 + A)B}$$

where in this case B represents R_f^*/R_f (and thus differs slightly from the case of the non-inverting amplifier). Thus

$$v_{out} = -Av_{in}^* = -\frac{Av_{in}}{1 + (1 + A)B}$$

The voltage gain is thus given by

$$v_{out}/v_{in} = -\frac{A}{1 + (1 + A)B}$$

$$= -\frac{A}{1 + AB} \qquad \textbf{6.6}$$

(since we reasonably can assume that $A \gg 1$)

or $v_{out}/v_{in} = -1/B$

(if we further assume that the feedback condition $AB \gg 1$ is satisfied). Once again the usual benefit of an amplifier gain dependent only on R_f and R_f^* obtains.

It is equally true in this case that the output impedance is decreased by the factor AB as compared with the non-fedback amplifier. We have to examine the case of the input impedance more carefully. It is certainly easy to show that the input impedance of the amplifier itself is increased by AB, in the sense that the current taken by it is reduced by this factor. However we are now asking the voltage source v_{in} to supply the current i which it did not have to do when no feedback existed, and this represents a *decrease* of the input impedance of the whole system as seen by v_{in}. We can calculate quite easily what the input impedance of the system now is if we remember that feedback reduces the voltage v_{in}^* to a very small value, and since $i = (v_{in} - v_{in}^*)/R_f^*$ we can write to a very good approximation $i = v_{in}/R_f^*$ or $v_{in}/i = R_f^*$. This equation tells us that the input impedance is simply R_f^*, a comparatively low value as we shall see later. (Note that this problem does not arise with the non-inverting amplifier, since here the amplifier itself supplies the current i, and the voltage source sees only the input resistance r_{in} increased by the factor AB.)

Before concluding this discussion one further advantage of negative feedback must be mentioned. This is the increase in bandwidth which follows its application. We can see in general why this should be so, since feedback acts to keep the overall gain of an amplifier constant even in the face of a drop in gain of its component transistors, and will of course act in a similar manner when this drop in gain is due to using signals of very high (or very low) frequencies. To place this result on a more formal basis we recall that at high frequencies the midband voltage gain, A_0, is multiplied by the factor $1/(1 + j\omega/\omega_U)$. Combining this with our result for, say, the inverting feedback amplifier that its gain $= -A/(1 + AB)$ we obtain, for high frequencies

$$\text{gain} = -\frac{A_0/(1 + j\omega/\omega_U)}{1 + \{A_0/(1 + j\omega/\omega_U)\}B}$$

$$\text{or gain} = -\frac{A_0}{1 + j\omega/\omega_U + A_0B} = -\frac{A_0}{1 + A_0B + j\omega/\omega_U}$$

which on dividing above and below by A_0B and rearranging

$$= -\frac{1/B}{1 + (1/A_0B) + (j\omega/A_0B\omega_U)} \quad \text{or finally}$$

$$\text{gain} \simeq -\frac{1/B}{1 + (j\omega/A_0B\omega_U)} \quad \text{since } A_0B \gg 1.$$

Comparing this with equation **2.22a** we see that it represents an amplifier with midband gain $1/B$ (as we expect) but with ω_U replaced by $A_0B\omega_U$. So the upper half power frequecy has been improved over the non-fedback case by the same feedback factor AB we have met so often before. An analagous reduction occurs in the lower half power frequency. We leave our discussion on feedback and frequency response at this point, but will return in Section 6.3 to discuss some pitfalls which arise in this connection.

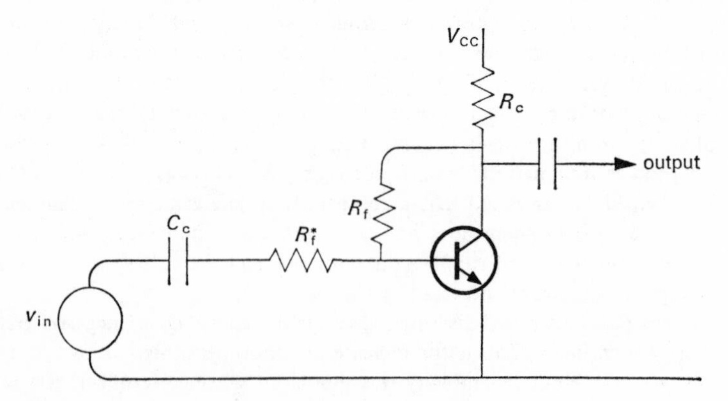

Figure 80. Single-stage, negative-feedback amplifier.

We turn now to some practical examples of feedback circuits. Figure 80 shows a simple single-stage transistor amplifier with the sort of feedback applied that we have discussed. R_f not only serves as a feedback resistor, but also sets the d.c. voltage on the base. The presence of the large input coupling capacitor C_c does not affect the feedback arrangements, and the gain is, as before, $-1/B = -R_f/R_f^*$. The output impedance is reduced by the usual factor. If even lower output impedance is required than this provides, we can incorporate an emitter follower into the circuit as shown in Figure 81. Once again R_f acts as part of the feedback network, and also with R_2 and R_{e_1}, sets the d.c. operating point for T_1. As R_2 is so much larger than R_f^*, we can ignore it in discussing the feedback ratio which is R_f^*/R_f as before. This corresponds to a gain of ten, which is typical or perhaps a little large for such a circuit. The feedback conditions, $AB \gg 1$ can be written $1/B \ll A$, which implies that the fedback gain must always be much less than the unfedback gain. If larger values are required, we must either cascade a number of circuits of the type of Figure 81, or else include further amplifying stages in the feedback loop.

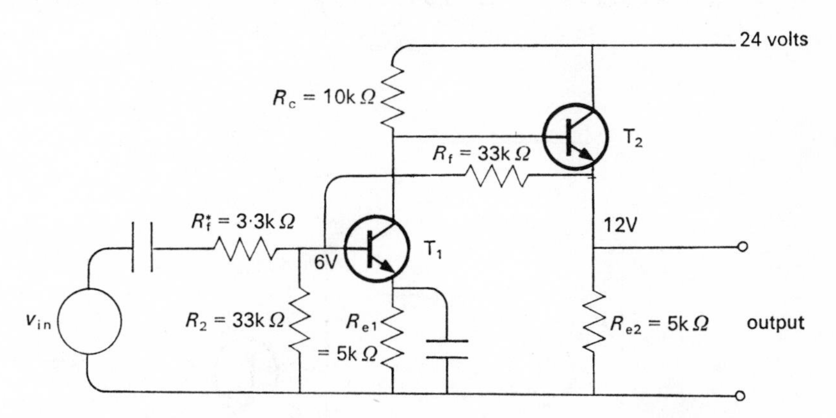

Figure 81. Single-stage amplifier with emitter follower included in feedback loop.

Before discussing the second solution we must mention one other important case of a feedback circuit with a single transistor, and one which we have met before in a different guise, the emitter (or source) follower. Considering this as a feedback device, we note that the *full* output voltage at the emitter opposes the input signal: this is the ultimate in feedback with $B = 1$. We thus expect a gain of $1/B = 1$, which indeed we obtain. The expected output impedance is $r_{out}/AB = R_e/(g_m R_e \times 1)$ or $1/g_m$, as before.

We turn now to feedback loops with two amplifying stages, an example of which is shown in Figure 82. As there is no reversal of signal polarity with such an amplifier (the gain is positive), a circuit analogous to Figure 79 must be used to ensure that the feedback is negative. In fact the feedback is effectively applied to the emitter of T_1, as the large decoupling capacitor across R_{e_1}

provides a short circuit for signals. R_f and R_f^* might typically be 5000 and 100 ohms respectively giving a gain of 51. The large capacitor C_c blocks off the d.c. potential at the collector of T_2 from the feedback network, but does not interfere with the feedback ratio. The rest of the circuit is just a straightforward two-stage amplifier, and should be self explanatory. As before, if very low output impedances are required, an emitter follower can be included in the loop as in Figure 83. This configuration is known as a 'ring of three' and is a very popular building block in amplifiers. In this case the feedback network is also the load for the emitter follower T_3, and has some effect on the biasing of T_1. The gain, of course, is the same as before, R_f/R_f^*, and typically would have the same numerical value.

Usually in circuits like those of Figures 82 and 83 there is a small adjustable capacitor, of the order of tens or even hundreds of picofarads in size, placed across R_f. This is to correct for the change in feedback ratio at high frequencies due to any stray capacity between the emitter of T_1 and ground. As R_{e_1} does not exist from the point of view of signals, this stray capacity is effectively across R_f^*. It thus effectively reduces the value of R_f^* at high frequencies, and hence in-

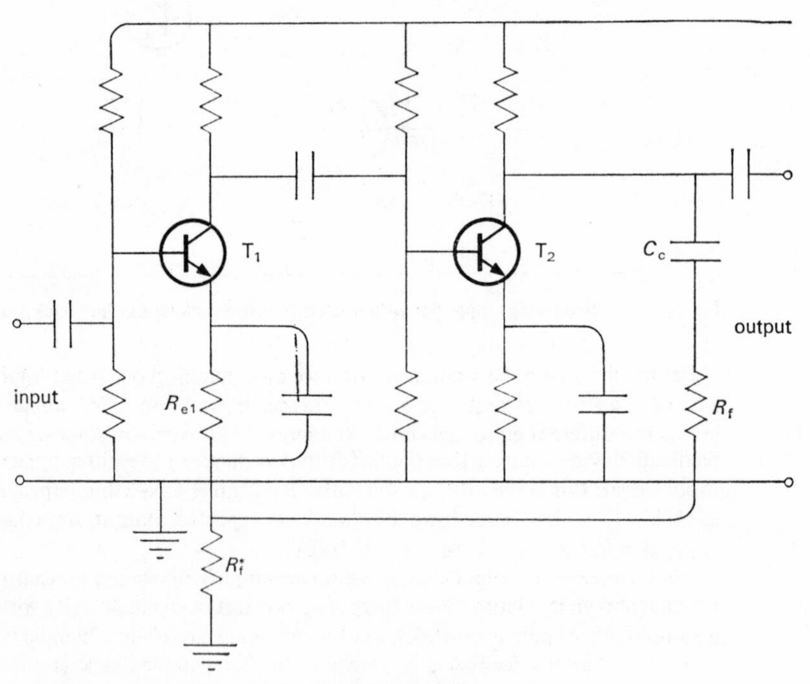

Figure 82. Two-stage, negative-feedback amplifier.

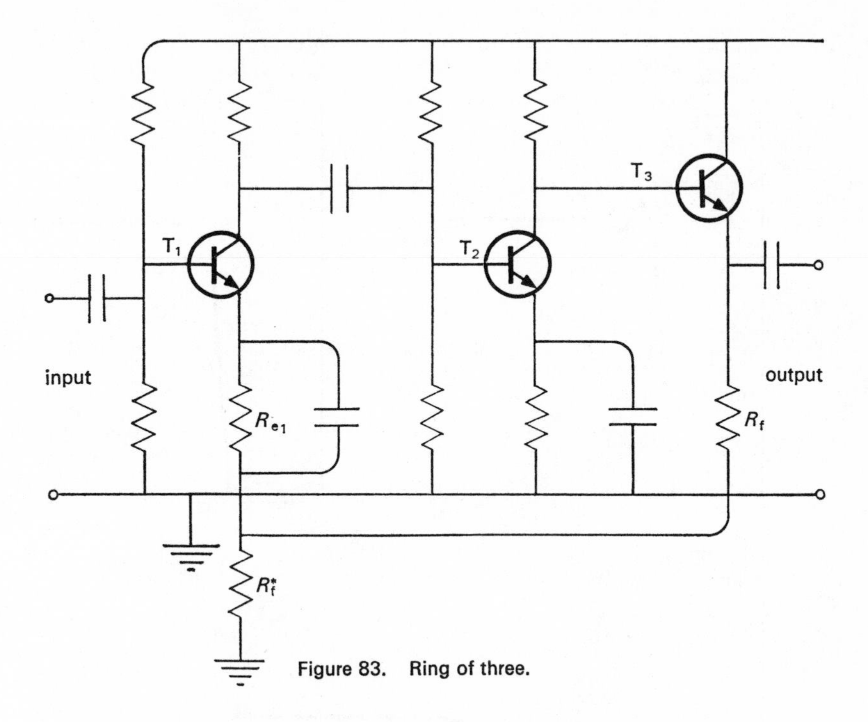

Figure 83. Ring of three.

creases the gain, R_f/R_f^*. To compensate for this R_f is correspondingly reduced by adding across it a small trimmer capacitor. (Although there is also stray capacity from the emitter of T_3 to ground, this is not important for our purpose, because it is across both R_f and R_f^*.) We can look at this in another way: if we have a potential divider such as R_f and R_f^* with stray capacities across them, it can easily be proved that the condition for the potential dividing ratio to be frequency independent is $R_f C_s = R_f^* C_s^*$ (with an obvious notation). C_s^* in this case is the stray capacity from the emitter of T_1 to ground, and C_s is the size of the trimmer capacitor which must be added across R_f. In practice C_s is not calculated accurately but its value is adjusted empirically with the aid of a test voltage step fed through the amplifier. Figure 84(a) shows what we should expect to see on an oscilloscope on the output if the trimmer capacitor were missing or adjusted to too small a value. The high-frequency response is too good, resulting in an overshoot on the rising edge of the step where the high frequency response shows up. This overshoot is followed by a short damped oscillation, the reason for which will be clearer when positive feedback is discussed. Figure 84(b) shows what would be expected with the trimmer capacitor adjusted to its optimum value – an almost perfect step. If the trimmer capacitor is made too large, the high frequency response becomes poor, and a step with bad rise time (Figure 84(c)) results.

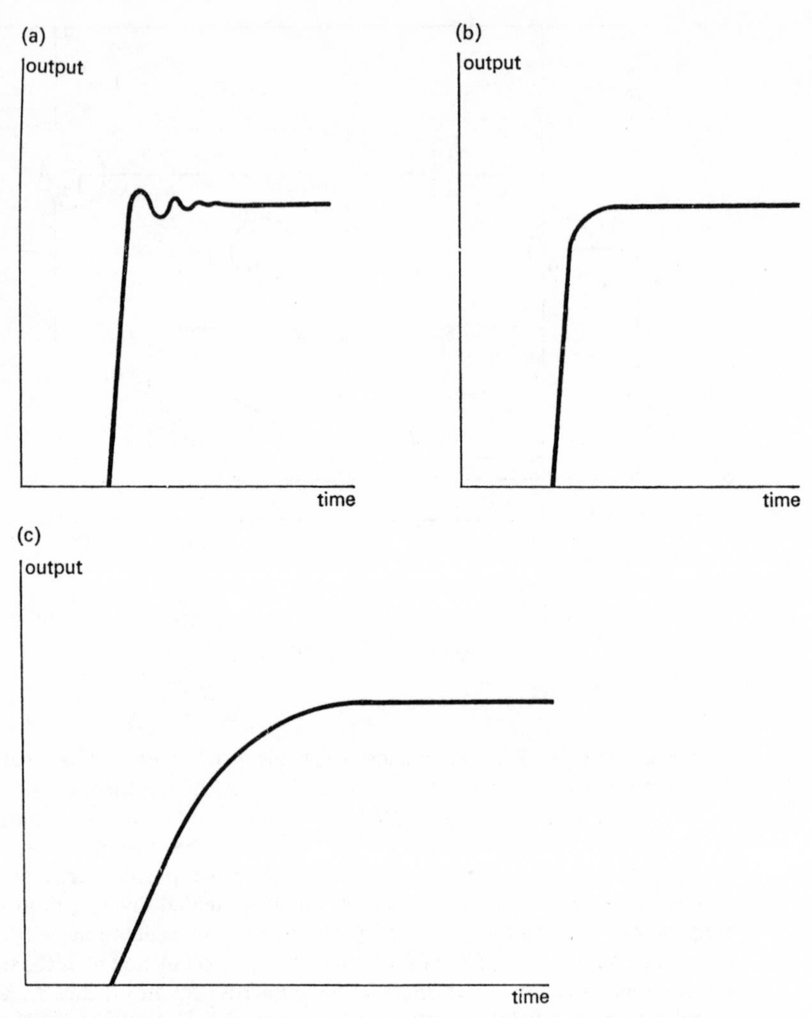

Figure 84. Influence of trimmer capacitor adjustment on output from feedback amplifier.

6·3 Instability in negative-feedback amplifiers

It might be thought that the process of adding more stages into our feedback loop, to produce amplifiers of even greater gain, could be continued indefinitely. In general this is not so. Every experimental worker must at some time or other have had the experience of constructing an amplifier, which, when the power was applied, simply oscillated: that is it produced an output signal with no apparent input. This is an example of *positive* feedback, where part of the output is fed back to the input so as to *increase* the signal already there. It can occur inadvertently via a poorly decoupled d.c. supply line, or even by too close a physical proximity between the output and input stages. Positive feedback is a technique of very great importance, to which the next chapter in this book is devoted. At the moment we merely wish to introduce enough about it to make intelligible some of its more unfortunate aspects in connexion with amplifiers. If, for example, the sign of the amplification A in Figure 78 is changed, positive feedback is obtained. An equation describing it which corresponds to equation **6.6** with A changed to $-A$ is

$$\text{voltage gain} = \frac{A}{1 - AB} \tag{6.7}$$

The full implications of this equation will not be discussed at this stage, merely that if $AB = 1$, the voltage gain becomes infinite. This corresponds to a finite output with zero input, or oscillation. At first it is not easy to see how an amplifier designed to operate with *negative* feedback can ever reach this condition, but it becomes more understandable when we remember that at high frequencies the midband voltage gain, A_0, is multiplied by the factor $1/(1 + j\omega/\omega_U)$, where ω_U corresponds to the upper half-power frequency (see equation **2.22a**). A full discussion of this problem of the transition from negative to positive feedback usually requires the introduction of a 'Nyquist plot' showing the magnitude and phase of AB as a function of frequency. We shall content ourselves with demonstrating how unwanted oscillation can arise in the particular case of a three-stage negative-feedback amplifier. We assume that each of the three stages has the same gain of A, to produce an overall gain before feedback of $A' = A^3$. The gain G with feedback, is given by equation **6.5** as

$$G = -\frac{A'}{1 + A'B} = -\frac{A^3}{1 + A^3 B}$$

or

$$G = -\frac{A_0^3\left(1 + \dfrac{j\omega}{\omega_U}\right)^{-3}}{1 + A_0^3 B\left(1 + \dfrac{j\omega}{\omega_U}\right)^{-3}} \tag{6.8}$$

If the denominator of the expression in equation **6.8** became zero, G would become infinite, and oscillations would be produced. For this to be so

$$1 + A_0^3 B\left(1 + \frac{j\omega}{\omega_U}\right)^{-3} = 0$$

or

$$1 + BA_0^3 + \frac{3j\omega}{\omega_U} - 3\left(\frac{\omega}{\omega_U}\right)^2 - j\left(\frac{\omega}{\omega_U}\right)^3 = 0 \tag{6.9}$$

Equating the imaginary parts to zero we obtain

$$3\left(\frac{\omega}{\omega_U}\right)-\left(\frac{\omega}{\omega_U}\right)^3=0$$

which (apart from the solution $\omega=0$) gives

$$\omega=\sqrt{3}\omega_U \tag{6.10}$$

Equating the real parts of **6.9** to zero we have

$$1+BA_0^3-3\left(\frac{\omega}{\omega_U}\right)^2=0$$

which on putting in the value for ω from equation **6.10** becomes

$$BA_0^3=8$$

(The other solution, $\omega=0$, gives $BA_0^3=-1$, which is impossible, as A and B are both positive quantities.)

Thus the value of $BA_0^3=A'B$ must not exceed eight, or oscillation will occur. As one of the conditions we seek to fulfil in feedback amplifiers is $A'B \gg 1$, it is clear that the present three-stage amplifier is not suitable for feedback applications. Repeating the preceding calculations for the one- or two-stage case it is found that the difficulty previously encountered cannot arise (that is there is no value of ω which makes G infinite). Consequently, if large values of gain are required a cascaded series of one- or two-stage feedback amplifiers must be used. Amplifiers with three or more stages in the feedback loop will not be suitable, unless special precautions which need not concern us here, are taken.

6·4 Other forms of negative feedback

In all our discussion up to now we have concentrated on one particular form of feedback, in fact that appropriate to the type of amplifier we have also concentrated on – the voltage amplifier. But there are, of course, other types of amplifier and we will now list them briefly in order that we can then discuss the form of feedback appropriate to each one.

First to recapitulate on the voltage amplifier we already know: it is fed from a voltage source (see Figure 85(a)) and on the output side it itself also acts as a voltage source. In order to ensure that most of the voltage from the source of signals appears at the input of the amplifier we arrange for r_{in} to be large (with respect to r_s, the internal resistance of the source): similarly we arrange to have the output impedance of the amplifier, r_{out}, small (with respect to any load resistor R_L we consider feeding) so that most of the output voltage appears across this load. The voltage gain of the amplifier is given by $A=v_{out}/v_{in}$.

Now let us look at the other amplifier configurations. In Chapter 1, when discussing nuclear radiation detectors we mentioned that sometimes we are interested in the current signals rather than the voltage signals from a detector.

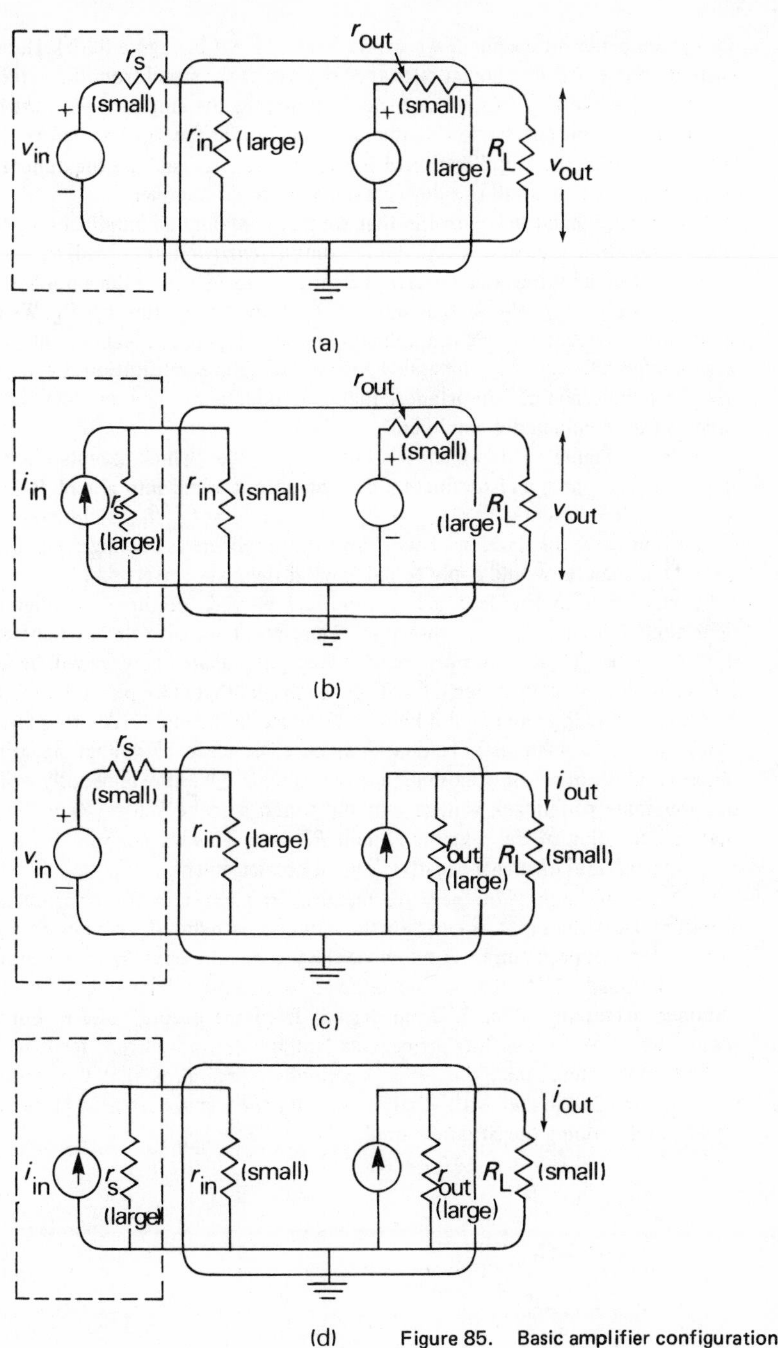

Figure 85. Basic amplifier configurations.

In this case the arrangement we would use is shown in Figure 85(b). Here the current source and its parallel resistance represents the signal from the detector: in order that most of this current should flow into the amplifier we must have r_{in} *small.* The output section of the amplifier can be the same as before. Note that in this case the amplification A is given by v_{out}/i_{in}, and has the dimensions of resistance: we thus call this device a transresistance amplifier.

On the other hand it is possible that we might require an amplifier to be fed from a voltage source and to provide an output *current* into a load R_L. In this case it will be advantageous to have the output impedance of the amplifier very large with respect to R_L, so that most of the current flows through R_L. We thus show the output section of the amplifier (Figure 85 (c)) as an ideal current source with a large resistance r_{out} in parallel with it. Here the amplification $A = i_{out}/v_{in}$ has the dimensions of 1/resistance, that is conductance. The amplifier is thus called a transconductance amplifier.

Finally in Figure 85(d) we show a current amplifier, which accepts a current signal from a source and produces a current output signal into a load R_L. Here we should have $r_s \gg r_{in}$ and $r_{out} \gg R_L$. The gain $A = i_{out}/i_{in}$ is dimensionless. (In all four cases discussed we have drawn the amplifiers as non-inverting, but of course our remarks would apply equally well if they were inverting.)

We now turn to the feedback appropriate to each of these amplifier configurations. The first of these, shown in Figure 86(a), we have already met earlier in this chapter. There we simply called it voltage feedback: now we will be more precise and call it 'voltage derived voltage feedback', that is we pick off a fraction of the output *voltage* and feed it back as a *voltage* to the input. (Another terminology calls it 'voltage series feedback' because the feedback voltage signal is in series with, but of course opposing, the original signal.) We have previously worked out the value of fedback voltage gain and found it to be $1/B = (R_f + R_f^*)/R_f^*$, that is a function of the resistors R_f and R_f^* only. This we saw was one of the attractive features of such a negative feedback arrangement.

Figure 86(b) shows how negative feedback is applied to the transresistance amplifier. Here the input *current* to the amplifier is reduced by allowing a part of it, i_f, the feedback current, to flow via the feedback resistor R_f to the negative output terminal. (Note that we are using an inverting amplifier.) In fact we have obtained a current feedback signal derived from the output voltage. For this reason we refer to this arrangement as 'voltage derived current feedback' or 'voltage shunt feedback', the latter terminology indicating that the feedback current flows in parallel with (that is, shunts) the signal current. The relevant equations describing the situation are:

$$i_{in}^* = i_{in} - i_f \qquad \textbf{6.11}$$

$$i_f = Ai_{in}^*/R_f \qquad \textbf{6.12}$$

$$v_{out} = -Ai_{in}^* \qquad \textbf{6.13}$$

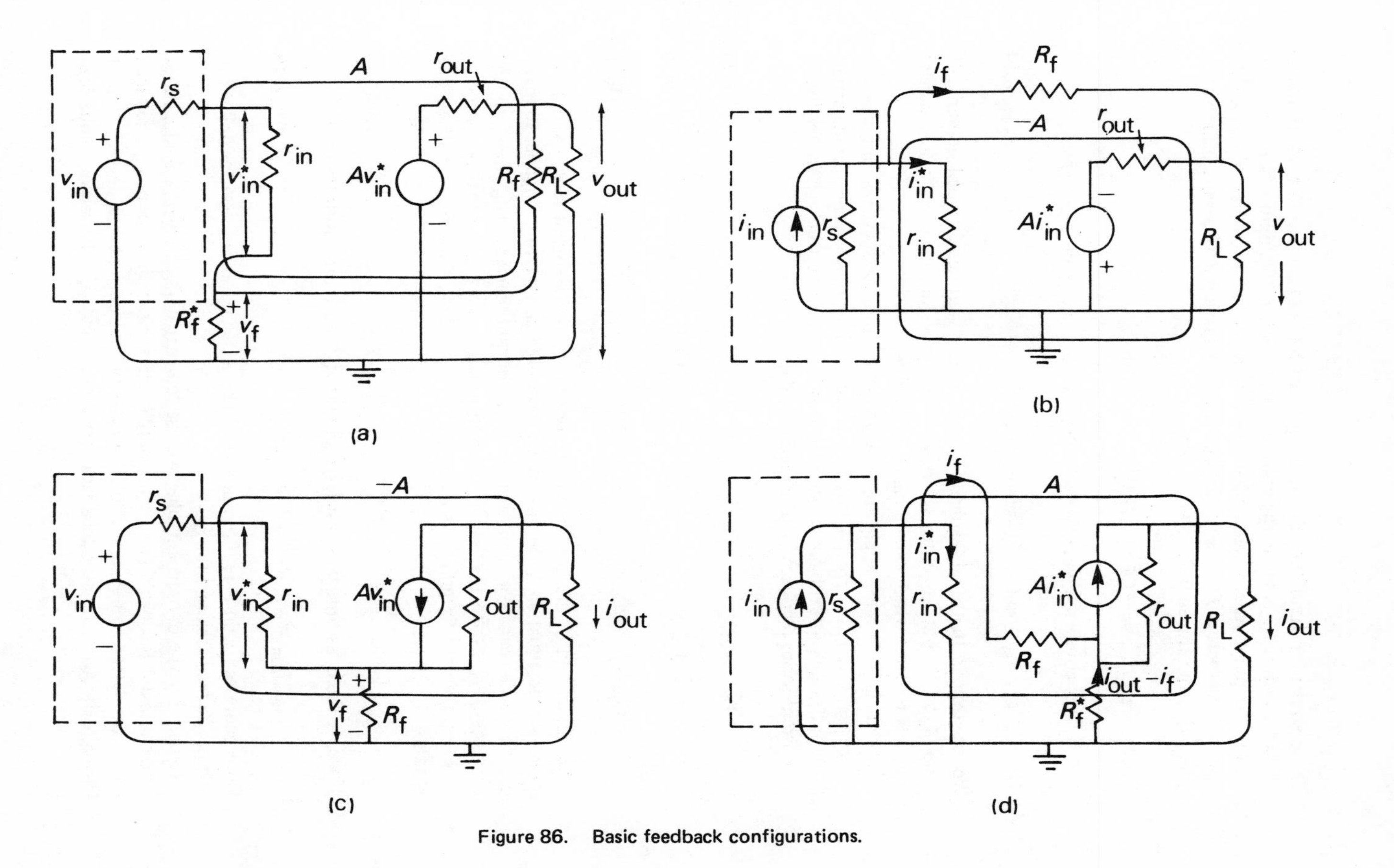

Figure 86. Basic feedback configurations.

These equations should be self explanatory with the exception of **6.12**. Here one notes that since r_{in} is small, the voltage is also small at the input of the amplifier. This is the point where the left-hand end of R_f is connected, and therefore the total voltage across R_f is very closely Ai^*_{in}, as equation **6.12** asserts. Eliminating i_f and i^*_{in} from the equations we obtain:

$$v_{out}/i_{in} = -(A^{-1} + R_f^{-1})^{-1}$$

If A is large compared with R_f we see that the fedback gain is given by:

$$v_{out}/i_{in} \simeq -(R_f^{-1})^{-1} = -R_f \qquad \textbf{6.14}$$

We are thus in the usual feedback situation. The gain has dropped numerically from A to R_f but in return we now have a gain dependant only on a passive device, a resistor, which can be made very stable.

Similar calculations apply to the fedback transconductance amplifier of Figure 86(c). Here the feedback voltage is obtained by allowing the output current to flow through a feedback resistor R_f. This is 'current derived voltage feedback' or 'current series feedback'. Our equations read:

$$v^*_{in} = v_{in} - v_f \qquad \textbf{6.15}$$

$$v_f = Av^*_{in}R_f \qquad \textbf{6.16}$$

$$i_{out} = -Av^*_{in} \qquad \textbf{6.17}$$

(The minus sign in the last equation arises for the same sort of reason as did the minus sign in equation **6.13**. That is, we have used an inverting amplifier to make the feedback negative, and therefore the output current Av^*_{in} flows in the opposite direction to the way we conventionally draw i_{out} for the standard case of the non-inverting amplifier.) Eliminating v_f and v^*_{in} from our equations one obtains:

$$i_{out}/v_{in} = -(A^{-1} + R_f)^{-1}$$

If we, as usual, assume that A is large so that $A^{-1} \ll R_f$, our result becomes

$$i_{out}/v_{in} \simeq -1/R_f \qquad \textbf{6.18}$$

Once more our gain is reduced in size (from A to $1/R_f$) but again is defined in terms of a feedback resistor R_f.

Finally for the current amplifier we use 'current derived current feedback' or 'current shunt feedback' (see Figure 86(d)). Here the input current i_{in} is reduced by i_f, the latter being proportional to the output current and derived from it through the agency of the feedback resistors R_f and R^*_f. The relevant equations are:

$$i_{in}^* = i_{in} - i_f \qquad 6.19$$

$$i_f R_f = (i_{out} - i_f) R_f^* \qquad 6.20$$

$$i_{out} = A i_{in}^* \qquad 6.21$$

where, as previously we assume the voltage on the input is small. Solving as usual we obtain:

$$i_{out}/i_{in} = A/(1 + AB) \quad \text{where } B = R_f^*/(R_f + R_f^*)$$

$$\text{or } i_{out}/i_{in} \simeq 1/B = (R_f + R_f^*)/R_f^* \quad \text{when } A \text{ is large.}$$

The familiar feedback properties are again evident.

The removal of the dependance of amplifier gain on the properties of the transistors making up the amplifier is not of course the only virtue of negative feedback. We have already seen in our discussions in the early part of this chapter how the input and output impedances of an amplifier with voltage derived voltage feedback were improved by the application of feedback, and similar calculations can be made for the other three configurations. However an informal discussion will suffice. Let us first remind ourselves why the output impedance of the circuit of Figure 86(a) is lower than that of Figure 85(a). The output impedance r_{out} is defined as the open circuit voltage divided by the short circuit current. The open circuit voltage is calculated under feedback conditions, and is therefore relatively small, whereas the short circuit current is calculated in the absence of feedback (since short circuiting the output removes voltage derived feedback) and is therefore relatively large. The ratio of these, r_{out}, is therefore much smaller than without feedback. We can say informally that even if we do not go so far as to short circuit the output, but merely attempt to draw some more current from it (by reducing R_L) then voltage derived feedback provides a mechanism which tries to maintain the output voltage constant, that is make the output approximate to a perfect voltage generator, i.e., reduce r_{out}. These remarks apply equally to Figure 86(b) which also features voltage derived feedback. For Figures 86(c) and 86(d) with current derived feedback the reverse is the case. Here the normal situation is to have the output short circuited (or nearly so, since R_L is assumed small) and feedback is operative in this condition, making the short circuit current relatively small. The open circuit voltage, on the other hand, is relatively large, since with no output circuit there can be no current derived feedback. The ratio of open circuit voltage to short circuit current, r_{out}, is therefore much *larger* than without feedback. Informally we can say that if we attempt to obtain a higher voltage output (by increasing R_L) then current derived feedback acts to maintain the output current constant, that is to make the output approximate to a perfect current generator, i.e., increase r_{out}. In all cases feedback acts to improve the output characteristics of the amplifier.

Similar remarks apply to the input impedance. For Figures 86(a) and 86(c) where the signal is fed back as a voltage, the voltage actually appearing across the amplifier input, v_{in}^*, is smaller than v_{in}. Consequently the current flowing through r_{in} is reduced, or in other words r_{in} is effectively increased. This of course is to be desired, for an amplifier being fed from a voltage source (since it guarantees that practically all the source voltage appears across the amplifier input, with little across r_s). On the other hand for the case of current feedback (Figures 86(b) and 86(d)) the current in r_{in} is reduced, making the voltage across it also smaller. A smaller voltage means (since it is being fed from a current source) that r_{in} is effectively smaller. This again is a good thing for an amplifier being fed from such a current source (since it guarantees that practically all the current from the source flows into the amplifier and not through r_s, the source impedance). Again, in all cases, feedback acts to improve amplifier performance.

For practical examples of all four feedback types shown in Figure 86, the reader is referred to reference 2. We have already had many examples earlier in this chapter of voltage derived voltage feedback. The only other type of direct interest to us and which we will discuss further is the current derived voltage feedback type of Figure 86(c). Figure 87 shows how such feedback is applied in practice to a common emitter stage by introducing an unbypassed resistor R_e' in the emitter lead. This resistor plays the role R_f does in our circuit of Figure 86(c). In all other respects the circuit is that of a conventional common emitter stage except that we have drawn V_{cc} and R_c in a slightly unusual manner to underline the fact that R_c is effectively in parallel with the load R_L. Since we postulated for this type of amplifier that R_L be small, we assume that the signal current i_{out} flows almost entirely through R_L as shown, with very little through R_c. The role of R_c as regards signals can thus be ignored. (Note that in this diagram we have shown the direction of i_{out} as it actually is for this inverting

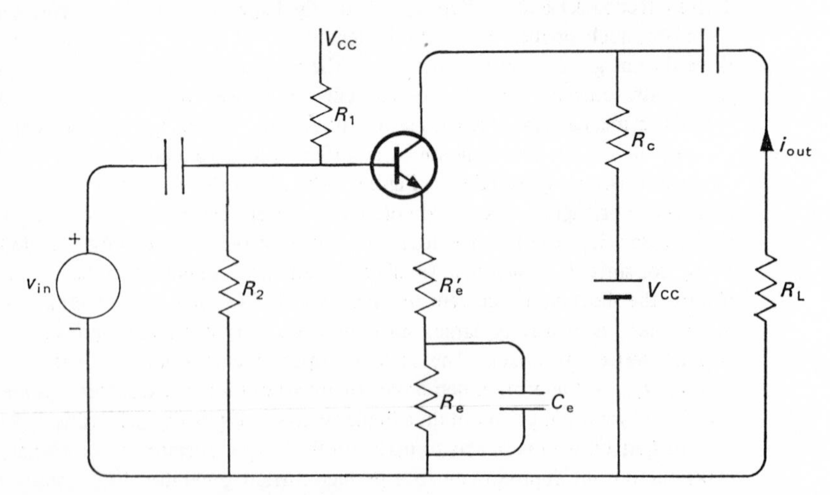

Figure 87. Current derived voltage feedback.

stage.) The current i_{out} delivered to the load will be, as we previously determined from out theory, v_{in}/R'_e, and thus independent both of the load and of any transistor parameters. The input and output impedances are both increased as we would expect for this type of amplifier, although the latter increase will not be obvious if we attempt a direct analysis of the circuit of Figure 87. This is so because our simple model of a transistor in the common emitter mode already assumes an output impedance $1/h_{oe}$ large enough to be ignored. (Compare, for example, Figure 60 and Figure 61.)

While the role of the circuit we are discussing is basically what we have already stated - to provide an output signal current depending only on the feedback resistor, we can, and often do, use it to provide output voltage signals. In this case R_L will be very large and the circuit will look like that of Figure 50, with however R'_e still present. (The fact that Figure 50 shows a *p-n-p* transistor is without significance.) In this case i_{out} must flow through R_c and the output voltage is given by:

$$v_{out} = -i_{out}R_c \quad \text{that is,}$$

$$v_{out} = -(v_{in}/R'_e)R_c \quad \text{or}$$

$$v_{out}/v_{in} = -(R_c/R'_e) \qquad 6.22$$

The voltage gain is thus defined by the ratio of the two resistors R_c and R'_e (and is of course negative, as we expect for an inverting stage). This is a result we have already noted in Section 6.1 and on page 88.

In retrospect one can arrive in a very simple fashion at equation **6.22**. If we apply a signal voltage v_{in} to the base of our transistor then the emitter will 'follow' by a similar amount. Consequently the signal current in the emitter lead will be v_{in}/R'_e. This current also flows through R_c, and the output thus drops by $(v_{in}/R'_e)R_c$, that is $v_{out} = -v_{in}(R'_e/R_c)$, which is the same as the previous result.

6·5 Operational amplifiers

In the discussion of the voltage-feedback amplifiers of Figures 82 and 83 we were at pains to ensure, by the use of a compensating external capacitor, that the feedback factor B was independent of frequency, over the frequency range of interest. It is clear, however, that interesting effects may be expected with feedback networks specifically designed to be frequency sensitive. For example a network which made the feedback factor large (and therefore the gain small) for all frequencies except those around a particular value would make a highly selective amplifier. An example of this sort of arrangement occurs in the next chapter. At the moment we shall investigate the effect of replacing one or both of the feedback resistors R_f and R_f^* in a voltage-feedback amplifier by frequency-sensitive elements, in particular capacitors. The resulting devices are known as 'operational amplifiers', because they are capable of performing accurate mathematical operations on the input signal such as differentiation or integration. Indeed, the voltage feedback amplifiers already discussed are operational amplifiers as they perform the operation of multiplying the input signal by the factor R_f/R_f^*.

Figure 88 shows an operational amplifier with the feedback resistor R_F^* replaced by a capacitor C_f: the triangular symbol for the amplifier itself is the usual one for operational amplifiers. One of the important characteristics we

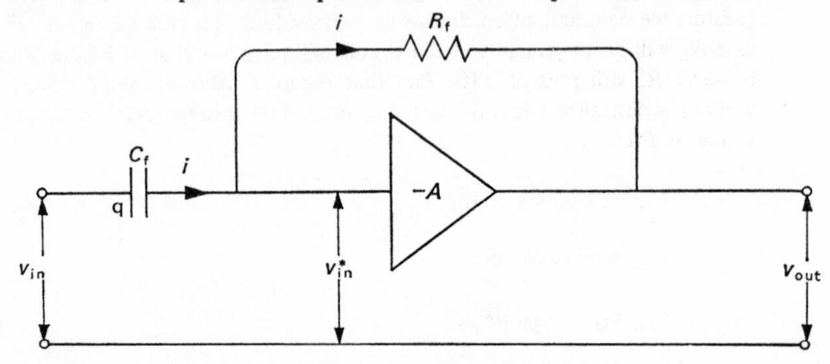

Figure 88. Operational amplifier for differentiation.

found for voltage-feedback amplifiers was the reduction of the voltage v_{in}^* across the actual input of the amplifier to a very small value, and we shall assume that this action also occurs in the present case. The equations governing the operation of the circuit are then as follows.

For the capacitor C_f

$$q = C_f v$$

where v is the voltage across it

or

$$i = C_f \frac{dv}{dt}$$

$$= C_f \frac{d}{dt}(v_{in} - v_{in}^*)$$

or

$$i \approx C_f \frac{dv_{in}}{dt}$$

(remembering the assumption about v_{in}^*).

Similarly for R_f

$$i = \frac{v_{in}^* - v_{out}}{R_f} \approx -\frac{v_{out}}{R_f}$$

Combining this with the previous equation

$$C_f \frac{dv_{in}}{dt} = -\frac{v_{out}}{R_f}$$

or

$$v_{out} = -R_f C_f \frac{dv_{in}}{dt}$$

Thus, apart from the constant $-R_f C_f$, the output is the exact differential with respect to time of the input, at least within the accuracy of the approximations used. The gain of the amplifier does not appear in the calculations, but it is there by implication, keeping the value of v_{in}^* to a low value.

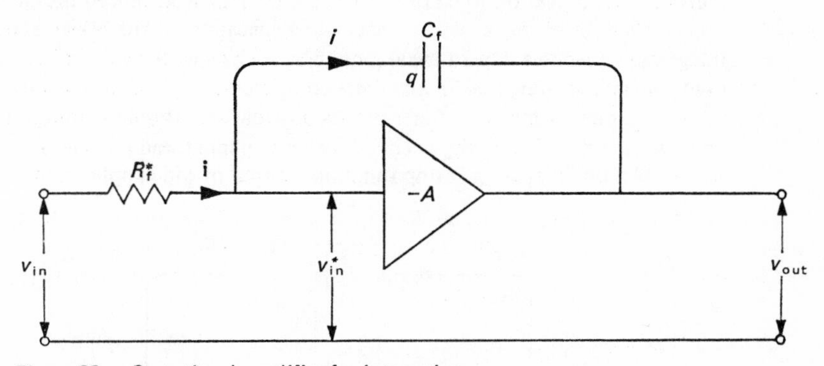

Figure 89. Operational amplifier for integration.

The circuit of Figure 89, where R_f is replaced by a capacitor C_f, performs integration, as can be seen from the corresponding equations

$$i = \frac{v_{in} - v_{in}^*}{R_f^*} \approx \frac{v_{in}}{R_f^*}$$

and

$$i = C_f \frac{d}{dt}(v_{in}^* - v_{out}) \approx -C_f \frac{dv_{out}}{dt}$$

thus

$$\frac{v_{in}}{R_f^*} = -C_f \frac{dv_{out}}{dt}$$

or

$$v_{\text{out}} = -\frac{1}{R_f^* C_f} \int v_{\text{in}}\, dt$$

which verifies our previous statement.

In both the preceding circuits the accuracy with which the relevant operation is performed depends on how closely v_{in}^* can be kept to zero, which in turn depends on the gain A. For perfect accuracy A would have to be infinite. A complete analysis for the practical case when A is large but finite, shows that for the circuit of Figure 89, for example, the quantity $R_f^* C_f$ is effectively multiplied by the large quantity A. Comparing this result with that of equation **3.6** for the simple integrating circuit of Figure 21, we see that for a step input, the output in the present case will be part of a rising exponential of very large time constant. It will thus be a very much better approximation to the straight line we expect for the perfect integrator, than will the output from the simple circuit at the same output voltage level. We have already met this sort of time constant multiplication when discussing the Miller effect; indeed the circuit of Figure 89 is an example of the deliberate application of the Miller effect for integration purposes. Operational amplifiers such as we have described can be used for obvious purposes in analogue computers; they can be used to shape pulses in a more accurate manner than is possible with simple shaping circuits, and the circuit of Figure 89 when fed with a step input, can provide, as we have seen, a steadily increasing 'ramp function' output of wide application.

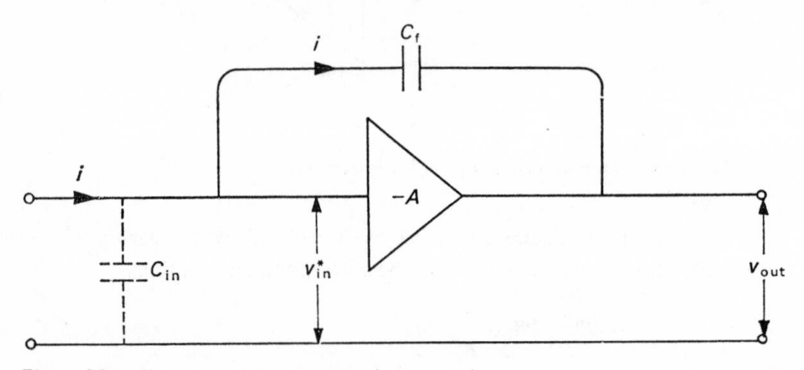

Figure 90. Charge-sensitive amplifier (schematic).

The last circuit of this general type which we shall discuss, is one of particular importance for nuclear physics instrumentation. It is the 'current integrator' or 'charge-sensitive amplifier' shown in Figure 90. In this case a feedback capacity C_f takes the place of R_f while the other capacity C_{in} represents the input capacity of the amplifier plus the additional capacity due to cables, etc, which may be present. We are particularly concerned with the input capacity in the present case, as the input to the amplifier is a current i, which would normally be expected to charge C_{in}. However, the action of C_f and C_{in} will be analogous to that of R_f and R_f^* in keeping v_{in}^* small and a more detailed analysis confirms this. Thus the capacity C_{in} is hardly charged at all, and the bulk of the current i flows to charge C_f. The usual equation for this capacitor is

$$i = C_f \frac{d}{dt}(v_{in}^* - v_{out}) \approx -C_f \frac{dv_{out}}{dt}.$$

Hence

$$v_{out} = -\frac{1}{C_f}\int i\, dt \qquad \textbf{6.23}$$

from which the name 'current integrator' is derived. If the source of current is a nuclear radiation detector, a current flows for a very short time only, and it will be more appropriate to write equation **6.23** as

$$v_{out} = -\frac{Q}{C_f}$$

where Q is the total charge produced by the detector, and is normally a measure of the energy of the particle or radiation which entered it. The reason for the name 'charge-sensitive amplifier' or 'charge-to-voltage converter' is now apparent. The conditions for correct functioning of the charge-sensitive amplifier may be determined by analogy with the resistor feedback case where $AB = AR_f^*/R \gg 1$. The impedance of a capacitor is inversely proportional to its capacitance, so we might deduce that the corresponding condition here would be $A(1/C_{in})/(1/C_f) \gg 1$ or $AC_f \gg C_{in}$, which closer analysis confirms. For $A = 100$ and $C_{in} = 10$ pF a suitable value of C_f would be 1 pF.

What are the advantages of the charge-sensitive amplifier over the ordinary voltage-sensitive type (that is, as in Figure 90, but with C_f removed). There, if a charge Q arrives from the detector and is deposited on C_{in}, the input voltage is Q/C_{in} and the output voltage of size $-AQ/C_{in}$. Comparing this with the value of $-Q/C_f$ for the charge-sensitive case where all the charge is drawn onto C_f we see, using typical values as indicated earlier, that the voltage-sensitive circuit has an output larger by a factor of ten. Usually some feedback would be applied to this amplifier to make it less sensitive to component changes and this would reduce its margin of superiority as regards size of output, but there is another important reason for preferring the charge-sensitive device. This is the fact that within the limits of the approximation used its output is independent of the input capacity. Suppose it were necessary to insert, between the detector and amplifier input, a cable of such length that its stray capacity to ground doubles C_{in} say. With the voltage-sensitive amplifier, the output signal corresponding to a nuclear particle with a certain energy (that is, to a particular input charge Q) would be halved; with the charge-sensitive device, the output signal is unchanged. Because of the nature of the experiments, and in the case of semiconductor detectors because of the nature of the detectors themselves, this property of the charge-sensitive arrangement makes it the preferred choice for detector pre-amplifiers in most cases.

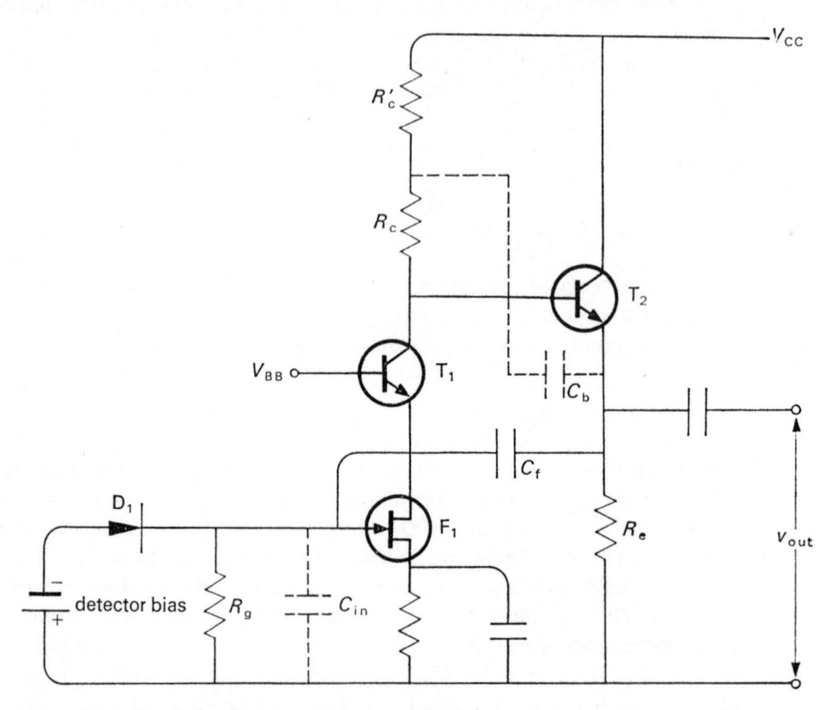

Figure 91. Circuit for charge-sensitive amplifier.

Figure 91 shows the circuit of a relatively simple charge-sensitive amplifier. Ignoring the function of the capacitor C_b for the moment, the action is as follows. The semiconductor detector D_1 which is basically a high-quality, reverse-biased diode, is normally non-conducting. When a nuclear particle enters the detector, a momentary current flows, which in the absence of feed-back action, would normally charge the stray capacity C_{in}. The FET F_1 and transistor T_1 form a cascode input stage (with the usual advantage of freedom from Miller effect) whose output goes via a direct coupled connexion to the base of the output emitter follower T_2. The feedback capacitor C_f, on which the charge to voltage conversion action depends, is connected from the emitter of T_2 to the gate of F_1, that is from output to input. The circuit will work as described, but its operation is improved by the addition of a coupling capacitor C_b connecting the emitter of T_2 back to the join of R_c and R_c'. The effect of this is as follows. Suppose that on the arrival of a signal, the collector of T_1 goes more positive. This signal is fed to the base of the emitter follower T_2; a signal almost as large appears by emitter-follower action on the emitter of T_2, and this in turn is fed to the top end of R_c. (R_c' is simply a load across which this signal can be developed: in its absence the output of T_2 would be shorted to ground for signals via the collector supply V_{CC}.) The voltage at the top of R_c thus follows almost exactly the voltage at the bottom, with the result that the current through it is reduced by a large factor over what one would normally expect to flow in it, as a result of the signal. In other words the size of R_c has been effectively increased for signals, and the gain correspondingly improved, without the unacceptably large d.c. drop associated with an actual resistor of equivalent size. The increase in gain improves the feedback action, by allowing the condition $AB \gg 1$ to be better fulfilled. This technique of 'bootstrapping' is a widely used one by no means confined to the type of circuit we are discussing. It can be used for example to effectively increase the size of base or gate resistors and thus improve the input impedance of amplifiers.

There is a final point about the charge-sensitive amplifier. Its action consists in drawing the incoming charge to the feedback capacitor and holding it there to produce an output step of size $-Q/C_f$. If the gain of the amplifier is infinite this voltage will continue to appear at the output for an indefinite period leading to 'pile-up' troubles with successive pulses. With a practical amplifier of finite gain A, some discharge of C_f will take place, and it can be shown that the output will fall with a time constant of $Ar_{in}C_f$, where r_{in} is the input impedance of the amplifier, that is, R_g for the case of the circuit of Figure 91. For reasons to be discussed in the chapter on noise, it may be necessary to fix R_g at a large value, in which case the time constant AR_gC_f may still be inconveniently large. Consequently it is usual to place directly across C_f a resistor R of such a size as to make the fall time, now simply RC_f, the desired value. This resistor R will now provide a d.c. link between input and output (since C_f with which it is in parallel is between input and output – see Figure 91), and we may wish to prevent this by placing a large capacitor in the line, whose only function will be to block this d.c. connexion. However with a little ingenuity we can in fact leave this d.c. connexion, and use it to provide negative d.c. feedback for the amplifier, with a resultant increase in stability. Indeed there is really no need now to have both R_g and R in the circuit, and we could have as easily achieved our purpose without R by connecting the bottom end of R_g, not as shown, but to the junction of C_f and R_e (that is put R_g directly across C_f), choosing at the same time a suitable value for the decay time constant R_gC_f. As a final sophistication it is even possible to eliminate R_g entirely, and discharge C_f by means of an optoelectronic coupling (i.e. a photodiode viewing a light emitting diode) across it. Eliminating the resistor in this fashion is obviously even better from the point of view of noise than simply making it large.

References and further reading

1. W. GOSLING, *Applied Electronics,* Macmillan, 1975.
2. J. MILLMAN and C. C. HALKIAS, *Integrated Electronics,* McGraw-Hill, 1972.

CHAPTER 7

Circuits with positive feedback

7·1 General principles of bistable circuits

Some of the unfortunate effects of positive feedback in amplifiers have already been encountered. In this chapter some valuable circuits which may be constructed using this form of feedback are discussed. They fall into two main classes, square-wave generators and trigger circuits on the one hand, and sinusoidal oscillators on the other; we shall deal with them in that order. The basic equation for an amplifier with positive voltage feedback (that is for an amplifier where a fraction of the output voltage is fed back to the input in order to increase the input signal already there) is, from equation **6.7**,

$$\text{voltage gain} = \frac{A}{1 - AB} \qquad \textbf{7.1}$$

If the feedback factor B is such that $AB < 1$ the amplifier behaves in a fairly normal way with a larger gain than for the unfedback case. But this gain is very dependent on the value of A, and thus on the characteristics of the transistors from which the amplifier is constructed. The output impedance will be larger, and the input impedance smaller than for the case with no feedback. However, there are applications in which positive-feedback amplifiers with large gains are useful. As AB approaches unity and the gain rises correspondingly, smaller and smaller input signals will drive the amplifier to the limit of its output. For $AB \geqslant 1$ equation **7.1** cannot be used to predict what will happen because the assumption made in deducing it, that a linear relation existed between input and output, no longer holds. We can see what happens as follows. Even without a signal the amplifier will now try to produce a large, indeed infinite output; but as the output increases the limit of the linear response of the amplifier is quickly reached, the gain A begins to fall, and with it the quantity AB. When the latter has fallen below unity a new position of equilibrium is possible, usually far removed from the normal conditions with no input. As B is often unity or close to it for the cases we shall discuss, the new position of equilibrium has A quite small. Indeed it is usual for the amplifier to swing right to the point where A is zero, and the amplifier inoperative. Clearly there are two directions in which the output can start to move when AB becomes greater than one, depending on a particular random fluctuation at the input, so there are consequently two possible equilibrium positions where A falls to zero, one with the

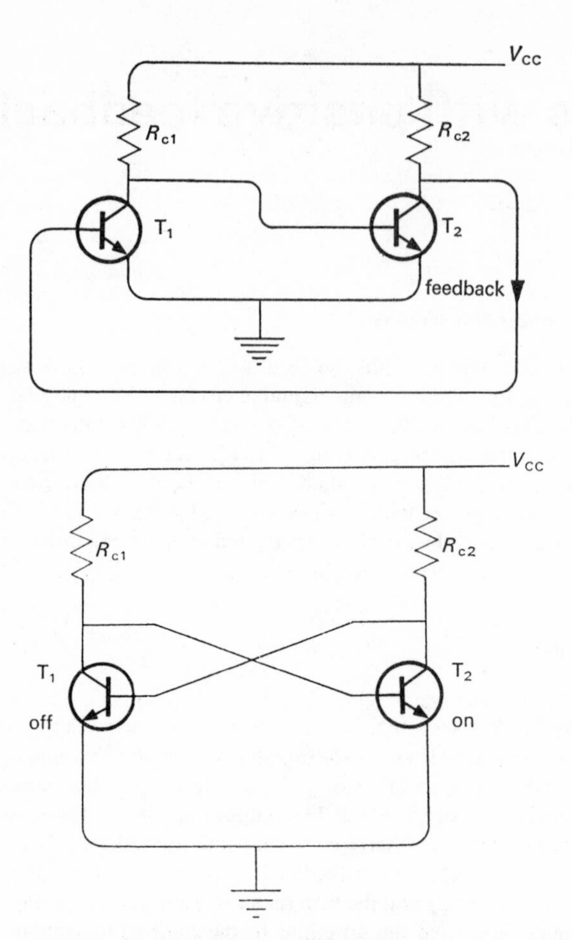

Figure 92. (a) Two-stage amplifier with positive feedback. (b) Same amplifier redrawn to emphasize symmetry of arrangement.

output transistor cut off, and the other with the output transistor fully on, and the preceding stage cut off.

These considerations will probably be clearer in the practical example in Figure 92(a). Here there is what looks like a two-stage transistor amplifier with d.c. coupling between the stages and the whole of the output fed back to the input. Figure 92(b) is the same circuit redrawn to show the symmetry of this particular arrangement. Because of this symmetry one might think it possible for the circuit to possess a state in which equal currents flowed in T_1 and T_2, and voltages at corresponding points of the two transistors were the same, and this in principle is true. However the slightest fluctuation in the voltage at the base of T_1 will be amplified by this transistor, fed to T_2, amplified further, and then fed back to the base of T_1 to augment the original fluctuation. As the full output from the collector of T_2 is fed back to the base of T_1, the factor B will be unity, and AB will be very much greater than one; the circuit will thus swing almost instantaneously into an equilibrium position at the end of the transistor's characteristics where AB has fallen to less than one, as discussed previously. Such an equilibrium position is indicated in Figure 92(b), where T_2 has been driven hard on, with a correspondingly low voltage on its collector. This voltage will be so low that it will be smaller than the value ($\approx 0{\cdot}6$ volts for silicon) required on the base of T_1, to which it is connected directly, to cause this transistor to conduct appreciably. T_1 is therefore effectively off, and the arrangement is stable.

(The transistor T_2 is in a rather curious state, with its collector voltage fallen below that of its base. It is said to be 'in saturation' and its operating point lies on the near vertical part of the characteristics on the extreme left of Figure 48(a). At first one might imagine that the current to the collector would cease, or even reverse itself under these conditions, but we can easily see that this is not so. In the case where the collector voltage has fallen to exactly that of the base (that is, when the transistor is driven on quite strongly), the base bias must be much higher than normal, and the base current correspondingly so. This means that there are a great number of electrons (for an *n-p-n* device) being injected from the emitter into the base region, and a large concentration of minority carriers there. Consequently many will diffuse across the base–collector junction, even without a voltage to collect them. So the collector will have to fall appreciably *below* the level of the base, say of the order of one volt, before this flow can be stopped, and the possibility of the situation which we envisaged in Figure 92(b) is thus assured.)

Another equilibrium position obviously exists with the roles of T_1 and T_2 interchanged, and hence this circuit is known as a 'bistable multivibrator' or simply a 'bistable'. It can be changed from one of these states to the other by the application of a suitable pulse to turn off the 'on' transistor, or alternatively by one which turns on the 'off' transistor, though the former is usually preferred, because we have the amplification of a conducting transistor to help us. Suppose a negative pulse is applied to the base of the *n-p-n* transistor T_2, to turn it off. As it starts to turn off, the voltage at its collector rises, which begins turning T_1 on. In principle if the circuit is brought back to the place where the loop gain AB is again greater than unity it will then move of its own accord to the other stable position with T_2 off. In practice, and particularly for a fast change over, a more detailed analysis shows that we may have to help the transition further on its way; in either case the pulse required to do this will only be of the order of one volt. Bistable multivibrators and other circuits to be described shortly belong to the family of 'regenerative-switching circuits' or simply 'trigger circuits'. 'Regenerative' refers to the use of positive feedback to switch the circuit from one stable state to the other, while the term 'trigger' implies that, like the firing of a gun, the input pulse is used only to start the process, which then proceeds of its own accord to its conclusion. Applications of the circuit of

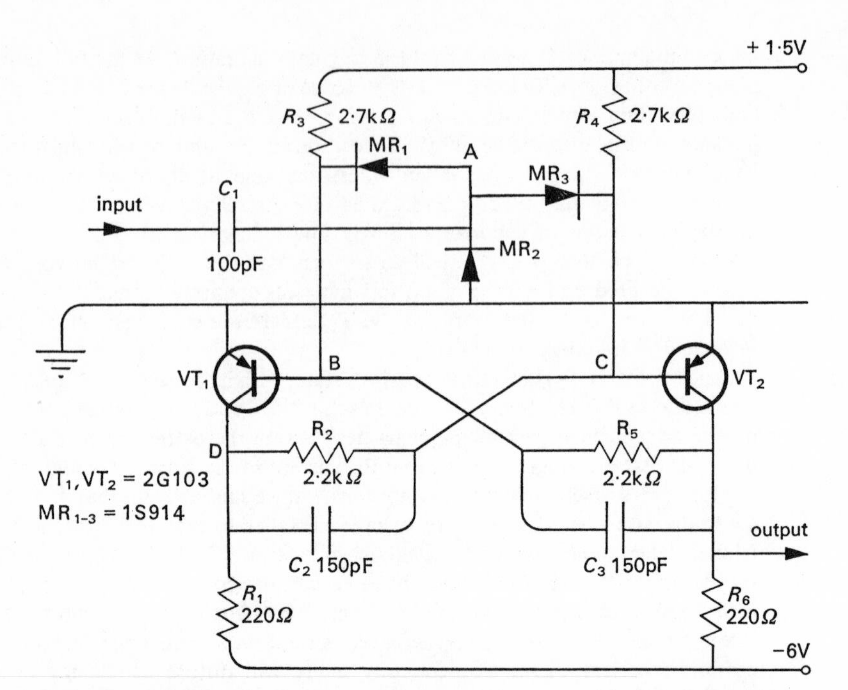

Figure 93. Bistable circuit with separate base-biasing supply (courtesy of Texas Instruments Ltd).

Figure 92(b), or similar circuits, as 'memories' and as counting devices will be met later.

The circuit we have just been discussing, although very simple, has some important disadvantages. In the first instance, the transistor labelled as 'off' is really very slightly 'on', and a preferable arrangement would be one which quite definitely reverse biased the base–emitter diode, and thus removed any possibility of trouble with I_{CO} at higher temperatures. Secondly, the difference between the voltage at the collector of a transistor when it is switched from the 'off' to the 'on' condition is very small, which means there is a very small output voltage from this arrangement. For these, and other reasons, a circuit of the kind shown in Figure 93 is preferable. This diagram may seem 'upside down' compared with that say of Figure 92(b), but as it is a commonly used convention to place the positive supply voltage at the top of the picture, and *p-n-p* transistors under this arrangement will always so appear, it is good practice to become familiar with them in this position. Ignoring for the moment the functions of the diodes MR_1, MR_2 and MR_3, as well as the small capacitors C_2 and C_3, let us concentrate on the d.c. potentials at various points. Let us assume also that it is the transistor VT_1 which is hard on: its collector voltage will then be close to that of its emitter, that is ground. The voltage on the base of VT_2, at the point C, can now be found from the values of R_2 and R_4, which form a potential divider between the point B (near ground potential) and the +1·5 volt bias supply. This turns out to be +0·7 volts approximately, which ensures that VT_2, a *p-n-p* device, is well turned off. Furthermore as the value of R_6 is so much less than that of R_5 and R_3, the voltage at the collector of the 'off' transistor, VT_2, is close to −6 volts, and thus very different from that of the conducting transistor. Thus by the addition of two potential divider networks and a bias supply we have met the objections mentioned in connexion with the previous simple circuit. (Indeed, the bias supply could be dispensed with at the expense of the voltage available across the collector resistor, by putting the two emitters to ground via a common bypassed resistor. This would provide an automatic negative bias for the emitters, so that the top ends of R_3 and R_4, previously at +1·5 volts, could now be grounded, yet still be positive with respect to the emitters as required.)

We now discuss the features of Figure 93 which were previously ignored, beginning with the diodes. The input arrangements shown are such as to allow the circuit to be used as a binary counter, or a 'divide by two' device in a way we shall discuss in detail in a later chapter. For this purpose we require that if an input pulse puts VT_1 off and VT_2 on, a further *identical* pulse should put VT_2 off and VT_1 on again, thus causing the bistable to revert to its original condition. This is achieved by the action of the so-called 'steering diodes' MR_1 and MR_3. First however, let us dispose of MR_2. This could be replaced by a resistor, on which the input pulse (which is positive) would be developed. For positive signals, for which it is reverse biased, it in fact acts as a high value resistor. For negative signals, or for undershoot on positive signals, the diode conducts, shorts the signal to ground, and thus clears the way for the arrival of a further genuine input signal. Returning to the 'steering diodes' we note that for the case we were discussing, where VT_2 was off and VT_1 on, the point C is slightly positive, and the point B, at the base of the 'on' transistor, slightly negative. Thus the diode MR_3 will be non-conducting, while MR_1 will be conducting, and will 'steer' the positive input signal to the base of the 'on' *p-n-p* transistor, VT_1, where it starts the regenerative cycle in which VT_1 is turned off, and VT_2 on. Under these new circumstances diode MR_1 will be non-conducting, and MR_3 conducting, and the succeeding input pulse will be steered to VT_2, where it will initiate the change back to the original state. In this way an input pulse will flip the bistable to its other state, irrespective of the state in which it finds it. For further details of the triggering mechanism, which is in fact more complicated than this simple approach suggests, reference 3 should be consulted.

The small capacitors C_2 and C_3 are known as 'commutating' or 'speed-up' capacitors. They have the same sort of function as the small capacitors placed across one of the resistors in a feedback network, that is they compensate for stray capacity and make the resistive networks frequency independent. In the present case C_3 compensates for the base–emitter capacity of VT_1, and C_2 plays a similar role for VT_2. These base–emitter capacities are mostly diffusion capacity, which represents the effect of charge storage in the base of the transistor, so the role of the speed-up capacitors could be alternatively considered as helping in the removal of this charge when a transition occurs.

The final question about this bistable circuit is how quickly it can be made to change from one state to another, that is how closely triggering pulses may follow one another. This resolution time will depend not only on how fast the transition between conducting and non-conducting states can take place, but also how long it takes for the various capacities in the circuits to settle down to their final conditions. The transition time will depend on the sort of considerations discussed in Chapter 4 in connexion with the high-frequency response of the common-emitter amplifier. Clearly, transistors with large values of ω_β, and low-load resistors should be used for best results. In fact, the resolution is usually limited by the settling time of the circuit and, in particular, the time for the speed-up capacitors to assume their final voltages. A full discussion of this process is lengthy and complicated by the fact that it depends on the size of the input signal. We should mention here that for the case of VT_1, originally conducting, being driven well off by an input signal, we would not expect any charging difficulties with C_3, as it is connected between the low impedance source of signals, and the collector of a transistor VT_2, which is coming on heavily. On the other hand although C_2 (Figure 93) has its right-hand side connected to a point of relatively low impedance (the base of a transistor coming on heavily), at its other side it must charge through R_1, as T_1 has now gone off. The limiting time constant in this case is thus $R_1 \times C_2$, or 33 nanoseconds for the numerical values given. Allowing a few time constants for settling, we expect that this particular circuit would trigger reliably on pulses separated by only 100 nanoseconds, that is at a rate of 10^7 pulses per second, and experimentally this is found to be so. We shall return to a discussion of the applications of bistable circuits, particularly in their integrated form, in Chapter 10.

7·2 The monostable multivibrator

The bistable multivibrator had two cross-coupling resistors R_2 and R_5 which helped to provide the system with two stable states. If this direct coupling is modified by replacing one or both of these resistors by capacitors, we obtain important classes of circuits with one or no permanently stable state. Figure 94 shows a circuit in which one of the cross-coupling resistors has been replaced by a fairly large capacitor C_2; the other resistor still remains, and with it its small speed-up capacitor C_1. The base of T_2 is connected to the supply voltage via the resistor R_{b_2}. This circuit is known as a 'monostable multivibrator', because, as we shall show shortly, it has only one permanently stable state. It is also known as a 'one-shot multivibrator' or 'univibrator'. Its action may be described as follows. As the base of T_2 is connected to the supply through the 3·9-kΩ resistor R_{b_2}, this transistor will normally be hard 'on' with its base a fraction of a volt positive, and drawing base current through R_{b_2}. Its collector will be close to ground potential, and hence from the divider action of R_k and R_{b_1} the base of T_1 is at about −2 volts. This *n-p-n* transistor is therefore well cut off, and the

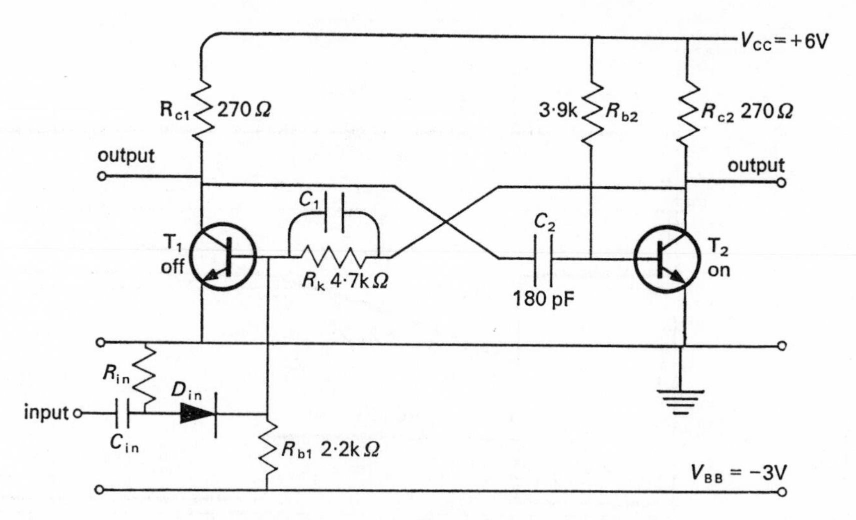

Figure 94. Monostable circuit (courtesy of Texas Instruments Ltd).

situation is perfectly stable. A change to a state where T_1 is on and T_2 off can be initiated by applying either a negative pulse to the base of the 'on' *n-p-n* transistor T_2, to turn it off, or (as shown) a positive pulse to the base of the 'off' transistor T_1 to turn it on. (The diode D_{in} through which the pulse is applied, although not always necessary, is a useful precaution, as, once the change-over has taken place, it becomes reverse biased in the manner of the steering diodes of Figure 93, thus disconnecting the monostable from the trigger source, and preventing any interaction between them during the time the circuit is in its other state.) Once T_1 has been turned on, regenerative-switching action completes the transition by turning T_2 off (and finishes the turning on of T_1 if the original input did not complete this). This action is in no way impaired by the existence of a capacitor C_2 in the feedback loop rather than a resistor, because this capacitor readily transmits signals from the collector of T_1 to the base of T_2.

This new state with T_1 on and T_2 off is not permanent, but only quasi-stable (that is the circuit will return spontaneously to its original stable state after a certain time). The easiest way to see this is to consider the voltage on the base of T_2. When the input is applied and T_1 goes on, a negative step of nearly six volts appears at the collector of this transistor and this is applied to the base of T_2 through the capacitor C_2. The action is very similar to that in Figure 24 with C_2 playing the role of C there, and R_{b_2} that of R. (The transistor T_2 is cut off when the step arrives, and so plays no part in the action.) The voltage at the base of T_2 must then decay exponentially with a time constant $\tau = R_{b_2} C_2$ as shown in Figure 95. This exponential decays towards the six-volt positive supply, because if T_2 were not there, C_2, being connected to this supply via R_{b_2}, would eventually reach this voltage. Before it can reach it however (in fact when it arrives at a small fraction of a volt above ground), T_2 comes on, and regeneration then

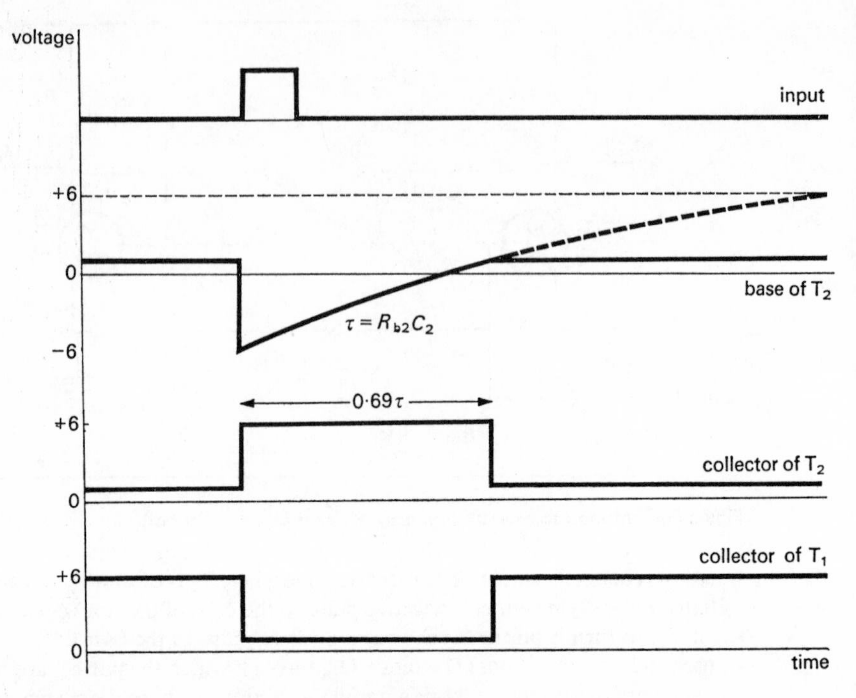

Figure 95. Waveforms in monostable multivibrator.

returns the monostable to its original stable state, where it will rest until another input pulse is applied. The voltage at the collector of T_2, where an output can be taken, rises sharply from the low collector-saturation voltage to almost the full six-volt supply voltage, when the monostable is triggered, and remains there until the voltage on the base of the transistor has decayed sufficiently for the reverse transition to take place. This output is thus a rectangular pulse, 6 volts high, which lasts for the time it takes the base, on its way from −6 volts to +6 volts, to reach a point just above ground (see Figure 95). The time to arrive at this half-way point on an exponential of time constant $R_{b_2}C_2$ is $R_{b_2}C_2 \log_e 2$ or $0{\cdot}69 R_{b_2}C_2$. For a capacitor C_2 of size 180 pF this works out at 0·5 μsec, but values from a small fraction of a microsecond to many seconds can be obtained with this type of circuit by suitable choice of capacitor. A rectangular pulse of the same magnitude but of opposite polarity may be obtained from the collector of T_1.

Small differences would be expected in practice from the idealized waveforms of Figure 95. For example, we know from the discussion about the bistable that the transition from one state to another cannot be expected to take place instantaneously, and consequently the outputs from either collector will have finite rise and fall times. The biggest of these will be $R_{c_1}C_2$ (analogous to

$R_1 \times C_2$ for the bistable) which occurs when T_1 switches off and the capacitor C_2 has to charge through R_{c_1}. Consequently the positive-going edge of the bottom pulse in Figure 95 will have a relatively poor rise time (although still small compared with its length which is controlled by $R_{b_2} C_2$). For this reason the output from the collector of T_2 is usually preferred.

The most important and useful fact about the monostable multivibrator is that it shows true trigger action, in that the output is independent both in amplitude and length of the input pulse. Provided that it only starts the transition going, the input pulse can have a variety of forms – it can be rectangular or exponential, and its height and length can vary widely without making any difference to the height of the output (approximately equal to the supply voltage) or its duration (defined by $R_{b_2} C_2$). The monostable thus acts as a 'discriminator' in producing zero output for input pulses below a certain size, and a standard output for pulses larger than this value. For example, such a circuit could be used to reject small spurious noise pulses from a nuclear detector, while standardizing in height and length the larger genuine signals.

The monostable can also be used to produce signal delays well beyond the range of the delay lines to be discussed in a later chapter. If the rectangular pulse from the collector of T_2 is passed through a differentiating network (Figure 24)

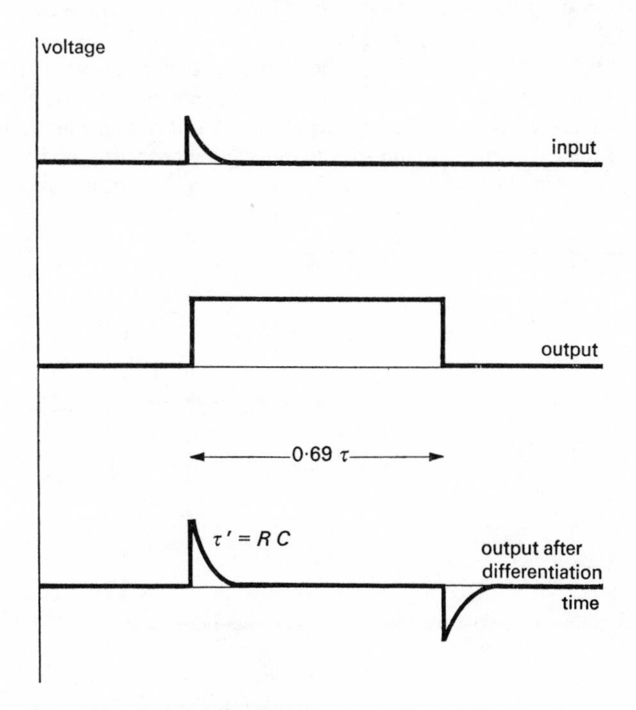

Figure 96. Delayed signal from monostable multivibrator.

with a time constant $\tau' = RC$ very much less than the length of the output pulse, we obtain the result shown in Figure 96. Consequently with this arrangement we obtain a signal delayed with respect to the original signal to the monostable by an amount $0{\cdot}69R_{b_2}C_2$, which can very easily be of the order of milliseconds.

The type of monostable discussed above should strictly be called the *collector-coupled* monostable, as the feedback coupling is from collector to base. There also exists an *emitter-coupled* monostable, whose general principles of operation are the same as for the previous type, but in which the resistive part of the feed-back is provided by an unbypassed emitter resistor, common to the two transistors. Emitter-coupling will be discussed in connexion with another trigger circuit in section 7·4.

7·3 The astable multivibrator

If the remaining resistive cross-coupling in the collector-coupled monostable is removed in favour of another capacitor, as shown in Figure 97, we have an 'astable multivibrator'. Its action and the waveforms shown in Figure 98 should be quite understandable from what has been said about the monostable multivibrator. In this case there is no stable state, but the circuit remains in quasi-stable states, with a particular transistor alternately off and on, for periods determined by the time constants $R_{b_1}C_1$ and $R_{b_2}C_2$. In the waveforms in Figure 98 the case of two nearly equal time constants is shown, and consequently the 'on' and 'off' times for a particular transitor are nearly equal, but it is possible to have them markedly different from one another. As in the monostable case one of the edges of the output pulse has a rise time limited by the product of the collector load and the cross-coupling capacitor, and this is shown in the diagram. The astable multivibrator has an obvious use as a free

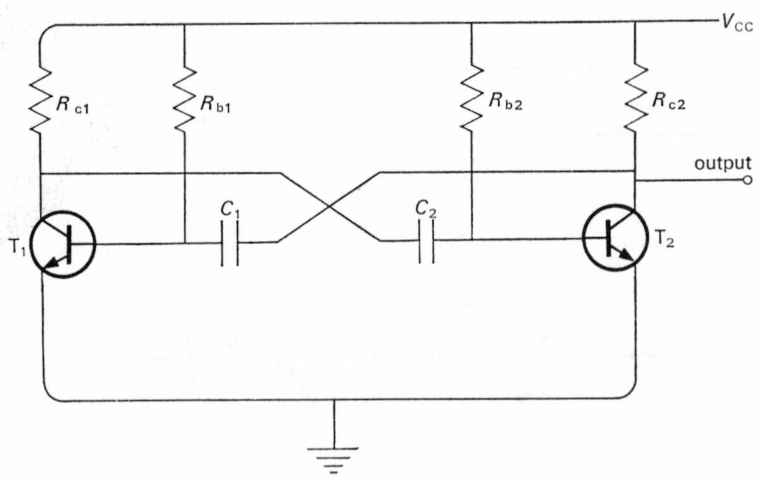

Figure 97. Astable circuit.

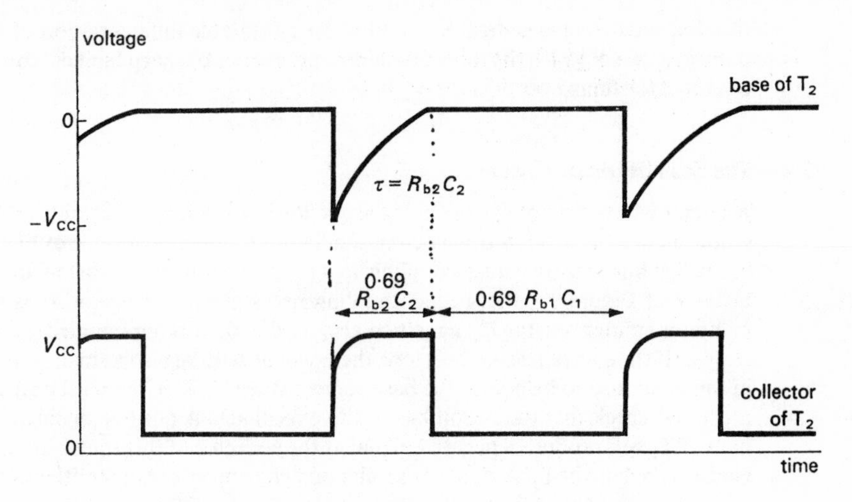

Figure 98. Waveforms in astable multivibrator.

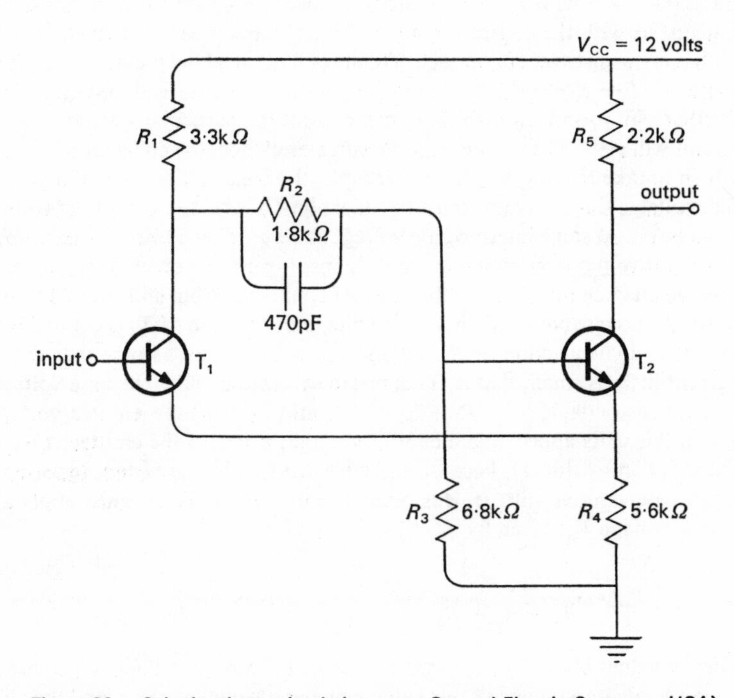

Figure 99. Schmitt trigger circuit (courtesy General Electric Company, USA).

running square wave generator; in addition by suitable differentiation of the output, as we did with the monostable circuit, a series of sharp 'spikes' can be obtained for timing purposes.

7·4 The Schmitt trigger circuit

A circuit which did not appear in the sequence from bistable to astable multivibrators was the Schmitt trigger circuit, or cathode-coupled bistable of Figure 99. It has one resistive cross-coupling R_2 (with speed-up capacitor) as in the bistable of Figure 93; the second coupling necessary for regeneration is the common-emitter resistor R_4, a method also used in the emitter-coupled monostable. (If the source resistance is low, there will usually be also a small resistor in the input lead to help limit the base current when T_1 is driven hard on.) We must first check that the circuit has positive feedback. A positive signal at the base of T_1 will produce a negative signal at the collector of this transistor, and hence at the base of T_2. A signal of similar polarity appears at the emitter of the latter transistor (since the emitter 'follows' the base) and this negative signal is fed back to the emitter of T_1. A negative signal on the emitter of a transistor adds to the effect of a positive signal on the base, so positive feedback exists, and regeneration is possible if the loop gain is sufficiently large. Of course, there is no stable state with both transistors conducting under these circumstances, but just as with the collector-coupled bistable, there will be two stable states with one transistor cut off in each. The special virtue of the present circuit is that there is a free electrode, the base of T_1, which is not directly involved in the feedback loop, and with which we can control the particular state in which the circuit will rest. (There is also another free electrode, the collector of T_2, from which to take the output.) If, for example, the base of T_1 is held at a very low d.c. voltage, this transistor must be off, while if it is held at a very high voltage it must be on. At some intermediate voltage as we gradually bring the base up, the regenerative flip over to the other state must occur. In order to calculate this voltage imagine that T_1 is on the point of conducting (but still cut off) while T_2 is on. As no current flows in T_1 the voltage of the base of T_2 is defined by the potential divider action of R_1, R_2 and R_3, provided we assume that the base current in T_2 is small, that is, T_2 is not in saturation. Thus the base voltage of this transistor is $V_{CC}R_3/(R_1+R_2+R_3)$, and as the base–emitter voltage is small, this is its approximate emitter voltage; it is also the emitter voltage of the input transistor T_1, because these electrodes are connected. Ignoring the small base–emitter voltage, it is clear that to turn on T_1 we must apply at its base a voltage V_{on} given by

$$V_{on} = \frac{V_{CC}\,R_3}{R_1+R_2+R_3}$$

For the values in the figure, this is 6·9 volts. If the input voltage is increased by a very small amount above this value, regeneration takes place, and the circuit switches over to the state with T_1 on and T_2 off. Quite a violent change has taken place in the circuit by increasing the input voltage through the last small amount, so it would not be surprising if, on reducing the base voltage again slightly, the circuit did not return to its original state. We shall show that we must reduce this voltage by an appreciable amount below V_{on} before such a change will occur. This means that the triggering point depends on whether the input voltage is being increased from a low level, or decreased from a high level. Such a 'hysteresis' effect occurs in all trigger circuits, although this is the first time we have studied it in detail. To determine this second triggering voltage, V_{off}, imagine that the base voltage of T_1 has been reduced from a high value to the point where T_2 is about to come on, but is not yet on. We are thus almost at the point of regeneration and a return to the other state. Figure 100 shows the

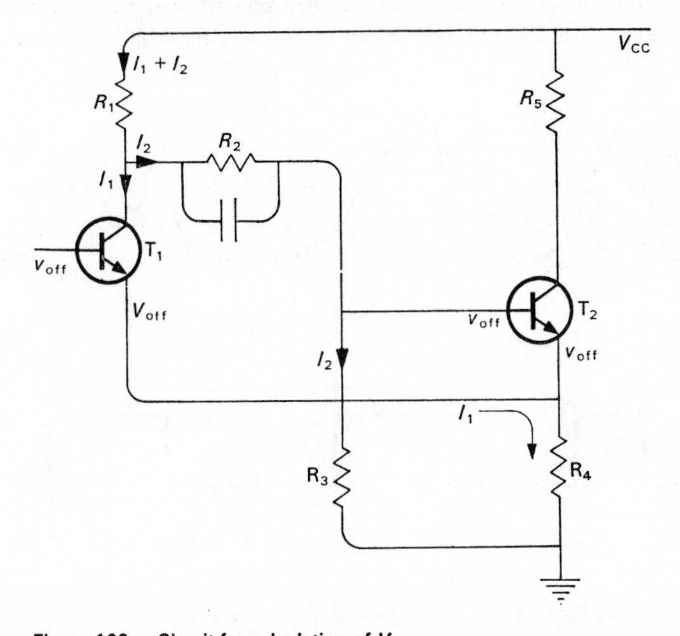

Figure 100. Circuit for calculation of V_{off}.

situation. As T_1 is on and T_2 is on the point of coming on, the bases of both these transistors are close to their respective emitters, which are connected together. Thus these four electrodes are all at approximately the same voltage V_{off}, as shown. Hence the current I_1 is V_{off}/R_4 and I_2 is V_{off}/R_3. The voltage drop across R_1 and R_2 is

$$V_{CC} - V_{off} = (I_1 + I_2) R_1 + I_2 R_2 \qquad 7.2$$

Substituting in equation 7.2 the values for I_1 and I_2 above, we have on solving for V_{off}

$$V_{off} = V_{CC}\left\{1+\frac{R_1+R_2}{R_3}+\frac{R_1}{R_4}\right\}^{-1}$$

which for the numerical values we are using is 5·1 volts. The hysteresis, $V_{on} - V_{off}$ ($=V_H$, say) is thus $6{\cdot}9 - 5{\cdot}1 = 1{\cdot}8$ volts for this circuit. We need not place too much confidence in the exact numerical value obtained, because we have neglected fractions of a volt here and there in our approximations. However the general order of magnitude is certainly right. The amount of hysteresis is clearly a function of the circuit design, and in particular of the loop gain AB. Indeed by adjusting the loop gain to be exactly unity we should be able to eliminate hysteresis entirely. We shall not consider methods of doing this, both because additional components, and further analysis would be needed, and because a certain amount of hysteresis is usually an advantage for applications of this circuit.

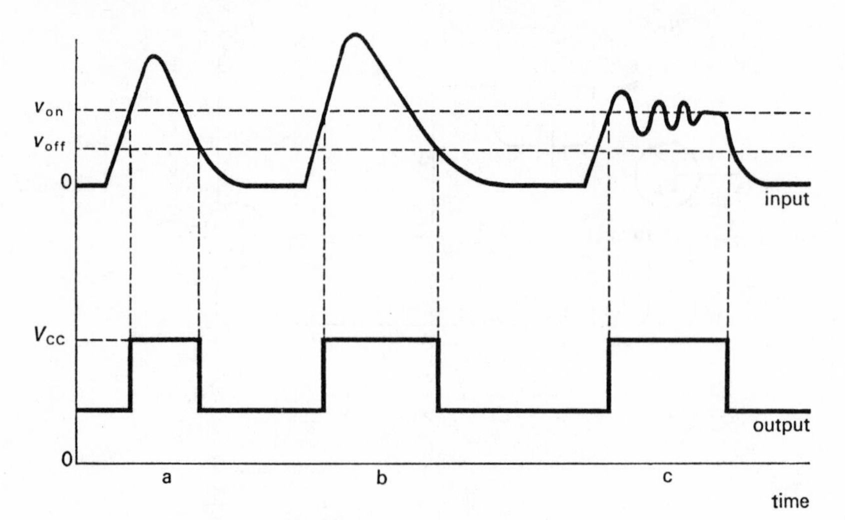

Figure 101. Output from Schmitt trigger circuit for various inputs.

The Schmitt trigger circuit is widely used as a discriminator. In Figure 101(a) we see that as soon as the input signal passes the level V_{on}, regeneration takes place, T_2 goes off and the level of the output voltage at the collector of this transistor rises sharply to the supply voltage. When the input level returns below V_{off}, the output drops to its original level. With the larger input pulse shown in Figure 101(b) the output is of the same magnitude as before, but lasting for a longer time; thus the output pulse is standard only in height. This is usually quite acceptable, but if variations in length are important it can be standardized by passing it through a monostable. Figure 101(c) shows how a certain amount of hysteresis can be helpful. The input pulse illustrated is one with some damped oscillation ('ringing') on the flat top; we do not want the discriminator going on and off a number of times, as would happen when the oscillations crossed the level V_{on} if the hysteresis were zero. All we need to know is that the pulse originally crossed the threshold, and the presence of hysteresis ensures just one single output to indicate this.

Because of its 'free' input electrode, the Schmitt circuit possesses the important advantage for a discriminator (and one not found in the collector-coupled monostable) that it is easy to arrange for the threshold level at which it triggers to be made variable. This is achieved, as in Figure 102, by connecting the base (via a resistor R on which the input signal can be developed) to a variable voltage V. This voltage may be quite simply obtained from a potentiometer across the supply V_{CC}. If, for example, a bias of +3 volts is placed on the base of T_1, an input pulse of size $V_{on} - 3 = 6{\cdot}9 - 3 = 3{\cdot}9$ volts, will trigger the circuit, and so by a simple adjustment of a potentiometer (which can be labelled to read the discriminator level directly) the required variability is obtained. However the maximum external bias applied must not exceed V_{off} or the circuit will not return to its normal state after the pulse disappears. The threshold can thus be varied from V_{on} (with no external bias) to $V_{on} - V_{off} = V_H$ (with the maximum allowable external bias V_{off}), or from 6·9 to 1·8 volts in the present case. The applications in pulse-height-analysis circuits, of the

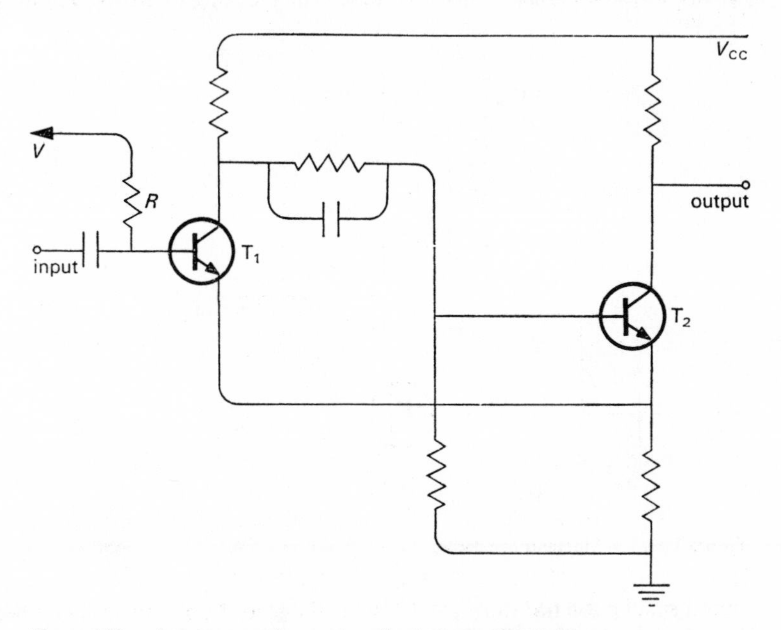

Figure 102. Arrangement for varying discriminator threshold.

Schmitt trigger circuit with variable threshold will be discussed in a later chapter.

Finally one should say that the sequence of circuits we have been treating is not the only way to produce monostable and astable multivibrators, and Schmitt trigger circuits. Basically we need only an amplifier with positive feedback, plus suitable timing and (where appropriate) triggering circuitry. The difference amplifier, with its two inputs, and the possibility, in many integrated forms, of an output which can swing both above and below ground, provides a very suitable mechanism for producing elegant and simple regenerative circuits. (See references 1 and 2.)

7·5 Regenerative circuits with a single active element: the blocking oscillator

So far circuits with two transistors have been discussed, in which the double inversion produced gave a signal which was in the same sense as the input, and thus provided positive feedback. However, two devices, the blocking oscillator and the tunnel diode, for very different reasons, require only a single active element. Both of these have been widely used in the past but neither is easily compatible with integrated circuit techniques, the former because it contains a transformer, and the latter because its manufacture requires rather unusual semiconductor material. Future developments may possibly, therefore, favour the types of circuits discussed in the previous sections, so we shall deal fairly briefly with the devices in question here. The blocking oscillator, Figure 103,

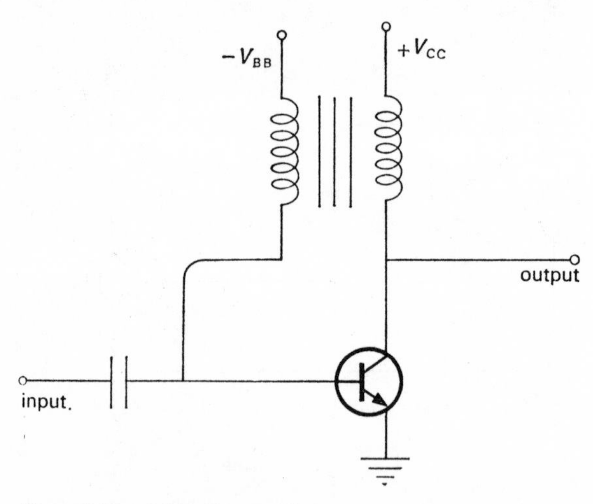

Figure 103. A blocking oscillator.

uses a small pulse transformer to provide the second polarity inversion necessary for positive feedback. In the present diagram it is arranged to act as the analogue of the monostable multivibrator. (The term 'blocking oscillator' arises because the oscillations are not continuous, but will 'block', or cut themselves off, after a single cycle.) The action of the circuit may be explained simply as follows. In the ordinary way the small negative bias applied to the base of the *n-p-n* transistor keeps it cut off in a perfectly stable manner. When a sufficiently large positive trigger pulse is applied to start the transistor conducting, a regenerative action begins. The collector voltage drops, and the sense of the transformer windings are such that the base is driven even further positive, very quickly driving the transistor well into saturation. The circuit is now in a quasi-stable state corresponding to a similar state in a circuit using two transistors. In this case, however, the duration of the state is not determined by a resistor–capacitor combination, but by the characteristics of the transformer itself. As the transistor is in saturation with its collector almost at ground, there is now a nearly constant voltage (equal to the supply voltage V_{CC}) across the primary of the transformer. A transformer is an a.c. device, so it would not be expected to be able to maintain a constant voltage output on the secondary with a steady voltage on the primary. Consequently the secondary voltage applied to the base of the transistor begins to drop. When the base voltage is insufficient to hold the transistor in saturation, the voltage at the collector begins to rise, and regenerative action comes in to help the circuit in this new direction. The transistor is thus quickly cut off, and the quasi-stable state ends. We expect the output at the collector to be a rectangular negative pulse, whose length is determined by the transformer characteristics, and whose height is practically equal to the supply voltage V_{CC}. In practice, because of the inductances in the circuit, the voltage at the collector will overshoot at the end of the pulse to a voltage higher than the supply voltage (see Figure 104). A diode is often connected from collector to supply, to eliminate this overshoot:

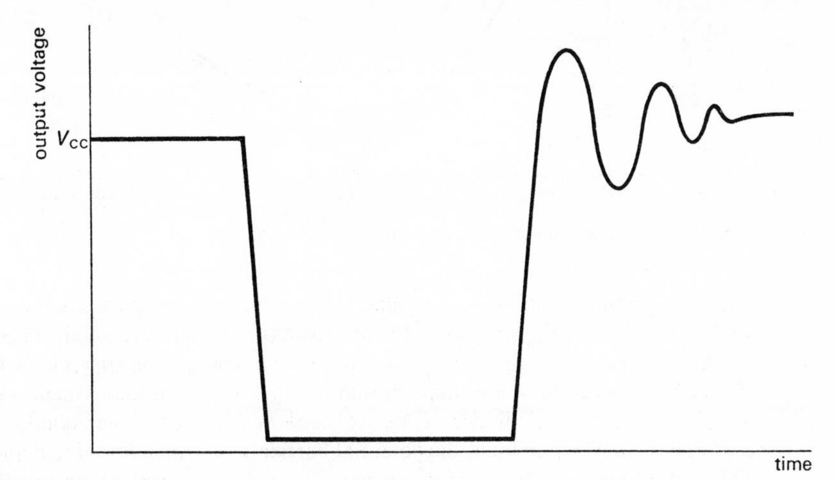

Figure 104. Output from blocking oscillator.

this both improves the output waveform, and also prevents possible damage to the transistor from overvoltage. The diode polarity is arranged so that it is normally non-conducting, but when the transistor collector tries to rise above the supply voltage, the diode conducts, and 'clamps' the collector to the supply voltage. Sometimes the output from a blocking oscillator is taken from a third winding on the transformer. Either end of this can be earthed so positive or negative output pulses can thus be obtained.

With altered bias conditions and possibly the addition of further components the circuit of Figure 103 can clearly be converted into an 'astable' or 'free-running' blocking oscillator. For obvious reasons however no blocking oscillator analogue of the bistable multivibrator exists.

7·6 The tunnel diode

The current–voltage characteristic for a tunnel (or Esaki) diode is shown in Figure 105. The part marked III on the curve is that due to the ordinary diode process. The anomalous rise in part I is due to what is known as 'tunnelling'.

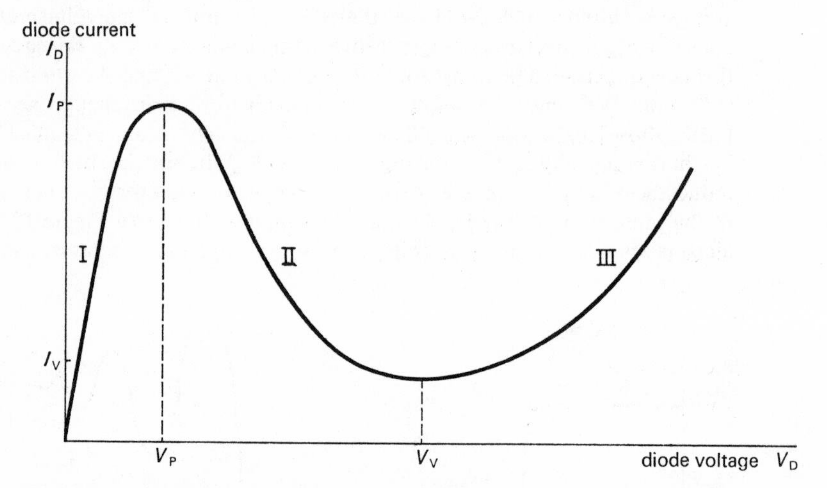

Figure 105. Tunnel-diode characteristic.

This is a quantum mechanical process whereby a particle has a small probability of surmounting a potential barrier, even though on classical grounds, it does not possess enough energy to do so. For tunnelling to be appreciable the depletion layer has to be less than one hundredth of the few microns mentioned as typical for normal diodes. Thus we need heavily doped semiconductor material from which to build the potential barrier, as we are allowed to deplete only a thin region to do so. This explains the previous remark about possible difficulties with tunnel diodes in integrated circuits. The tunnelling is at a maximum for a particular voltage V_P when the energy levels on either side of the junction are brought into a particular relationship, after which the curve drops to join the conventional diode characteristic. The part of the curve labelled II is of particular interest, since there the dynamic resistance of the diode, dV_D/dI_D is negative. The peak and valley voltages V_P and V_V, are small for all semiconductors, and for germanium are 50 mV and 300 mV, respectively. For the same material there is about one order of magnitude, or a little less, between the peak and valley currents I_P and I_V, and the value of I_P might typically lie in the one to ten milliampere range.

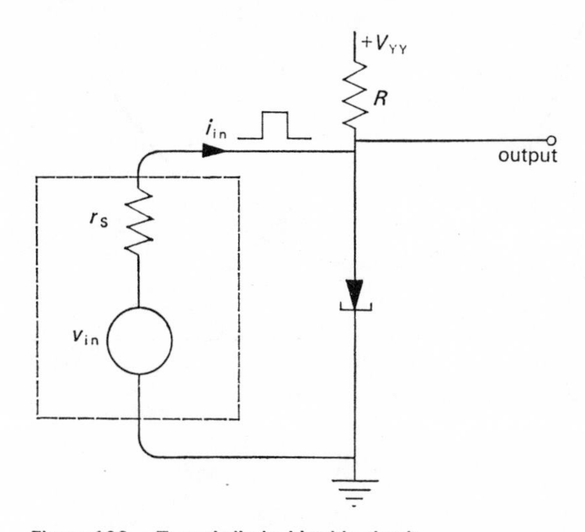

Figure 106. Tunnel-diode, bistable circuit.

Figure 106 shows a tunnel diode fed from a supply voltage V_{YY} through a load resistor R. (The symbol shown for the diode is the most common but not the only one in use.) Figure 107 shows the diode characteristic, crossed by the load line whose slope is, as usual, $-1/R$, and which intersects the voltage axis at V_{YY}. From the magnitudes given previously for the currents and voltages for tunnel diodes, V_{YY} will be less than a volt, and R of the order of 100 ohms, for the arrangement drawn. Possible operating points are given by the intersection of the diode characteristic and the load line, and for the particular values of V_{YY} and R chosen, there are obviously three, of which those of A and C are stable. We clearly have a bistable circuit which we can switch from one state to the other by a suitable signal. Suppose for example that the circuit is originally at the point A, and we feed in a pulse of current i_{in} (that is a voltage pulse from a source with internal resistance r_s which is much greater than R, in order to avoid loading the circuit with a low source impedance, see Figure 106). The new load line is shown dashed in Figure 108; it has the same slope as before, but intersects the diode characteristic only at C′. The circuit thus switches very rapidly to this new position, but as soon as the input pulse ends, it returns to the point C, and the switching action is complete. A similar pulse of reverse polarity will switch the circuit back to its original operating point at A.

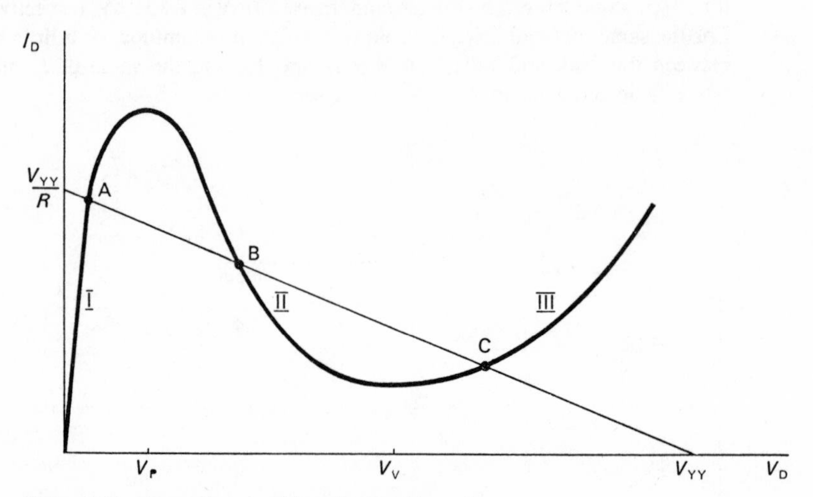

Figure 107. Operating conditions for bistable action.

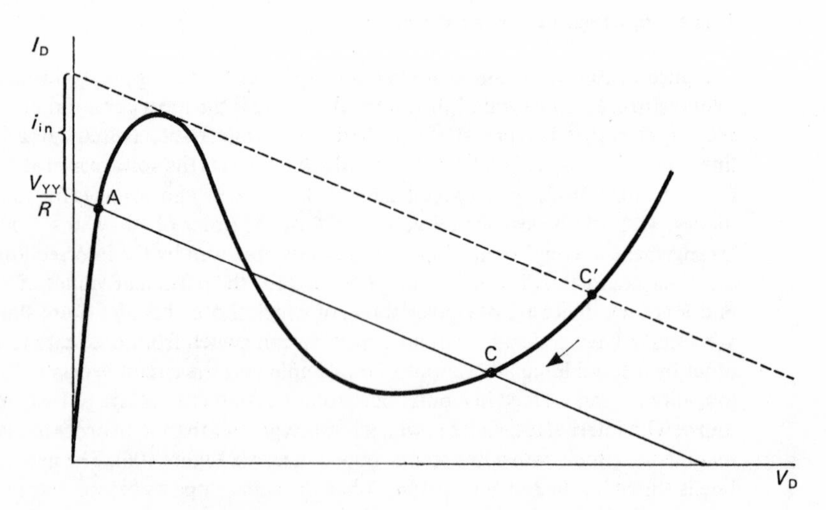

Figure 108. Switching the bistable.

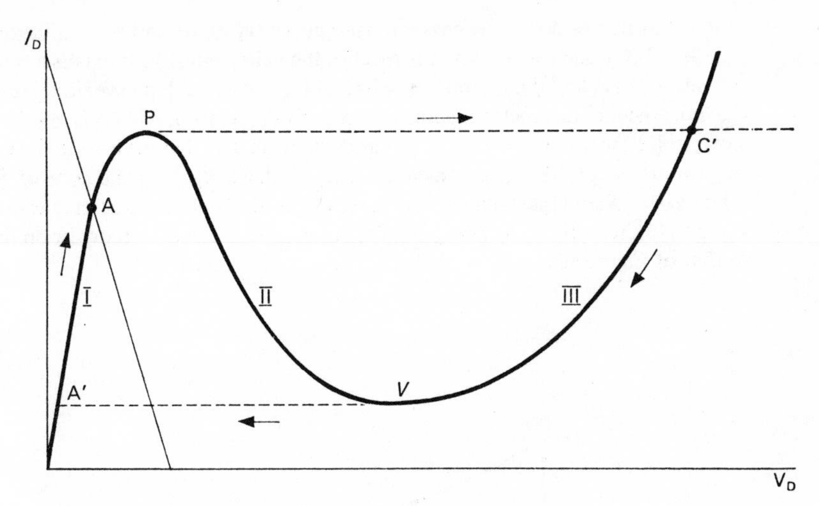

Figure 109. Operating conditions for monostable action.

To produce a monostable arrangement the supply voltage and load resistor must be adjusted to operate at a point A as shown in Figure 109, where only one stable state exists. If the circuit is triggered to part III of the curve, it will, finding no stable position there, return immediately to the stable point at A. To ensure that the monostable remains in its quasi-stable state for a finite time, a small inductance L is added in series with R, as shown in Figure 110. If as before we bring the operating point above the peak, the circuit jumps to a new operating point C′ (Figure 109). This point is at the same level as P, as the presence of an inductor means that the current cannot change instantaneously. The operating

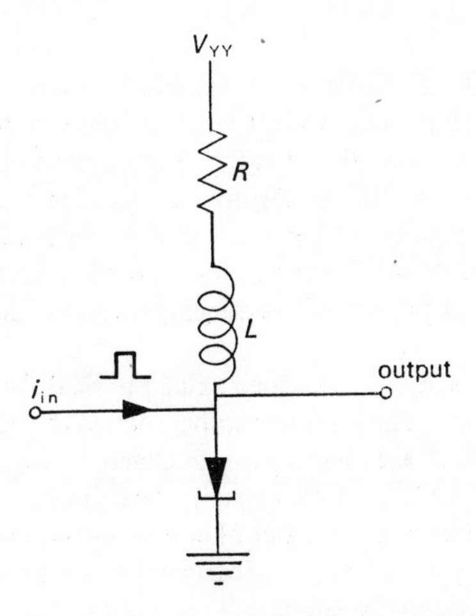

Figure 110. Tunnel-diode, monostable circuit.

point then moves down the characteristic curve trying to find an equilibrium position, but none exists. When it reaches the valley point V, it makes a rapid transition to A′, and thence returns to its stable position A. However in this case the movement of the operating point from C′ to V, and from A′ to A is relatively slow; indeed we know from Figure 27 and equation **3.11** that the time taken will be proportional to the time constant L/R_{total}, where R_{total} is the sum of the load resistor R and the dynamic resistance of the diode at the point in question. Figure 111 shows the sort of output pulse obtainable, the lettering corresponding to that of Figure 109.

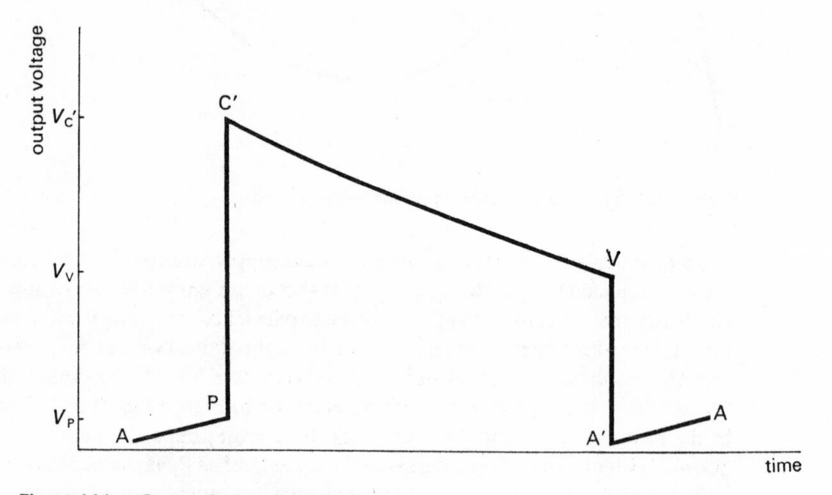

Figure 111. Output from monostable.

The arrangement for using the tunnel diode as an astable multivibrator should be fairly obvious from what we have said. In this case the supply voltage and load resistor are adjusted so that the load line cuts the characteristic once, at some point such as B in Figure 107 on the negatively sloping part of the characteristic. No stable position of equilibrium exists, but if we include an inductance in the circuit as in the monostable, the operating point will circulate continuously round the loop PC′VA′ of Figure 109, producing a succession of pulses of the type shown in Figure 111.

The main advantage of tunnel diodes, apart from their small physical size, is their extremely fast rate of switching from one state to another, due to the basic nature of the tunnelling effect itself. There are no problems of charge storage or transit time, the chief limitation being the stray capacity and inductance associated with the device. Switching times of the order of one nanosecond are easily obtainable. On the debit side, the output pulse is clearly of the same order of magnitude as the valley voltage so it must be small.

7·7 Sinusoidal oscillators

Oscillators which produce a sinusoidal output are essential elements in circuits for telecommunication. The physicist will usually meet them as timing devices, an example of which we shall see later in connexion with the multichannel analyser. Most sinusoidal oscillators have the same basic starting point as astable multivibrator oscillators – an amplifier with positive feedback, loop gain AB greater than one, and hence an output with no input. They differ from the astable multivibrator in that only one particular frequency is fed back with this large gain, either by having a filter circuit in the feedback loop (phase-shift oscillators) or else by arranging that the amplifier itself only produces a high gain for one particular frequency (resonant-feedback oscillators). In a third type of 'negative-resistance oscillator' a device with negative dynamic resistance cancels out the dissipative elements in a tuned circuit.

7·8 Phase-shift oscillators

Figure 112 shows, schematically, an example of a phase-shift oscillator, in which a fraction of the output of an amplifier, $Z_F^*/(Z_F + Z_F^*)$, is fed back to provide its own input. The terminology arises because for only one particular

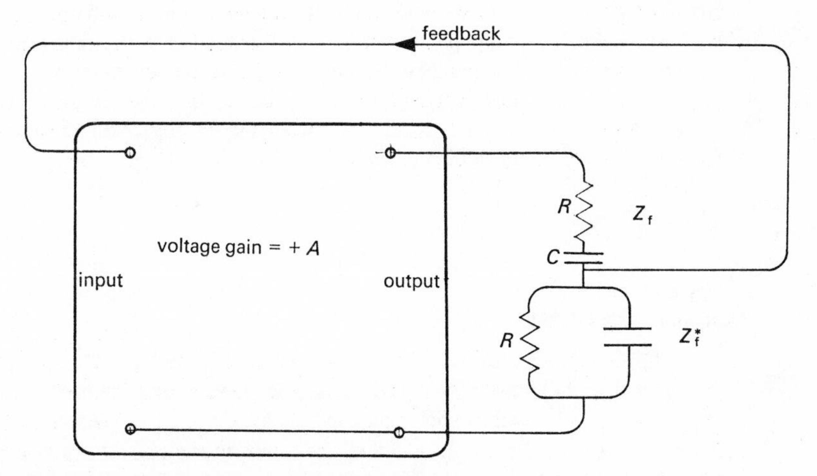

Figure 112. Phase-shift oscillator (schematic).

frequency will the phase be shifted by the correct amount for oscillation. This particular circuit is known as the 'Wien bridge' oscillator from the name associated with the filter arrangement on the output. The input impedance is assumed to be high enough to throw a negligible load on the output and we have no trouble, as in Figure 77 with the impedance of an external source of signals, as there is none. The fraction fed back is

$$B = \frac{\left\{\dfrac{1}{R} + j\omega C\right\}^{-1}}{\left\{\dfrac{1}{R} + j\omega C\right\}^{-1} + R + \dfrac{1}{j\omega C}}$$

which after simplification becomes

$$B = \frac{1}{3 + j(\omega RC - \dfrac{1}{\omega RC})}$$

For the particular frequency given by $RC\omega = 1$, or $\omega = 1/RC$, B is real and equal to $\frac{1}{3}$; this value of ω also makes the magnitude of B a maximum. So for a value of $\omega = 1/RC$, not only is there the maximum amount of feedback, but since the value of B has no imaginary part, no additional phase shift is introduced by the network, and the positive value of the amplifier gain (equivalent to zero phase shift) ensures that the possibility of oscillation exists. For oscillation to take place, $AB \geqslant 1$; therefore, in this case A must be at least three. For larger values of A the output will rise until limited by the non-linearity of the amplifier.

This type of oscillator is widely used at audio- and low radio-frequencies, where large inductors and capacitors would be needed for the resonant-feedback type. At higher frequencies the latter is favoured, because it is less affected by stray capacities, which can be incorporated into the tuning capacitance. For $R = 5000$ and $C = 0{\cdot}01\ \mu$F, the oscillation frequency for the circuit of Figure 112 would be about 3000 Hz.

7·9 Resonant-feedback oscillators

Figure 113(a) shows an amplifier of a type met in section 2·11. The effective collector load, and therefore the voltage gain, is large only for a frequency given by $\omega_0 = (LC)^{-\frac{1}{2}}$. Oscillation will therefore be obtained at this frequency if a sufficient fraction of the output is fed to the input, in the correct relation for positive feedback. We do this in Figure 113(b), by feeding back to the emitter from a tapping point on the inductor through a coupling capacitor C_c, the base being to ground for signals because of the large capacitor C_b. (Feedback to the base, rather than the emitter, would, as in Chapter 6, be in the negative mode.) This tapped-inductor arrangement is known as the Hartley oscillator. If instead, a capacitative divider is used, as in Figure 113(c), we obtain what is known as the Colpitts circuit. The output in either case can be taken from a number of places – from the collector, from the emitter, or, by making the inductor one winding of a transformer, from a secondary winding.

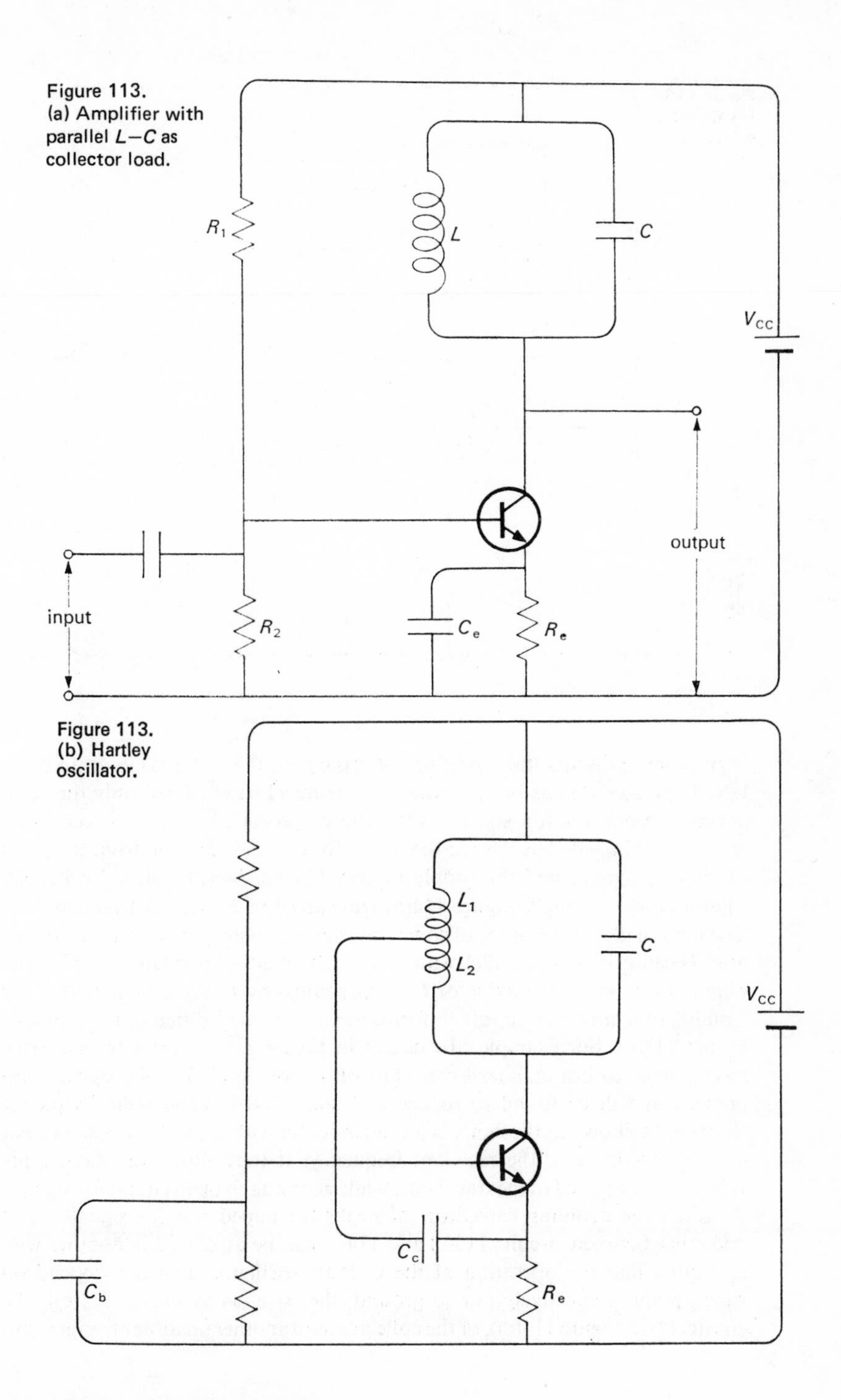

Figure 113.
(a) Amplifier with parallel $L-C$ as collector load.

Figure 113.
(b) Hartley oscillator.

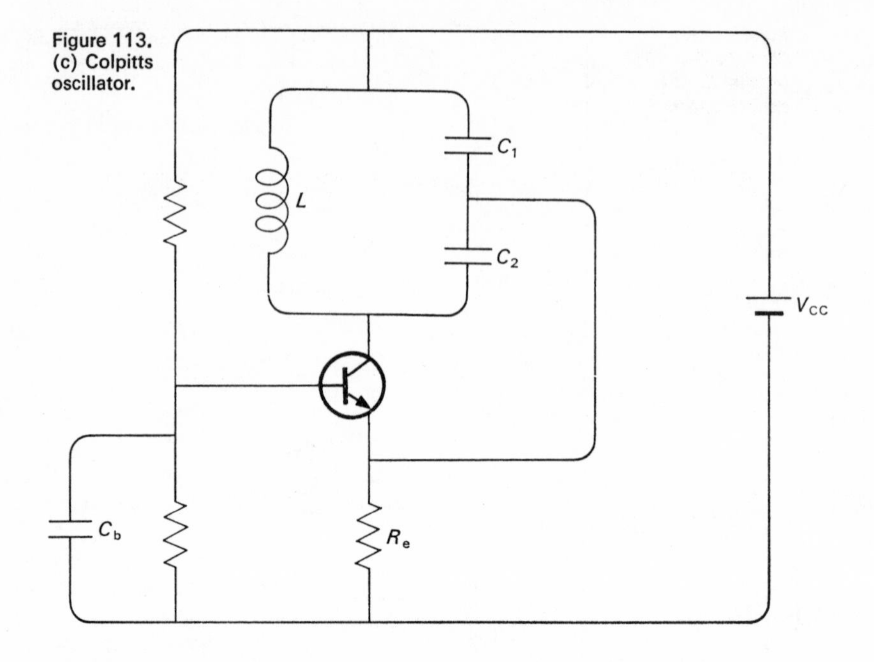

Figure 113. (c) Colpitts oscillator.

In order to discuss the conditions necessary for the Colpitts circuit to oscillate, Figure 113(c) has been redrawn in Figure 114(a) to show only the components important for signals, with those associated only with d.c. level settings eliminated. The base of the transistor is thus to ground from the point of view of signals, and the supply battery V_{CC} has been replaced by a short. Figure 114(b) is simply Figure 114(a) rearranged in a more convenient way. The resistor R_e, of the order of some thousands of ohms, has also been eliminated because it is in parallel with the much smaller impedance of C_1. The ohmic resistance of the inductor L is represented by R. We note here that the Colpitts oscillator can appear in forms apparently very different from that of Figure 113(c). For example, the circuit in Figure 114(c) seems to bear little resemblance to our standard case, but on removing all but the signal components, it will be found to reduce to Figure 114(b). (The radio-frequency (RF) choke shown in the figure is just an inductor with a low d.c. resistance but a high impedance at the resonant frequency. It thus allows the d.c. supply voltage to be applied to the transistor, while acting as an open circuit for signals. C_c is a large coupling capacitor, of negligible impedance for signals.) The difference between circuits 113(c) and 114(c) can be expressed in another way by noting that the operation of the Colpitts oscillator does not depend on which point is effectively put to ground, the base (as in Figure 113(c)), the emitter (as in Figure 114(c)), or the collector as in another possible arrangement.

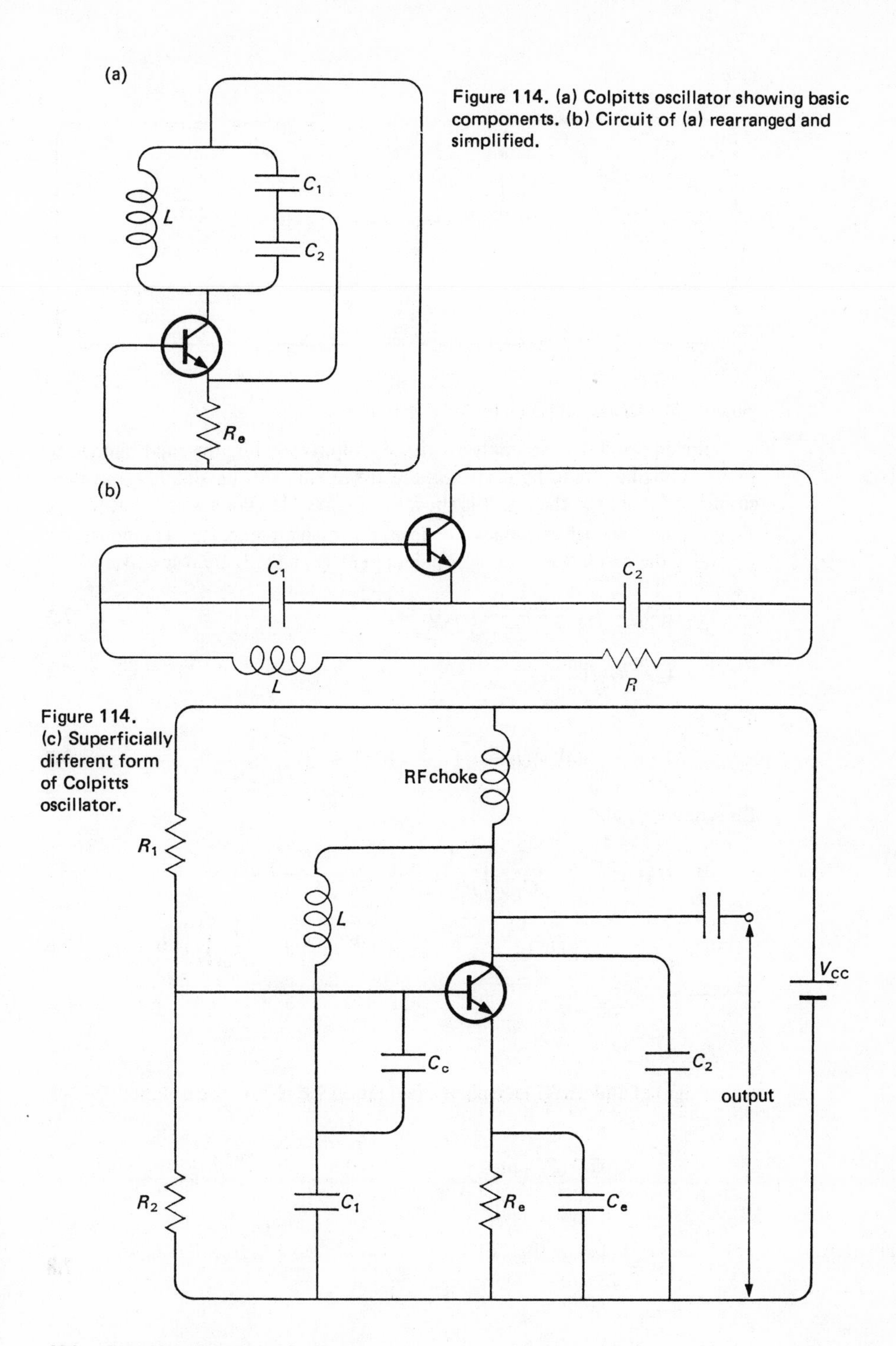

Figure 114. (a) Colpitts oscillator showing basic components. (b) Circuit of (a) rearranged and simplified.

Figure 114. (c) Superficially different form of Colpitts oscillator.

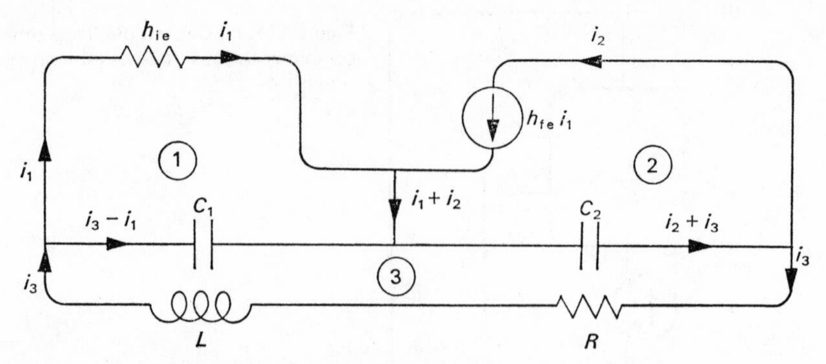

Figure 115. Equivalent circuit for Colpitts oscillator.

Returning now to the analysis of the Colpitts circuit, the final change in Figure 114(b) is made by replacing the transistor with its simple equivalent circuit of Figure 61: the result is shown in Figure 115. We assume that we are applying just enough feedback to keep the oscillation going. The equations governing the currents and voltages in loops (1), (2) and (3) are respectively

$$i_1 h_{ie} - (i_3 - i_1)\frac{1}{jC_1\,\omega} = 0 \tag{7.3}$$

$$i_2 = h_{fe}\, i_1 \tag{7.4}$$

and

$$i_3(R + jL\omega) + (i_3 - i_1)\frac{1}{jC_1\,\omega} + (i_3 + i_2)\frac{1}{jC_2\,\omega} = 0 \tag{7.5}$$

Eliminating i_2 and i_3

$$i_1\left\{\left(\frac{h_{fe}}{jC_2\,\omega} - \frac{1}{jC_1\,\omega}\right)\frac{1}{jC_1\,\omega} + \left(h_{ie} + \frac{1}{jC_1\,\omega}\right) \times \right.$$
$$\left. \times\left(R + jL\omega + \frac{1}{jC\omega}\right)\right\} = 0 \tag{7.6}$$

where

$$\frac{1}{C} = \frac{1}{C_1} + \frac{1}{C_2}.$$

Equating real and imaginary parts of equation **7.6** to zero we obtain

$$\omega^2 = \frac{1}{LC} + \frac{R}{LC_1\,h_{ie}} \tag{7.7}$$

or

$$\omega^2 \approx \frac{1}{LC} \tag{7.8}$$

and

$$h_{fe} = \frac{C_2}{C_1} + h_{ie} RC_1 C_2 \omega^2 + C_2 \left(L\omega^2 - \frac{1}{C}\right) \qquad 7.9$$

which becomes, using the approximate value of ω^2 from equation **7.8**

$$h_{fe} = \frac{C_2}{C_1} + \frac{h_{ie} R(C_1 + C_2)}{L} \qquad 7.10$$

Equation **7.8** tells us that the natural frequency of oscillation, given by $\omega^2 = 1/LC$ is not appreciably changed by the feedback. Equation **7.10** gives the minimum value of h_{fe} needed to keep the oscillation going, for given values of the other components. An important approximation to equation **7.10** is obtained by neglecting the second term on the right with its small factor R/L to give

$$h_{fe} \geqslant \frac{C_2}{C_1} \qquad 7.11$$

for oscillation to occur. Equation **7.11** together with equation **7.8** may alternatively be used to determine C_1 and C_2, if one is given h_{fe}, L and the frequency of oscillation required. For $h_{fe} = 50$, $L = 10^{-4}$ henrys, and $\omega = 10^7$ sec^{-1}, one obtains $C_1 = 100$ pF and $C_2 = 5000$ pF for oscillation to just occur. But since equation **7.11** can be written $C_1 \geqslant C_2/h_{fe}$ a larger value of C_1 would leave a margin of safety, particularly if the circuit has an appreciable external load. However this is not the only limitation on the value of C_1: there is another solution to equation **7.10** with $C_2 \ll C_1$. In this case, neglecting the term C_2/C_1, and also C_2 with respect to C_1 in the second term on the right, we have

$$C_1 \leqslant \frac{L h_{fe}}{R h_{ie}} \qquad 7.12$$

Using the same values as before, and in addition $h_{ie} = 2000$ ohms and $R = 20$ ohms, we obtain $C_1 = 0{\cdot}125$ μF and $C_2 = 100$ pF. Oscillation is possible between the limits defined by equations **7.11** and **7.12**.

The latter limit is interesting because it is easily interpretable in terms of voltage feedback. The voltage gain for a transistor amplifier is $g_m R_c$ where g_m is h_{fe}/h_{ie}, and in this case the collector load R_c is that of a tuned circuit at resonance, L/RC (see section 2·11). The fraction fed back from across C_1 is $(1/C_1)/(1/C_1 + 1/C_2) \approx C_2/C_1$ assuming $C_1 \gg C_2$. So the value of AB = gain × feedback factor is given by

$$AB = \frac{h_{fe}}{h_{ie}} \frac{L}{RC} \frac{C_2}{C_1} \qquad 7.13$$

or as $$\frac{1}{C}=\frac{1}{C_1}+\frac{1}{C_2}\approx\frac{1}{C_2}$$

$$AB=\frac{h_{fe}}{h_{ie}}\frac{L}{R}\frac{1}{C_1} \qquad \textbf{7.14}$$

For oscillation $AB=1$, hence solving equation **7.14** for C_1 gives

$$C_1=\frac{L\,h_{fe}}{R\,h_{ie}}$$

which is identical with the condition for oscillation given by equation **7.12**.

For the Hartley oscillator we might expect the same results as we have obtained for the Colpitts case with $L_1\,\omega$ replacing $1/C_1\,\omega$ (and similarly for L_2). The relations will actually be more complicated since mutual inductance will exist between the parts L_1 and L_2 of the coil.

7·10 Crystal oscillators

It is important in oscillators used as the basis of a timing process for the frequency of oscillation to be extremely stable. In general the methods used for stabilization lie outside the scope of this book, but a few points can be made. We see from equation **7.7** that the value of ω depends largely on the values of L and C, but, because of the small second term, also to a minor extent on h_{ie} (and in a more complicated analysis than this, on other transistor parameters as well). These parameters will vary, for example, with changes in the supply voltage or in temperature, and this variation will be reflected in a change in the frequency of oscillation. For the most critical applications it is important to reduce these changes to a minimum, and fortunately, our simple analysis indicates how this may be done. Equation **7.7** tells us that the more the factor R/L is reduced, that is the larger we make the Q of the circuit, the nearer the resonance frequency will approach the value of $(LC)^{-\frac{1}{2}}$, and the less dependent it will be on transistor parameters. Resonant circuits formed from quartz crystals provide Qs of the order of 10^5, that is about 1000 times better than conventional circuits.

Quartz crystal resonators depend on the piezo-electric effect: if a thin slice of quartz, say about 0·5 cm thick, has electrodes evaporated onto either side of it, a voltage applied to the electrodes will cause the slice to deform mechanically. Conversely a mechanical deformation of the slice will cause charges of opposite signs to appear on the electrodes. At a particular frequency corresponding to the natural mechanical resonance of the slice, an ideal crystal once set into oscillation, would continue to oscillate indefinitely. We can compare this with the series-resonant LC circuit of Figure 19(b). Here also if oscillation at the resonant frequency $\omega_0=(LC)^{-\frac{1}{2}}$ is initiated by an external voltage source, it will, in the ideal case, continue when the external voltage is reduced to zero, because the impedance of the circuit for currents at the resonant frequency is zero. In practice some power from a transistor would have to be supplied to make good the losses due to dissipation in the non-zero resistance R of the inductor, and the same would be true, to a lesser extent, with the crystal. From the electrical point of view, the crystal resonator may be regarded as equivalent to a series-tuned resonance circuit as in Figure 116. The position is complicated by the fact that we must also take account of the ordinary capacity between the crystal electrodes which is indicated by the capacitor C'. Of course a conventional tuned circuit may also have stray capacity across it, but in the case of the crystal, the relative sizes of C and C' make the result markedly different from the tuned-circuit case.

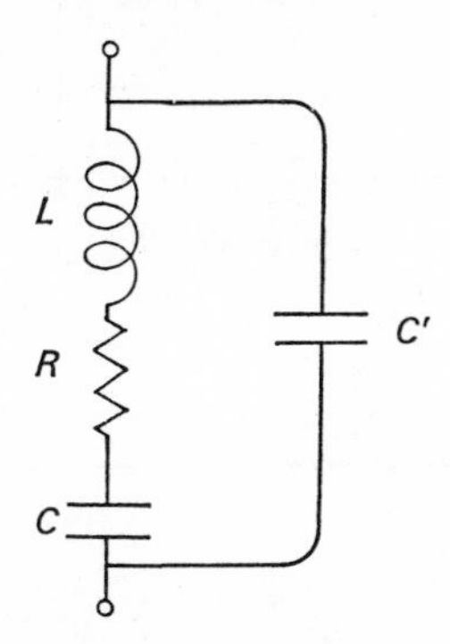

Figure 116. Equivalent circuit for crystal.

The impedance Z of the crystal of Figure 116 is simply that of the two arms in parallel, and neglecting R, is given by

$$Z^{-1} = \left(jL\omega + \frac{1}{jC\omega}\right)^{-1} + \left(\frac{1}{jC'\omega}\right)^{-1}$$

which on simplification becomes

$$Z = \frac{1 - \omega^2 LC}{j\omega(C + C')\left(1 - \dfrac{\omega^2 LCC'}{C + C'}\right)} \qquad 7.15$$

This equation defines two important frequencies. The first is the ordinary series resonance, which occurs when the left-hand branch in Figure 116 has zero impedance, that is when $Z = 0$ and $\omega = (LC)^{-\frac{1}{2}}$. The second is when Z is infinite and a parallel resonance is given by $\omega^2 = (C + C')/LCC'$ or $\omega^2 = 1/LC''$, where C'' is equal to C and C' in series, that is, $1/C'' = 1/C + 1/C'$. For the crystal C will usually be much less than one per cent of C', so to a close approximation $C'' = C$, and the parallel and series resonances almost coincide, the former occurring at the slightly higher frequency. Because of the very wide range of crystal characteristics available, it is difficult to quote typical values,

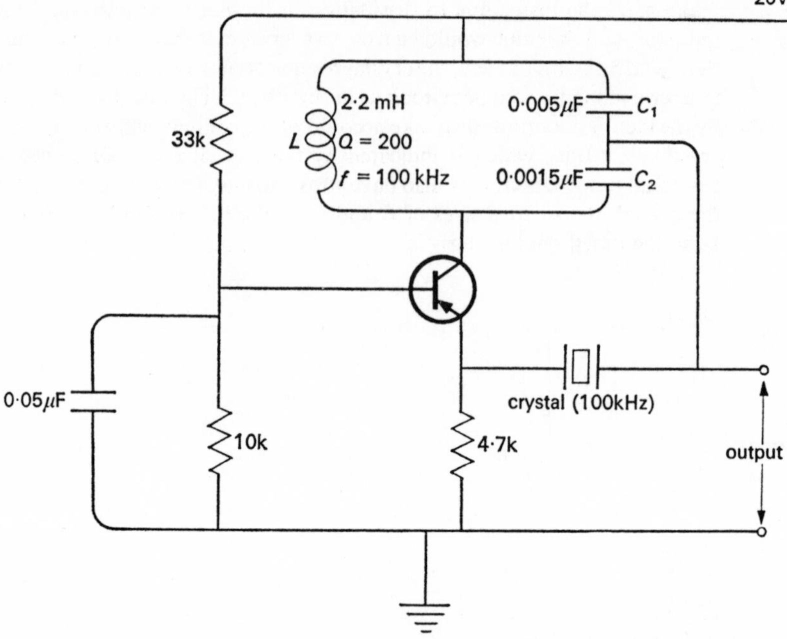

Figure 117. Series-mode crystal oscillator (courtesy U. S. Department of Defense).

but the following figures (from reference 5) will give some idea of the numerical quantities involved. For $L = 31{\cdot}2$ henrys, $C = 0{\cdot}081$ pF, $R = 130$ ohms, and $C' = 34$ pF, we obtain a Q of 150,000, a series resonance frequency of 100 kHz, and a parallel resonance frequency 100 Hz higher.

In practical oscillators, the crystal may be used in either its series or parallel resonance mode. Figure 117 shows the former case. Oscillation occurs at a frequency where the impedance in the feedback line to the emitter is a minimum, that is at the series resonance of the crystal. The tuned-anode circuit (peaking at the same frequency as the crystal) still remains to provide the necessary gain and impedance relations for feedback, but it is the much narrower resonance of the crystal which controls the frequency of operation. Figure 118 shows the crystal used in its parallel mode in the 'Pierce oscillator'. It is basically the same circuit as Figure 114(c), with the crystal replacing the tuned circuit. The external capacitors C_1 and C_2 must still be retained to provide the potential divider action: it is clear from the relation between C'', C and C', that the addition of such external capacity will not appreciably change the resonance frequency.

Crystal oscillators can be expected to hold their frequency constant to about one part in a million in ordinary circumstances. With careful control of the

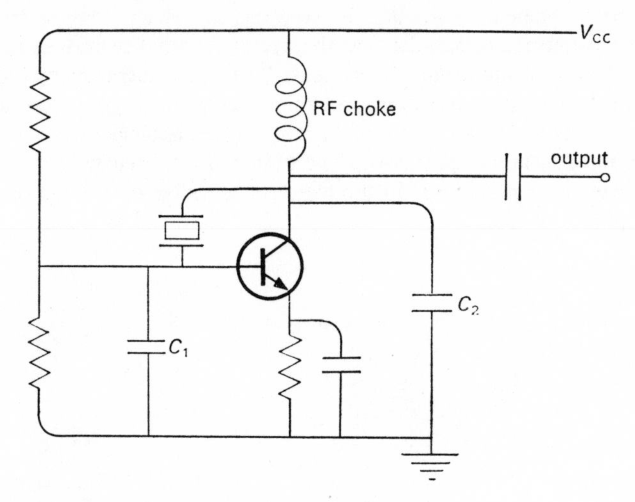

Figure 118. Parallel-mode crystal oscillator.

supply voltage and crystal temperature, constancy of the order of one part in 10^9 can be obtained.

7·11 Negative-resistance oscillators

As mentioned in the previous section, oscillations once started in an idealized resonance circuit, would continue indefinitely. In practice, because of the series resistance of the inductor (labelled R_s in Figure 119(a), previously called simply R) even if an oscillation were started, it would quickly die away due to the power losses in R_s. We can think of the role of the transistor in Figure 113(c) for example as being to make good the power lost in this way. Figure 119(b) shows another situation in which oscillations would die away unless maintained by an

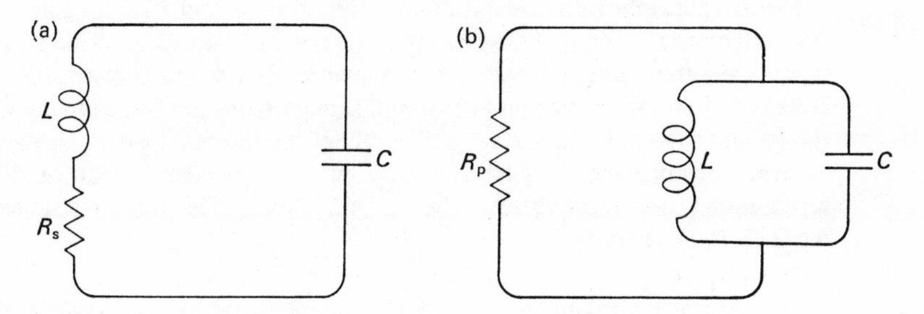

Figure 119. Losses in a resonant circuit.

outside source – where a resistor R_p, representing, say, an external load on the oscillator, is connected in parallel with the circuit. In practice both series and parallel resistance would usually be present. These resistances abstract signal power from the circuit: if however, we can add enough 'negative resistance' (that is something which will add signal power to the circuit) we can return to the condition of sustained oscillation. The tunnel diode discussed previously, will do this for us if it is biased in the region of negative dynamic resistance (region II of Figure 105).

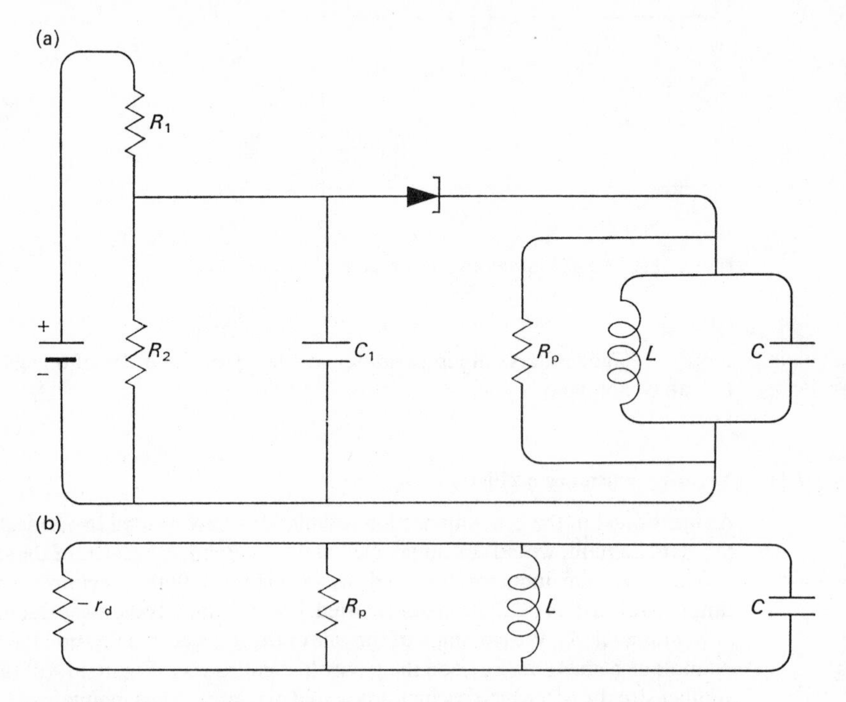

Figure 120. Tunnel diode oscillator (a) actual circuit (b) equivalent circuit.

Figure 120(a) shows a possible circuit. Resistors R_1 and R_2 form a potential divider across the supply battery to provide the d.c. bias, (adjustable if necessary), while the decoupling capacitor C_1 ensures that this biasing arrangement is a short circuit from the point of view of signals. Consider first the case where the inductor is an ideal one (that is $R_s = 0$) but the external load R_p in parallel is an important factor. Figure 120(b) shows the equivalent circuit, with $-r_d$ representing the negative resistance of the diode. The total resistance in parallel, R_t, is given by

$$\frac{1}{R_t} = \frac{1}{R_p} + \left(\frac{-1}{r_d}\right)$$

If $r_d = R_p$, then $1/R_t$ is zero, R_t is infinite, and no load is thrown on the tuned circuit. Sustained oscillation is thus possible. If R_p is less than r_d, then R_t is positive, and no sustained oscillation is possible. If R_p is greater than r_d, the amplitude of the oscillation will build up until a part of the tunnel diode curve is reached where the negative resistance has increased to such an extent that the build-up process stabilizes.

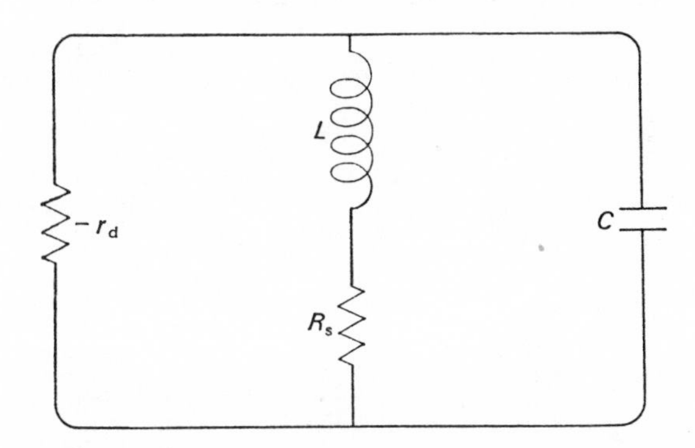

Figure 121. Equivalent circuit for the case of non-ideal inductor.

The equivalent circuit in the case of a tuned circuit with an inductor of non-negligible resistance R_s, but with negligible external load, is shown in Figure 121. The mathematical treatment is not quite so easy here, but we can obtain the result in a plausible way as follows. From the point of view of damping the oscillations, a small series resistance like R_s of the present figure, would be equivalent to a large resistance in parallel with the circuit. At the end of Chapter 2, it was shown that the impedance of a parallel LC circuit at resonance was approximately L/RC. (The R of that equation is the R_s here.) This is another way of saying that a resistance of R_s in series is equivalent to one of L/R_sC in parallel. The condition for oscillation in the circuit of Figure 121 is therefore $r_d = L/R_sC$. If both R_s and R_p are present the condition for oscillation is obtained by adding R_p and L/R_sC in parallel, and equating the result to r_d. For a germanium tunnel diode r_d might typically be 50 ohms.

Of course, when dealing with negative-resistance devices, we are in no sense 'getting something for nothing'. Although the d.c. supply does not appear in any of the equivalent circuits, it is from there that the power to sustain the oscillations is drawn. Tunnel diodes are attractive as oscillators for the same reasons that made them useful in multivibrator type circuits – small physical size, and high speed of operation. Oscillation at frequencies as high as 10 kMHz (10^{10} Hz) is possible.

References and further reading

1. W. GOSLING, *Applied Electronics,* Macmillan, 1975.
2. J. MILLMAN and C. C. HALKIAS, *Integrated Electronics,* McGraw-Hill, 1972.
3. *Texas Instruments Application Note* no. 5. The 2G103 in saturating binary and decade counters. [This Application Note deals in greater detail with the circuit of Figure 93.]
4. S. AHMED, 'Astable and monostable multivibrators', *Texas Instruments Semiconductor Application Report,* vol. 1, no, 9, 1966.
5. P. VIGOUREUX and C. F. BOOTH, *Quartz Vibrators and their Applications,* H.M.S.O., 1950.

CHAPTER 8

Delay lines

8·1 Introduction

The string of inductors and capacitors shown in Figure 122 is known as a 'lumped-parameter delay line' or simply a 'lumped delay line', the first word in the expression implying that it is made up of discrete components. It is interesting, not only for its own sake, but because it is a reasonably good model of a

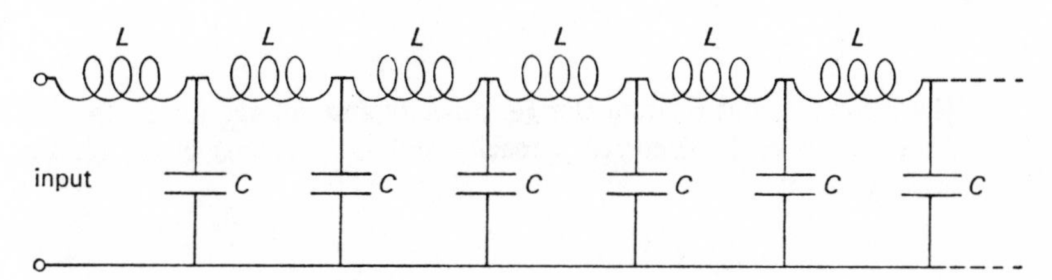

Figure 122. Lumped delay line.

coaxial cable, or other type of transmission line. In this case L and C will represent the inductance and shunt capacitance respectively, per unit length. The model is a reasonable one as L and C are usually more important than the other parameters involved – the ohmic resistance and leakage of the cable.

There are two important attributes of such delay lines, whether 'lumped' or 'distributed'. If at the input of a line, a voltage step is applied this step progresses down the line undistorted (in the ideal case) and with a finite and constant velocity. The delay time per unit length (or per section for the lumped line), that is the time taken for the step to move through unit length (or one section), is given by $T_d = (LC)^{\frac{1}{2}}$. The other important property of a line is the load it throws on the source of signals connected to it. For an infinitely long line this load is purely resistive and of size $(L/C)^{\frac{1}{2}}$. This is known as the 'characteristic impedance' of the line, and written Z_0 (to cover the more general case where it may not be purely resistive in character). A satisfactory proof of the facts just quoted, using the method of the Laplace Transform, may be found in references 1 and 2 at the end of this chapter. We shall discuss the same topic in a simpler way.

Figure 123 shows a small length of a distributed delay line with a signal present. Two equations describe the situation. The first of these is that con-

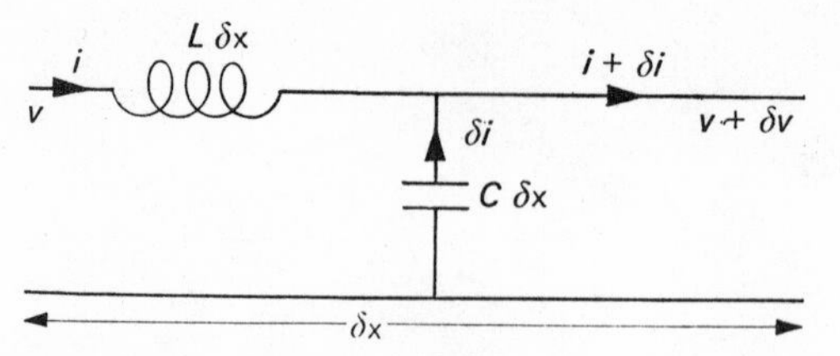

Figure 123. Delay-line element.

necting the voltage across an inductor with the rate of change of current through it, and for the present case is

$$-\delta v = (L\delta x)\frac{\partial i}{\partial t}$$

or

$$\frac{\partial v}{\partial x} = -L\frac{\partial i}{\partial t} \qquad \textbf{8.1}$$

The second is that relating charge, capacity and voltage for a capacitor, or what is the same thing, current, capacity, and rate of change of voltage. For the present circuit

$$-\delta i = (C\delta x)\frac{\partial v}{\partial t}$$

or

$$\frac{\partial i}{\partial x} = -C\frac{\partial v}{\partial t} \qquad \textbf{8.2}$$

By differentiating equation **8.1** with respect to x, and equation **8.2** with respect to t, and equating the values for $\partial^2 i/\partial x\,\partial t$ and $\partial^2 i/\partial t\,\partial x$, we obtain

$$\frac{\partial^2 v}{\partial x^2} = LC\frac{\partial^2 v}{\partial t^2} \qquad \textbf{8.3}$$

which is the well-known wave equation. A similar equation can be obtained for the current. A solution of equation **8.3** (as can be verified directly by substitution) is $v = f(x - Ut)$, where $U = (LC)^{-\frac{1}{2}}$, This represents a wave moving to the right with a velocity $U = (LC)^{-\frac{1}{2}}$, independent of frequency. It is not hard to believe that a voltage step, which we can consider built up, by the Fourier method, from basic sinusoidal components, will also travel with this velocity $(LC)^{-\frac{1}{2}}$. This is equivalent to the previous statement that the time delay per unit length, T_d, is $(LC)^{\frac{1}{2}}$.

A further solution of equation **8.3** is given by $v = f(x + Ut)$, where U has the same value as before. This represents a wave moving to the left along the line, and in general such a wave will be present, even with a signal generator on the left-hand end of the line only, due to reflections at the other end, as we shall later discuss. For the moment, however, confining our attention to an infinitely long line so that these reflections do not occur, let us consider the particular sinusoidal solution $v = v_0 \sin(x - Ut)$, and the corresponding solution for the current, $i = i_0 \sin(x - Ut)$. We can verify that these satisfy our basic equations **8.1** and **8.2**, provided the constants v_0 and i_0 are connected by the simple relation, $v_0 = i_0(L/C)^{\frac{1}{2}}$. Note there is no phase difference between v and i at any point on the line, and in particular at the origin of signals. Thus $v/i = v_0/i_0 = (L/C)^{\frac{1}{2}}$ (which has the dimensions of resistance), and our previous statement that the line acts as a resistance of size $Z_0 = (L/C)^{\frac{1}{2}}$ is thus confirmed.

From the last fact we can make an interesting deduction. Suppose an infinite line as in Figure 124(a) were cut across at AA′, and the part of the line to the right of AA′ removed and replaced by a resistor equal to Z_0, as in Figure 124(b).

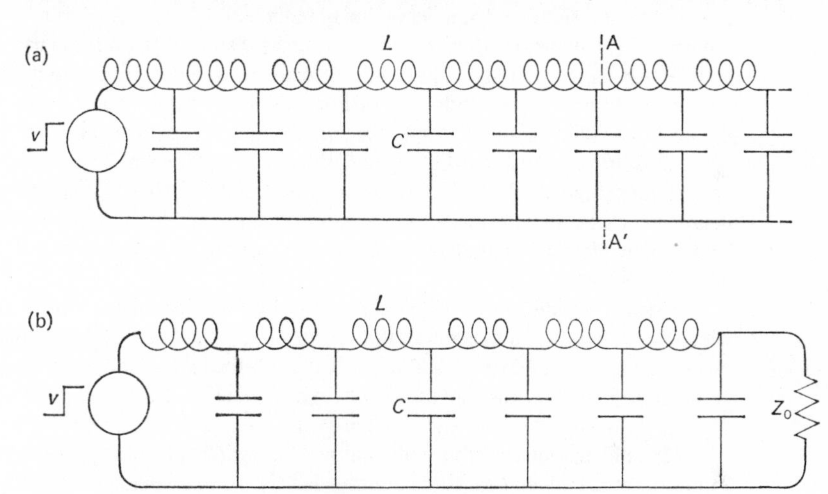

Figure 124. Equivalence of infinite line and correctly terminated finite line.

From the point of view of the generator, nothing has changed, because the resistor is a perfect replacement for the infinite line to the right of AA′. In other words, a lumped delay line of a finite number of sections, or a coaxial cable of finite length, will act as an infinite line if terminated in its characteristic resistance. Any voltage step sent down an infinitely long line will never return, and we can now deduce that the same is true for a line correctly terminated: in other words all the energy of the applied signal is dissipated in the resistor. We contrast this with the case where the terminating resistor is of any other size than Z_0, where reflection of the signal can be shown to occur. (It may be helpful in this connexion to think of the analogy of sound waves in organ pipes.) We shall discuss later, in more detail, reflections for two particular cases in which we are especially interested, that of the short-circuited line (terminating resistor zero), and that of the open-circuit line (terminating resistor infinite). Practical delay lines will differ from the idealized ones we have discussed both in the attenuation of signals travelling down them, and in the degradation of the rise time of voltage steps. The latter is particularly bad with lumped delay lines, as we shall see.

8·2 Delay application

The first, and most obvious application of delay lines is to delay signals. Suppose for example that we wish to record a signal from a nuclear radiation detector characterizing a certain nuclear event, provided a signal from another detector, indicating the occurrence of a related nuclear event, is obtained within a certain time. (Alternatively we might wish to veto the recording of the first signal if the second arrived within a stated period.) In either case the solution is to pass the first signal down a delay line of sufficient delay to allow us to verify the arrival (or non-arrival) of the second signal. Depending on the particular problem in hand, we may wish to produce delays ranging from some nanoseconds to a few microseconds. The ordinary general-purpose coaxial cable, of the order of half a centimetre or less in outer diameter, has distributed capacitance and inductance such that Z_0 lies in the 50 to 100 ohms range, and the velocity of propagation of signals is about $\frac{2}{3}$ that of light. This corresponds to a delay per metre of approximately 5 nanoseconds, which for a manageable length of cable, means that this type is only suitable for the shorter delays envisaged above.

For delays of the order of microseconds lumped parameter lines can be used. For example if a line were required to produce a delay of 2 μsec with a characteristic impedance of 200 ohms, then arbitrarily assuming the number of sections n to be 8, the equations for finding L and C become $T_d = T_D/8 = 2 \times 10^{-6}/8 = (LC)^{\frac{1}{2}}$ and $Z_0 = 200 = (L/C)^{\frac{1}{2}}$. This gives $L = 50$ μH and $C = 1250$ pF. In fact our choice of n cannot be entirely arbitrary, because it is found that for a given overall delay time T_D, the 10 to 90 per cent rise time of a step, T_R, decreases with increasing number of sections – which explains in general the better performance of continuous lines. One empirical formula states that

$$\frac{T_R}{T_D} \approx n^{-\frac{2}{3}} \quad (3 < n < 30)$$

In our case this would correspond to a ratio of rise time to delay time of $\frac{1}{4}$, which means that more sections are called for if the application is at all critical as regards rise time.

The other method of obtaining microsecond delays with reasonable physical sizes is to use cables in which the inductance per unit length, and thus the delay, has been increased by using an inner conductor in the form of a helix wound on a core of magnetic material. Cables with delays of the order of one microsecond per metre, and characteristic impedances of the order of one thousand ohms (increased also by the increased value of the inductance) are commercially available. Rise times of the order of 10 per cent of the delay time occur here also.

8·3 Pulse shaping with delay lines

In Chapter 3 we discussed some of the advantages and disadvantages of delay-line shaped pulses. Figure 125 illustrates a possible method of producing

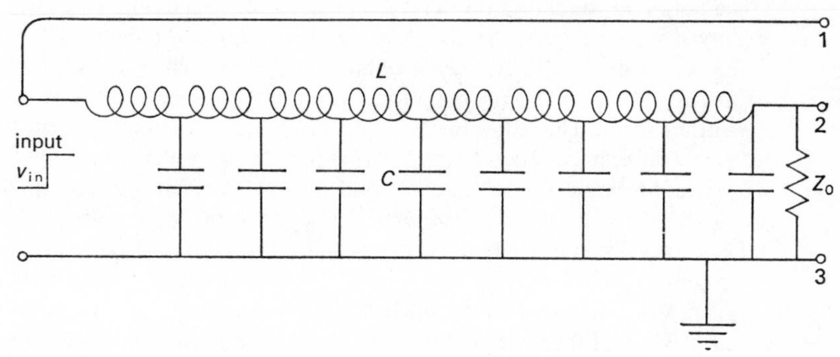

Figure 125. Circuit for delay-line pulse shaping.

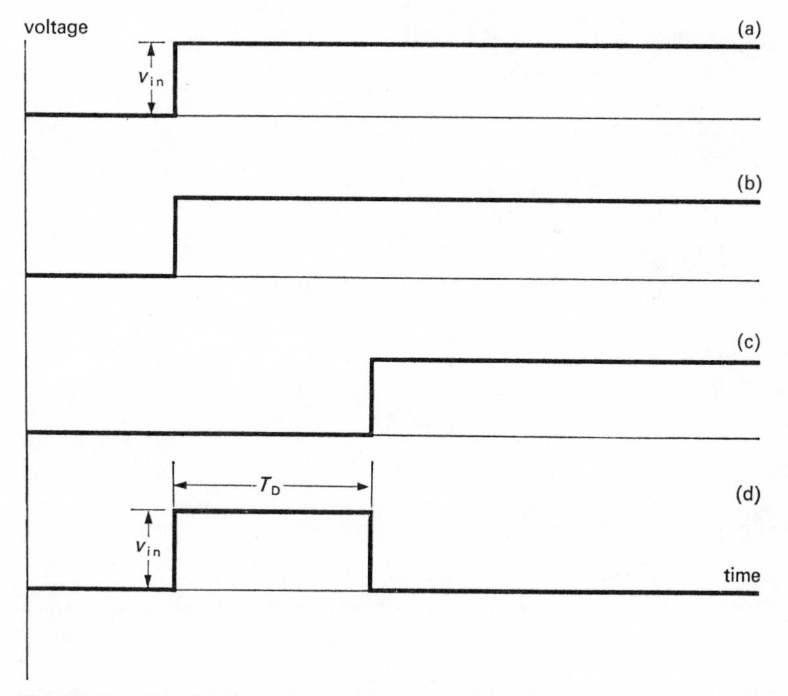

Figure 126. Waveforms in delay-line-shaping circuit.

such pulses from a basic voltage step coming from a detector. Here the input step is applied at the input of a delay line, whose output terminals, labelled (2) and (3), are correctly terminated with its characteristic resistor Z_0. At the same time the input is applied between the top wire of the diagram and ground, and we should expect this corresponding output to appear at terminal (1) with no delay. Figure 126(a) shows the input step, and Figure 126(b) shows the output between terminal (1) and ground, appearing, as we supposed, with no time lag. The output at the end of the delay line, that is between terminals (2) and (3) is delayed by the time T_D and is shown in section (c) of the figure. The voltage between terminals (1) and (2) is obtained by substracting the waveforms in sections (b) and (c) of the diagram, and is shown in section (d). It is of the required form. The same result could have been obtained by considering directly the voltage between terminals (1) and (2). When the input is applied the voltage at terminal (1) rises at once, while that at (2) remains at ground. After a time T_D, the voltage of (2) rises by the amount of the input voltage, reducing the voltage difference between (1) and (2) to zero, and completing the process.

Figure 127 shows an alternative, and more commonly used method. It specifically illustrates a delay line in the form of a cable, and is convenient in that one side of the input and one side of the output, as well as the case of the cable, can all be at ground. There is a resistor of size Z_0 at the *input* of the cable,

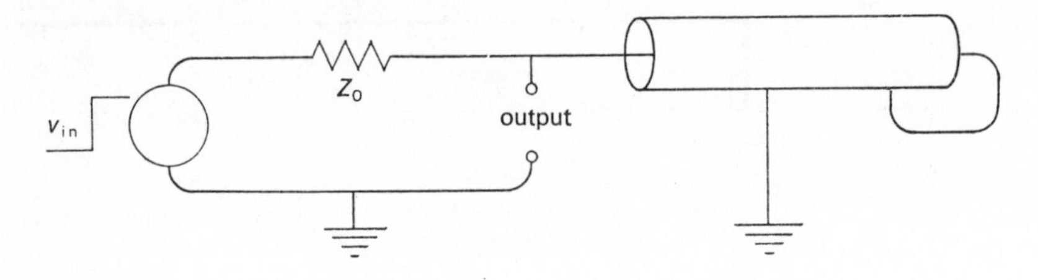

Figure 127. More convenient arrangement for delay-line pulse shaping.

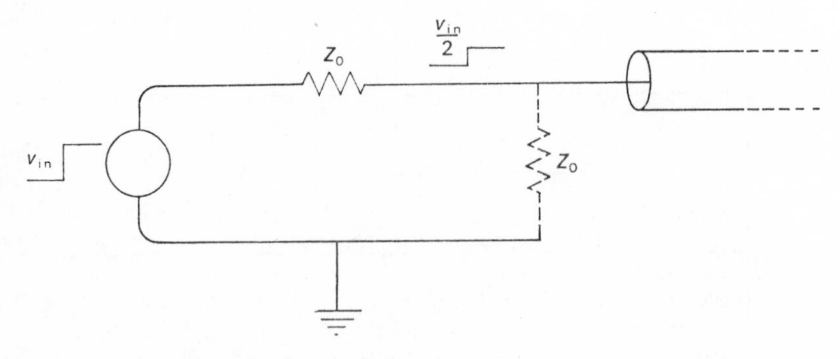

Figure 128. Equivalent circuit when step starts down line.

and a short at the far end. The output is taken between the input of the cable and ground. We shall examine the action in stages. First of all, as the signal sets off down the line, it has no way of 'knowing' that the line is not an infinite one, and hence the latter will appear at that moment to be an impedance of size Z_0, as in Figure 128. A dividing action takes place, and only half the original input appears at the input of the line. Thus a voltage step of size $v_{in}/2$ moves down the line, charging the distributed capacity as it goes, and arriving at the far end after a time T_D. As this end is shorted, there can never be any voltage there. So when the step arrives at this point it must be reflected with reversed polarity, that is an inverted step of size $-v_{in}/2$ must set off backwards down the line to preserve zero voltage at the shorted end. (The analogy with the closed end of an organ pipe is apparent.) Finally the inverted step travels backwards along the line, discharging the distributed capacity as it goes, and after a further time T_D reaches the input end. Assuming the voltage source is perfect, with zero resistance, the signal sees the line correctly terminated with its characteristic impedance Z_0, no reflection take place, and the process ends. The output (Figure 129), is a rectangular pulse of height $v_{in}/2$ and length $2T_D$, as the voltage there rises to half the input signal as soon as this is applied, and remains there until the reflected signal returns after a time $2T_D$, when it reduces to zero. In practice it is unnecessary for the voltage source to have zero resistance: for example if the characteristic impedance of the line were 75 ohms, and the internal resistance of the source 25 ohms, we would merely use a resistor of 50 ohms in series with the source to provide the requisite value of Z_0. But the source resistance cannot exceed Z_0 (as we cannot add on negative resistance!), which means that an emitter follower, but not a source follower, must be used for driving delay lines of this sort of characteristic impedance.

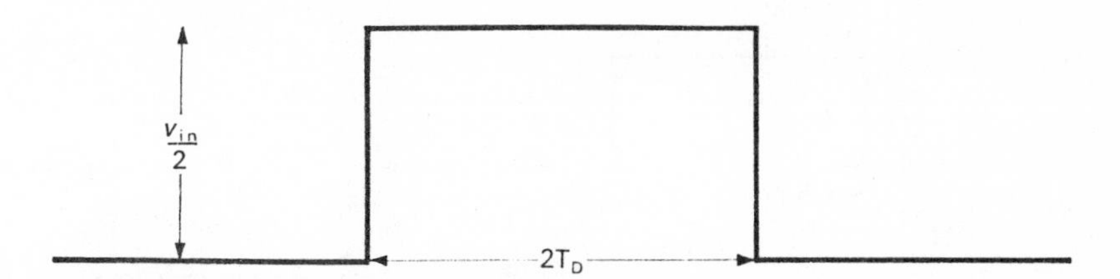

Figure 129. Output from shaping circuit of Figure 127 (idealized).

As mentioned earlier, every real delay line attenuates the step as it passes along it. Thus the returning step will be smaller than when it set out, and will not be able to completely eliminate the voltage at the output when it arrives back there. The output signal will consequently have a 'pedestal' on it as shown in Figure 130(a). The remedy for this is to have a gentle slope on the top of the original signal as shown in Figure 130(b), so that the voltage due to the generator at the output slowly declines. The returning step, although still

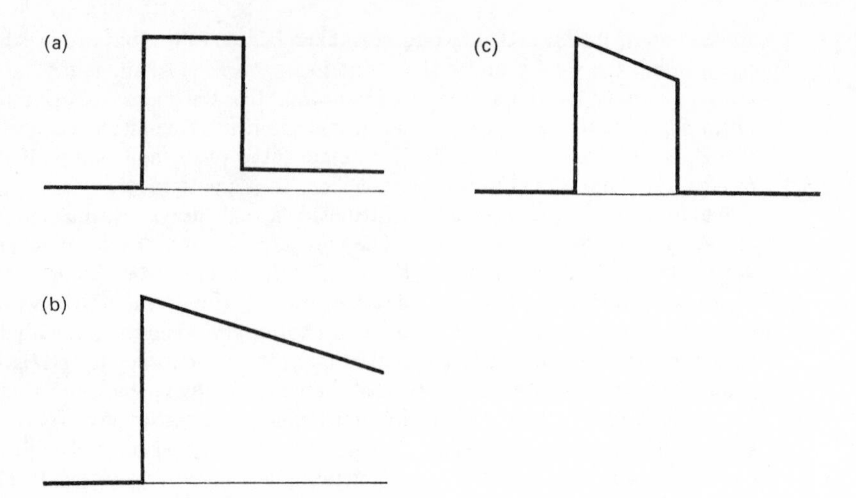

Figure 130. Output obtained in practice with delay-line shaping, and its correction.

attenuated in height, finds it now has a rather smaller signal to cancel than was there originally, and a little empirical adjustment makes exact cancellation possible (see Figure 130(c)). Fortunately, slope of the signal top occurs naturally, either due to the time constant of the detector, or to a later coupling network, so it is only a question of adjusting one of these appropriately. The signal from any real delay line will also have a finite rise time, and as the trailing edge is just the rising edge inverted, this will have a finite fall time as well.

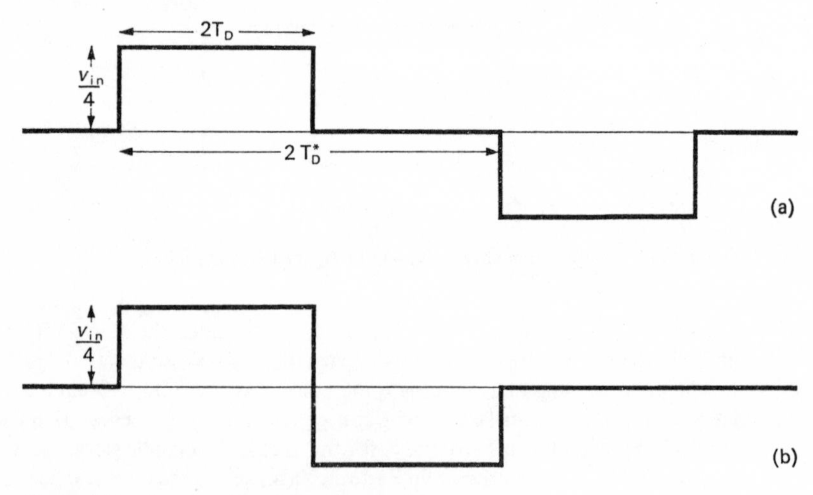

Figure 131. Double-delay-line pulse shaping.

If the delay-line shaped pulse of Figure 129 is passed down a further delay line with a one way travel time of T_D^*, but in a circuit otherwise similar to that of Figure 127, a little consideration will show that the output signal is as in Figure 131(a). Finally if T_D^* is made equal to T_D, the 'double delay line shaped' pulse whose virtues were discussed in an earlier chapter appears (see Figure 131(b)). Note however that its height has again been halved.

8·4 Termination of signal cables

Any length of coaxial cable carrying a signal, behaves as a delay line, even when there is no specific need for such delay. It is necessary, therefore, to be aware of the presence of this delay, and if necessary to make allowance for it, and also, in principle, to terminate the cable correctly to avoid unwanted reflections. We say 'in principle', because we shall discuss in a moment exactly when we can avoid doing this, but we shall first consider techniques for correct termination.

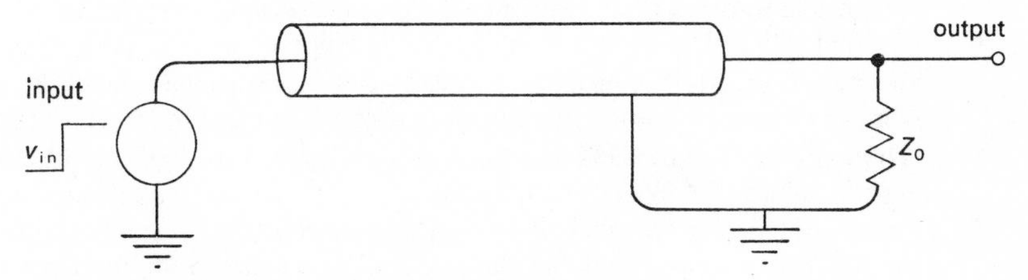

Figure 132. Signal cable terminated at the receiving end.

Figure 132 shows a cable terminated at the receiving end in a manner discussed at the beginning of this chapter. The signal goes down the line and is completely absorbed at the far end in the characteristic impedance Z_0. We have assumed that the source of signals has zero internal resistance: if this is not so only a certain fraction of the input will be transmitted down the line, for reasons similar to those in Figure 128. The output signal (which will still of course be terminated correctly in Z_0) will drop correspondingly in size. This loss can be avoided by the use of the circuit in Figure 133 where the characteristic resistor is placed at the input. If the generator resistance is not zero, an appropriate resistor is used to make the total Z_0. The signal transmitted down the line is now

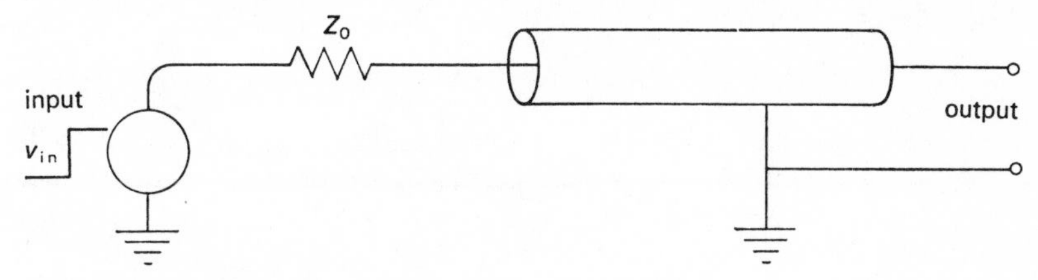

Figure 133. Signal cable terminated at the sending end.

$v_{in}/2$ from the usual divider action. The considerations when this step reaches the open-circuited receiving end are the opposite of those already discussed for the short-circuited case. Now there is an anti-node for voltage, that is reflection without inversion, and a node or zero for current, as no current can flow when a circuit is open. The step of size $v_{in}/2$ thus 'stands on its own head' to make the total voltage at the open end v_{in}. This step then returns along the line, leaving the distributed capacity charged to the full voltage v_{in} as it goes, and is finally absorbed in the characteristic impedance at the sending end. No further action occurs, and we have at the output end the full input voltage. In practice the receiving end cannot be exactly open circuited, because the function of the cable is to transmit the signal to, say, the input of an amplifier with finite input impedance. But the action described will be basically unimpaired if the load on the output is very large compared with Z_0. The foregoing discussion was for a step of voltage, but it also applies to a rectangular pulse, as this is merely two successive steps of opposite polarity, and by a further extension clearly applies to any shape of pulse.

Consideration of what happens when a step is sent down an improperly terminated line will help to decide when we need not bother in practice, to terminate cables correctly. Suppose that we send a step v_{in} down the line of Figure 132 from an ideal source, but with the termination at the receiving end a resistor of arbitrary size Z. Figure 134 lists the sort of reflections we expect to obtain for various values of Z. We have already discussed the cases at the extreme top and bottom of the diagram where $Z = \infty$ and $Z = 0$ respectively, as well as the case for $Z = Z_0$. It is reasonable to assume that for a terminating resistor of size between Z_0 and ∞, the reflected signal will be non-inverted but reduced in size from the open-circuit case, while that for Z between Z_0 and zero will be inverted

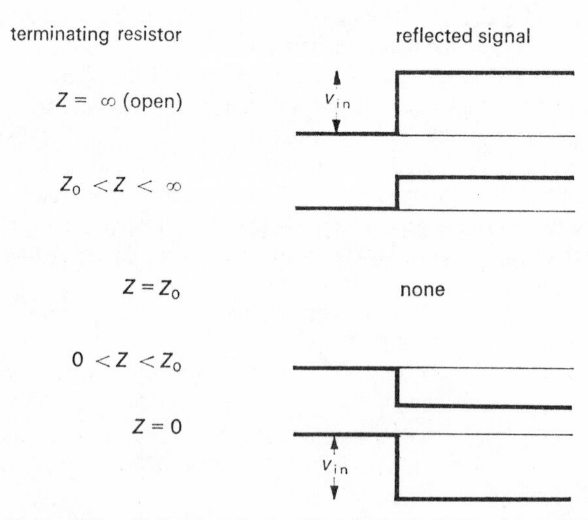

Figure 134. Reflections from cable end for various values of terminating resistor.

and similarly reduced. Let us suppose that we have chosen a resistor less than Z_0 and of size such that an inverted pulse of half the incoming height is reflected. The action starts with a signal of size v_{in} moving down the line to the output end, charging the line capacity as it goes. When it arrives at the output, half of it, in our example, is reflected as an inverted signal, and returns along the line reducing the charge on the distributed capacitance to $v_{in}/2$ as it moves along. The output voltage is thus, for the moment, $v_{in}/2$. As the source resistance is assumed to be zero, the inverted signal which returns there is reflected with polarity reversed, and its full height $v_{in}/2$. This positive signal travels once more to the output end, where, as on the first occasion, half of it is reflected as an inverted signal of size $v_{in}/4$. The signal at the output just after a time $3T_D$ has thus risen to a height of $v_{in}/2 + v_{in}/4$ or $3v_{in}/4$ as shown in Figure 135. The rest of the process should now be clear. The signal at the output continues to rise, at times separated by $2T_D$, by amounts which are half of the preceding rise

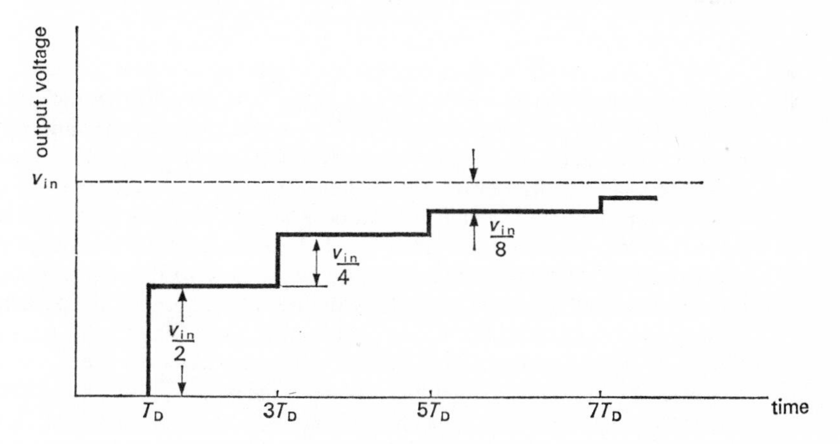

Figure 135. Output from incorrectly terminated line fed with step signal ($Z < Z_0$).

to a final value v_{in}. If the terminating resistor is below Z_0, but of different size from the one we have chosen, the pattern will, in general, be the same, but the fraction reflected at each stage will be different. But if the terminating resistor is greater than Z_0, and of size such that a pulse of half the incoming height is reflected at the end, a corresponding analysis will show that the output first rises *above* the value v_{in} by an amount $v_{in}/2$, then after a further time $2T_D$ falls below v_{in} by an amount $v_{in}/4$, arriving at a final value v_{in} in a series of steps of diminishing size, which makes the output appear alternately above and below v_{in} (Figure 136). The output voltage in this and the preceding case, both reach the same value v_{in} which we should expect to obtain if there were no delay line present between input and output, but only after a time which is long compared with T_D, and in different ways for the two sizes of termination.

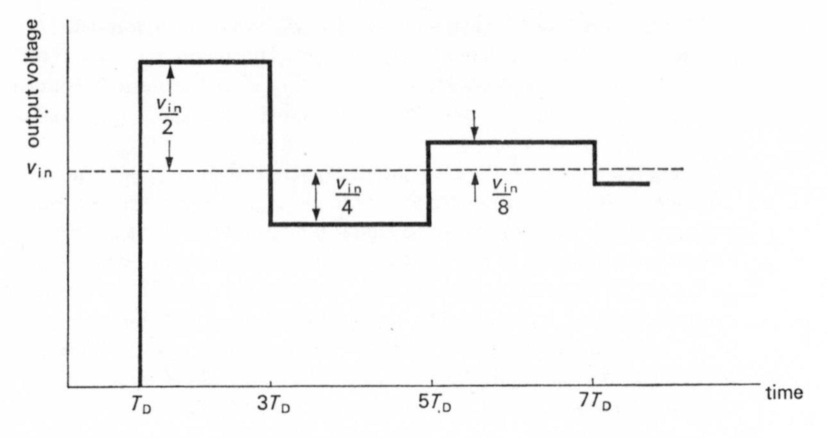

Figure 136. Output from incorrectly terminated line fed with step signal ($Z > Z_0$).

This gives a clue as to when we must terminate signal cables correctly, and when we can afford to ignore the problem. We note two things in connexion with the previous discussion. Firstly we have assumed that the rise time of the original input signal to the line was very small (we have actually drawn it infinitely small) and certainly very much less than T_D; secondly the pattern of steps in both Figures 135 and 136 will have disappeared and the signal level stabilized at v_{in} after a time of say $10T_D$. (Indeed we might think of the pattern of Figure 135 as having a 'rise time' of a few times T_D.) If we have an input signal with a finite rise time which is more than a few times T_D, the pattern of steps produced by the cable will be completely obscured by the inherent rise time of the signal, and there is no point in bothering to terminate the line with its characteristic impedance. But if the rise time of the signal is comparable with or less than the cable delay time, correct termination is imperative to avoid distortion of the leading edge of the signal with spurious reflections. From figures previously given it is clear that unless we are dealing with very long lengths of ordinary coaxial cable, correct termination is not essential for signals with microsecond rise times. For example five metres of coaxial cable might have a delay time of about twenty-five nanoseconds. Thus if the signal we are dealing with has a one-microsecond rise time correct termination is superfluous, but with a ten-nanosecond rise time signal, or if we increased the cable length to fifty metres, then it would become necessary.

8·5 Delay lines in pulse generators

Delay lines are used in pulse generating equipment to produce rectangular pulses for test purposes. They have the advantage over the multivibrator type of generator that the pulse length is determined solely by a static element, the delay line, and does not depend on the stability of transistor characteristics.

The principles whereby a voltage step is converted into a rectangular pulse have already been discussed in connexion with pulse shaping and nothing further need be said. The voltage step in the present case is produced by opening (or closing) a switch connected to a well-regulated power supply, the latter defining the pulse height. The switch is usually of a special type with mercury contacts in order to produce a fast rising step.

References and further reading

1. A. T. STARR, *Electronics*, Pitman, 1959.
2. G. E. OWEN and P. W. KEATON, *Fundamentals of Electronics*, volume 1, Harper & Row, 1966.
3. I. A. D. LEWIS and F. H. WELLS, *Millimicrosecond Pulse Techniques*, Pergamon Press, 1959.

CHAPTER 9

Power supplies

9·1 Basic rectifying circuits

Direct current power for electronic equipment may be supplied from batteries, and for portable equipment this is nearly always the case. For fixed laboratory apparatus, particularly where considerable current is drawn, d.c. supplies are normally obtained by rectifying and smoothing the mains. We discuss, in this chapter, circuits for such supplies.

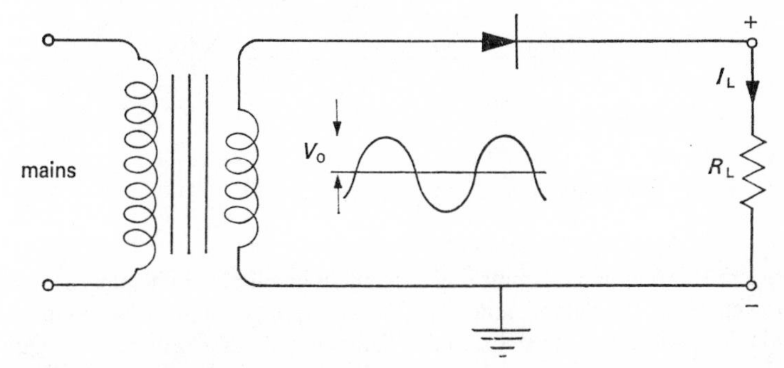

Figure 137. Simple rectifying circuit.

The simplest arrangement is shown in Figure 137. Here the transformer both isolates the mains from the equipment, and also steps down the voltage to a suitable value – say 12 volts r.m.s. typically. R_L represents the load on the power supply. For the sake of simplicity we assume in what follows that the semiconductor diode used is a perfect rectifier – that is it has zero resistance in the forward direction, and an infinite resistance in the backward direction – an assumption which we have seen is not too far from the truth. We assume also that the transformer is perfect, and in particular that its secondary winding has a negligible d.c. resistance. The current flowing in the load resistance R_L is then as shown in Figure 138(b), since the diode can only conduct on the positive half cycles of the input voltage of Figure 138(a). We have thus obtained a current which is unidirectional, and the combination of transformer and diode can be considered as a d.c. generator with polarity as indicated in Figure 137. (If negative voltages with respect to ground are required, the diode can be reversed.) Clearly however, the large variations in the amplitude of the output in

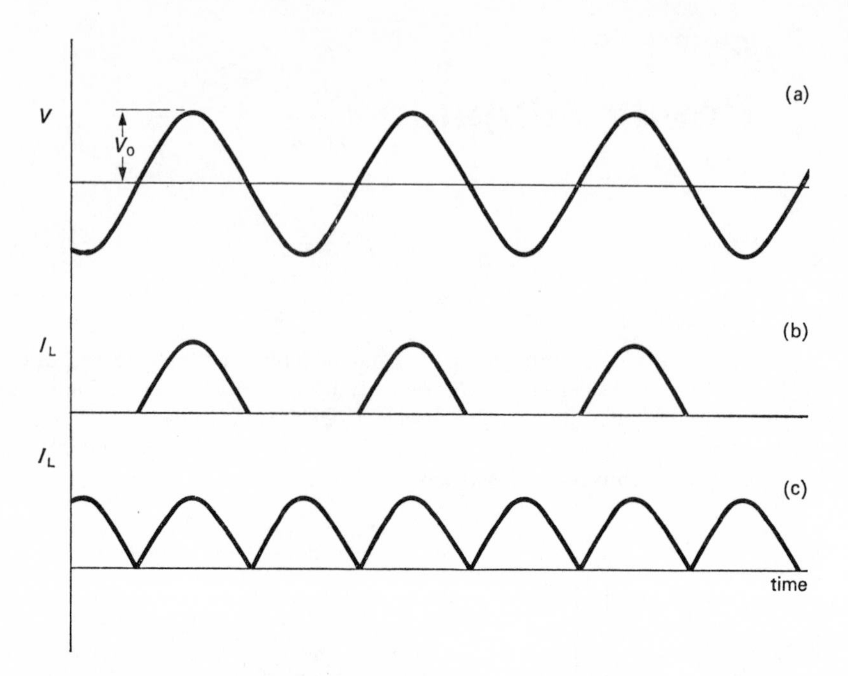

Figure 138. Waveforms in rectifying circuits.

Figure 138(b) make this simple arrangement unsuitable in its present form as a power supply for electronic devices. We can improve matters by changing from this 'half-wave' arrangement to the 'full-wave' rectifying circuit of Figure 139 which uses a centre tapped transformer. This is effectively two circuits of the type shown in Figure 137 joined together, so that the diodes conduct on successive half cycles of the input wave form to produce a current through the

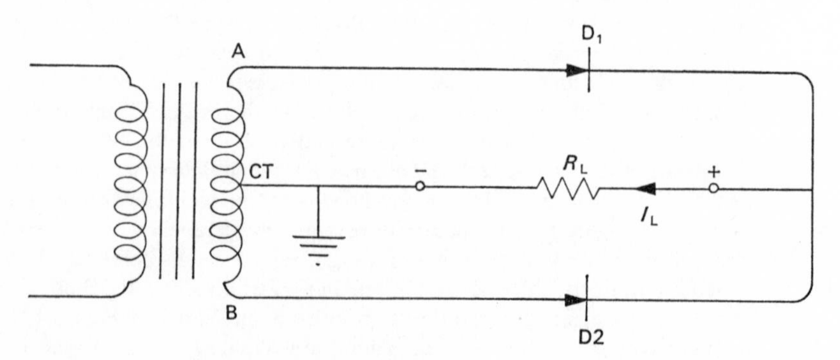

Figure 139. Full-wave rectifying circuit.

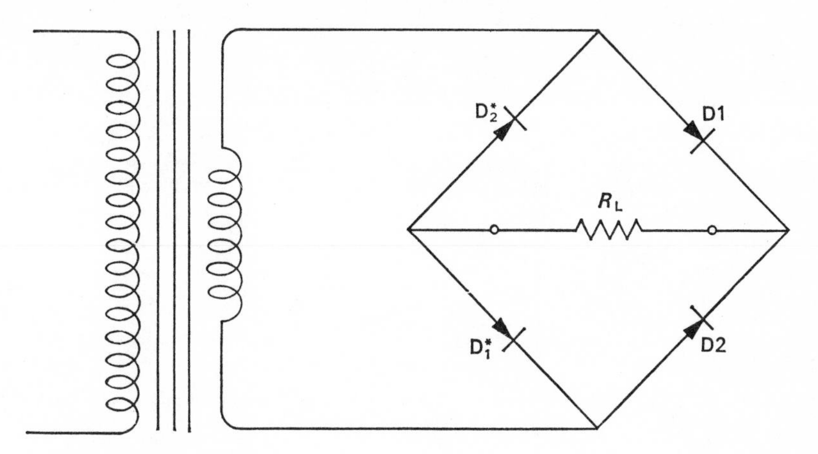

Figure 140. Bridge circuit.

load as shown in Figure 138(c). We can see how this arises by thinking of the voltages of the points A and B with respect to the centre tap CT. When A is positive with respect to CT, B is negative with respect to that point, and hence diode D_1 conducts and D_2 is cut off. On the other half of the cycle conditions are reversed and D_2 conducts. The current through the load fluctuates less wildly than for the half-wave case, reaching zero only momentarily, twice per cycle. As the transformer is centre tapped we shall need a secondary giving twice as much voltage between its extremities compared with the half-wave case, to obtain the same peak current through the load. The 'bridge circuit' shown in Figure 140 has the smoother output of the full-wave circuit without the need for centre tapping, at the expense of using two additional diodes. In this arrangement diodes D_1 and D_1^* conduct on one half cycle, and diodes D_2 and D_2^* on the other.

9·2 Smoothing circuits

To produce an output approximating better to a steady d.c. voltage than the full-wave rectifier, a 'smoothing' or 'filter' circuit is used on the output. In the simplest case this may be either a large capacitor across the output as in Figure 141, or a large inductor in series with the load as in Figure 142. The action of these two types of filter should be obvious in a general way. In the capacitor filter, the capacitor tries to store charge when the voltage is at a maximum, and thus smooth out fluctuations in output voltage, while in the inductor filter, the inductor tries to prevent changing currents through the load and thus produce a steadier current flow. We shall now examine the action of these circuits more closely.

Figure 143 shows the details of the output waveform with a full-wave

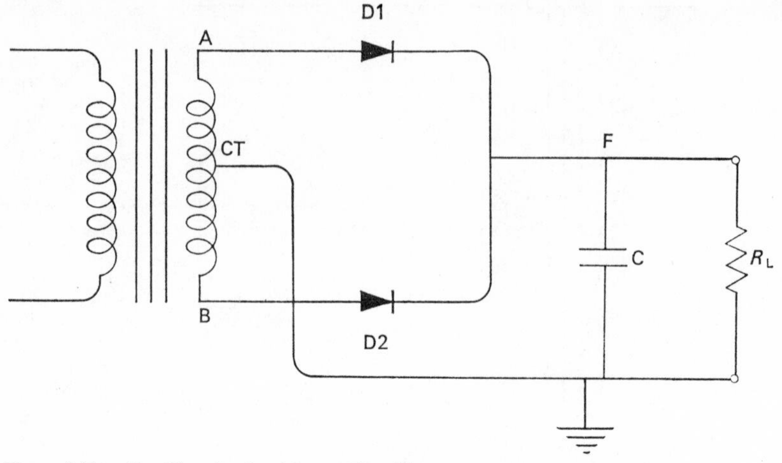

Figure 141. Rectifier circuit with capacitor filter.

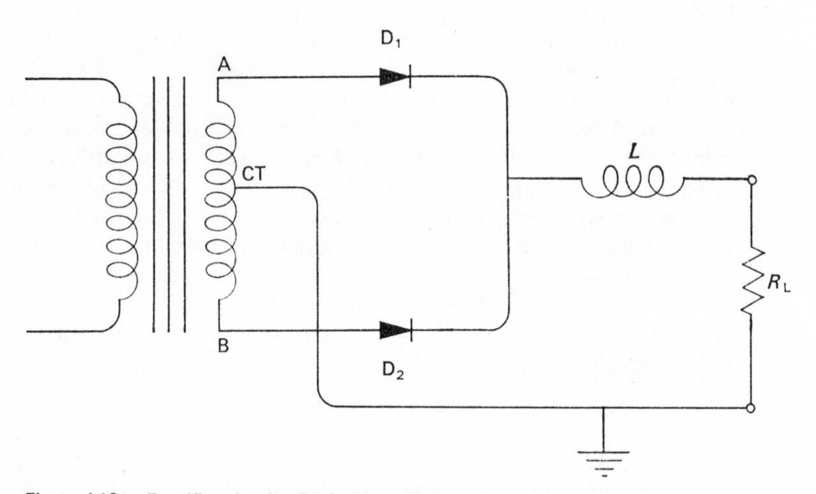

Figure 142. Rectifier circuit with inductor filter.

rectifier and a capacitor filter. Because of the storage action of the capacitor, the voltage never reaches zero. At the point X, for example, one of the diodes would normally be supplying current to the load, the other being cut off. Because of the stored charge on the capacitor, its voltage is actually higher than that which would be provided at that moment by the transformer and the 'on' diode, so this diode too is cut off. The current to the load is now being supplied from the charge stored in the capacitor, and therefore the voltage must decay exponentially with time constant $R_L C$. At the point Y the other diode would normally

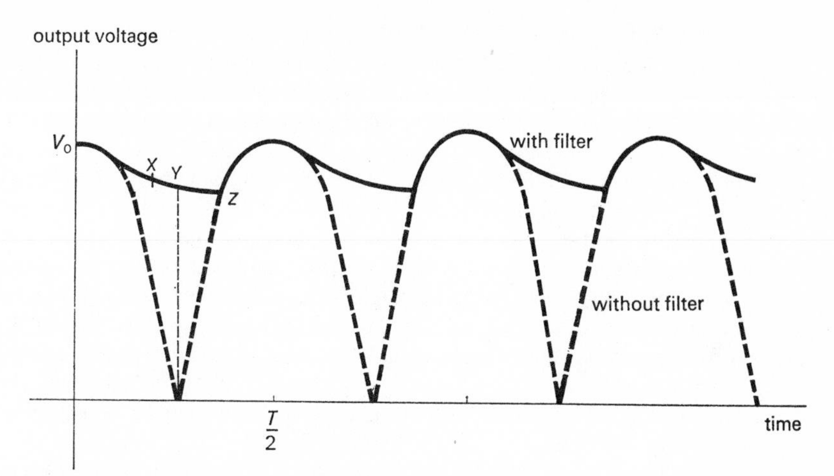

Figure 143. Output waveform with capacitor filter.

take over, but here again because of the higher capacitor voltage, it supplies no current initially. However at the point Z, the transformer–diode voltage reaches that of the capacitor, and from here to the end of the half cycle, the capacitor is recharged from the supply. The whole process then repeats itself with the roles of the diodes interchanged. The efficiency of the filtering system may be characterized by calculating the 'droop' in voltage at its lowest point as a fraction of peak voltage, and we shall do this for the particular case where the time constant $R_L C$ is large compared with $T/2$, the time between successive maxima of the unfiltered output. This is equivalent to assuming that the 'droop' is going to be small.

The voltage at a typical point X is given by $V = V_0 \exp(-t/R_L C)$, which with our approximation becomes $V_0(1 - t/R_L C)$. Another consequence of this approximation is that the point Z of Figure 143 will not be far from the peak of the cycle, and thus at Z we can put $t \approx T/2$. The voltage at this point in the cycle is thus $V_Z = V_0(1 - T/2R_L C)$, and the fractional 'droop', $(V_0 - V_Z)/V_0$ is given by $T/2R_L C = 1/2fR_L C$, where f is the supply frequency. For a given load R_L, the 'droop' or ripple improves as C is increased. There is however, an upper limit to the size we can make C because of the possible damage to the diodes from large currents during the recharging of the capacitor between the point Z and the peak of the cycle. The energy supplied by the capacitor to the load between recharging is $(V_0^2/R_L)T/2$ approximately, and this must be restored during charging. If by increasing C we reduce the time between the point Z and the peak of the cycle, higher charging currents would be expected and in fact an approximate analysis shows that the charging current rises as the square root of C. Often the manufacturer will advise directly the largest value of capacitor which may be used with a given diode.

At this point it is appropriate to mention a few other characteristics and restrictions for rectifier diodes. Let us suppose that each half of the transformer secondary in the full-wave circuit of Figure 141 produces a peak voltage of 20 volts. Assuming as before that the time constant $R_L C$ is large with respect to $T/2$, there is across the capacitor an almost steady voltage of 20 volts. The point F in the diagram is thus at +20 volts with respect to ground. Since the points A and B, at different points of the cycle can both reach –20 volts with respect to the grounded centre tap, it is clear that the diodes can momentarily have across them 40 volts, that is twice the output voltage. The diode in a full-wave rectifying circuit must thus be able to stand twice the output voltage without breaking down in the manner discussed in connexion with Figure 6. We thus need to check that the 'peak-inverse (or reverse) voltage' quoted for a diode is larger than twice the d.c. voltage we intend to use it to produce. Diodes can be connected in series to provide higher peak inverse voltages, and commercially available 'stacks' with up to a dozen diodes may have peak inverse voltages of 6000 volts.

The other piece of information we need for a design is the greatest average forward current that the diode will supply – that is, forgetting the intermittent nature of this current, which we have allowed for otherwise, the largest current, averaged over a few cycles, which we can draw through the diode for our load. The manufacturer will also provide information on how much the diode actually diverges from the ideal diode with zero forward resistance and infinite back resistance, which we have assumed up to now. The reverse current at a specified voltage will indicate the departure from ideal reverse characteristics, and the voltage drop in the diode at a given forward current will specify how far it departs from the ideal zero resistance forward characteristics. In view of the tremendous range of rectifier diodes available, it is impossible to quote 'typical' values for these quantities, but the following data for a Mullard type *BY*100 silicon rectifier diode will at least indicate orders of magnitude. In this case the maximum average forward current is 550 mA (averaged over 50 msec), and the peak inverse voltage 800 volts. A 200-μF capacitor can be connected directly across the diode provided a 5-ohm current-limiting resistance is included in series with it. As regards departures from ideal characteristics, the reverse current at a reverse voltage of 1·25 kV (much higher than normally used) is only 10 μA, while there is a voltage drop in the diode of only 1·5 volts at a forward current of 5 amp – the latter being much greater than the allowable average forward current, and indeed equal to the maximum forward current which may be drawn instantaneously.

Returning now to the value of $1/2fR_L C$ for the ripple in a capacitor filtered supply, we see that for a given value of capacitor, the ripple is inversely proportional to R_L. Indeed we can write the equation $V_Z = V_0(1 - T/2R_L C)$ as

$$V_Z = V_0 - \frac{I_0 T}{2C}$$

$$= V_0 - \frac{I_0}{2fC} \qquad \textbf{9.1}$$

since the ripple is small and the mean current through the load I_0 is approximately V_0/R_L. This type of filter thus works best when feeding small currents into large loads: under these circumstances the ripple is very small and the output voltage is almost equal to the peak alternating voltage. The term 'regulation' is used in connexion with power supplies to describe how the output voltage decreases with load, and is usually characterized by the quantity $\delta V/\delta I$, the drop in output voltage with output current – in fact the output impedance. In the present case the output voltage varies with time between V_0 and V_Z, with an average value of $(V_0 + V_Z)/2 = V_0 - I_0/(4fC)$. Thus $\delta V_{av}/\delta I_0$ is, apart from sign, $1/(4fC)$. For a capacitor of 200 μF and a frequency of 50 Hz the output impedance will be 25 ohms, even assuming, as we have been doing, that the diodes and transformer are ideal. We shall see later how poorly this compares with electronically regulated supplies.

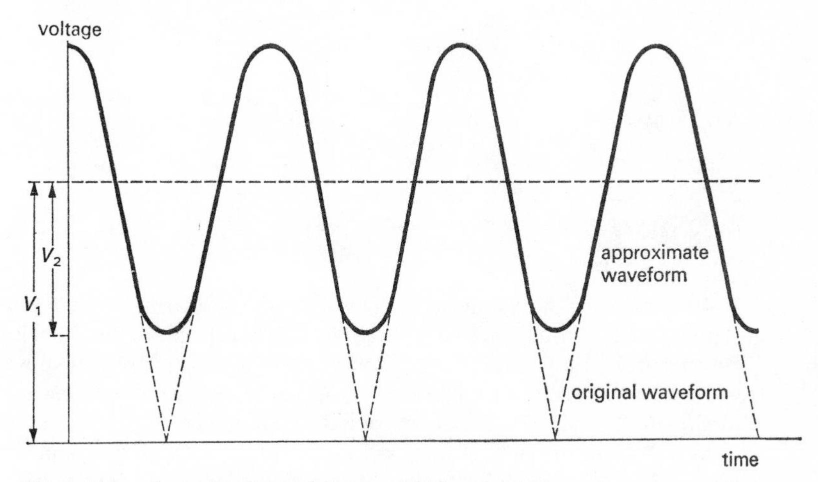

Figure 144. Approximation to full-wave rectifier waveform.

We shall treat the inductor filter of Figure 142 in a different way from the capacitor filter. In this case we start with an approximation to the output from a full-wave rectifier (Figure 144) made up of a steady d.c. voltage V_1, on which is superimposed an alternating voltage of peak value V_2 and frequency twice that of the original unrectified voltage of Figure 138(a). This is the first stage of a 'Fourier analysis' of the rectified wave form into a whole series of sinusoidal voltages of frequencies $2f$, $4f$, $6f$, etc., and gradually diminishing amplitudes. The mathematical analysis tells us that V_2 should be $\frac{2}{3}$ of V_1 in size, and the amplitude of the next component should be only $\frac{1}{5}$ of V_2. It is therefore a reasonably good approximation to consider only V_1 and V_2, particularly as it will later be clear that the higher frequency components are more easily removed by filtering. We may now calculate the percentage ripple appearing across the resistance R_L of Figure 142. For an ideal inductor of zero d.c. resistance, the whole d.c. component V_1 appears across R_L. The alternating voltage of frequency $2f$ appearing across R_L is $V_2 R_L/\{R_L + jL(2\omega)\}$. If we assume that L is large enough for $L(2\omega)$ to be very much greater than R_L this reduces to $V_2 R_L/2L\omega$ in magnitude. We can take as a measure of the percentage ripple on the output, this quantity divided by the d.c. component V_1; as $V_2 = (2/3)\,V_1$ this reduces to $R_L/3L\omega$. Thus, in contrast to the capacitor filter, the inductor filter works best for small values of the load, that is for large currents. Unfortunately, as R_L approaches zero and the filtering improves, the magnitude of the output drops to V_1, which is considerably less than the original peak value of $V_1 + V_2$. The fact that the characteristics of the inductor filter with changing load are complementary to those of the capacitor filter makes the composite filter of Figure 145 – the L-section filter as it is called, an attractive proposition.

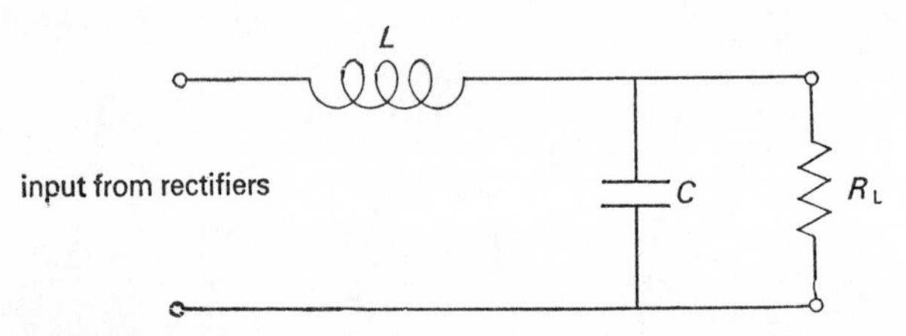

Figure 145. L-section filter.

Before we deal with it mathematically it is worth considering why the Fourier analysis method could not have been applied to the capacitor filter – that is why the output of the full-wave rectifier could not have been considered as equivalent to a generator producing a direct voltage of size V_1 and an alternating voltage of size V_2. The answer is that, as we have seen, the current supplied to the capacitor is an intermittent one, and it is difficult to think of a model of a generator which will simulate this. Such a difficulty does not arise with the inductor filter. We make these remarks at this stage because it is possible that such troubles could arise with the L-section filter of Figure 145: clearly if the size of the inductor is small, the filter approximates to a capacitor filter, and there will be intermittent current flow. In what follows we assume that the value of the inductor is sufficiently large to avoid this difficulty, and we shall later obtain an approximate condition to ensure such a situation.

Using the previous model for the rectifier system output, we note (Figure 145) that the direct component V_1 appears in full across R_L. If we make the reasonable assumption that the impedance of the capacitor for the alternating frequency in question is small compared with both that of the inductor and that of R_L, the alternating voltage across the output is simply $V_2(1/jC2\omega)/(jL2\omega)$ or $V_2/4LC\omega^2$ in magnitude. A measure of the ripple is obtained by dividing this quantity by V_1, to give $1/(6LC\omega^2)$, remembering the relation between V_1 and V_2. In contrast with the two previous types of filter, the ripple is independent of the load, so we may quote a value for it without further qualification. For a frequency of 50 Hz and values of L and C of 10 henrys and 200 μF respectively, the ripple is less than 0·1 per cent. Like the inductor filter, the L-section filter has a d.c. output V_1 appreciably less than the peak value of the input to it $(V_1 + V_2)$.

The condition for avoidance of intermittent current is obtained from the fact that the alternating current flow is largely controlled by the inductor, and thus is of magnitude $V_2/(L2\omega)$, while the direct current is controlled by R_L and is of magnitude V_1/R_L. If the overall current is just to reach zero, these two components must be equal, that is $V_2/L2\omega = V_1/R_L$, which reduces to $L\omega = R_L/3$. As this treatment is approximate we should make $L\omega \gg R_L/3$ for safety. This condition is in addition to the two previous conditions – that the impedance of C must be much less than that of both L and R_L for the frequency in question; however, there is nothing incompatible in the three requirements.

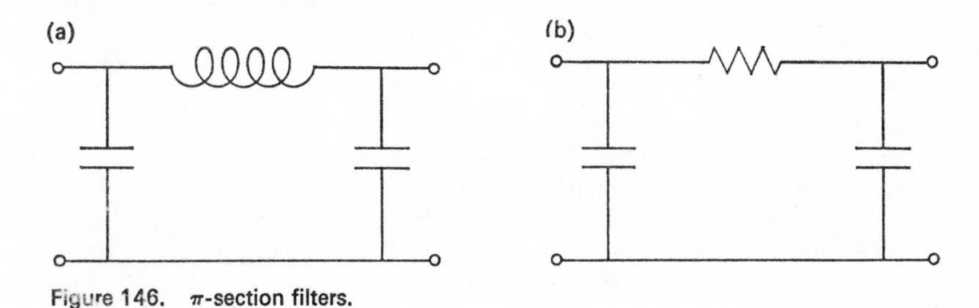

Figure 146. π-section filters.

Further types of filter can be built up from those we have discussed. For example, we may cascade a number of L-section filters, or we may combine a capacitor filter with an L-section filter to give the π-section filter of Figure 146(a). The latter combines the good ripple qualities of the L-section filter with the property of high output voltage under low load possessed by the capacitor filter. Sometimes the bulky inductor in the π-section filter is replaced by a resistor, as in Figure 146(b). The theory of its filtering action is the same as that of the circuit in Figure 146(a), with the impedance of the inductor replaced by that of the resistor. But as the resistor, unlike the inductor, also acts on the d.c. component, there will be a worsening in regulation – in fact the d.c. output impedance of the supply will be increased by the amount of the resistor. Further electronic control of the output, which we shall discuss shortly, can be used to offset this.

9·3 Other forms of power supply

Two rather different forms of power supply remain to be discussed. These are both suitable for providing a high voltage at a small current, as would be required for example, for the operation of a geiger counter. The first, shown in

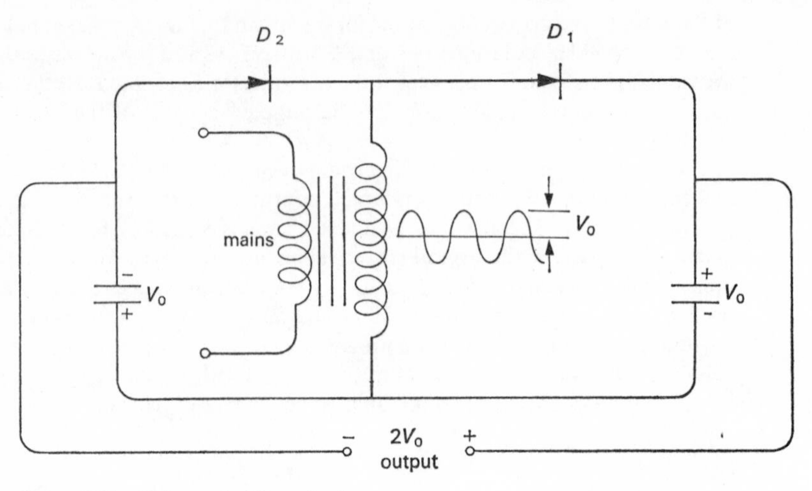

Figure 147. Voltage doubling circuit.

Figure 147, is the 'voltage doubler' which can be thought of as two half-wave rectifier circuits placed back to back in order to add their outputs. Each has a capacitor filter on its output and these capacitors are charged on alternate half cycles to a voltage, which if the load is small, is close to the peak V_0. The output voltage is thus $2V_0$, justifying the name 'voltage doubler'. Because of the basic half-wave nature of the arrangement, the ripple at a given load current will be worse than for the circuits previously discussed. Voltage treblers, quadruplers, etc., can also be constructed, on the same general lines as for the voltage doubler.

The second type is the r.f. (for 'radio-frequency') power supply. In discussing sinusoidal oscillators in Chapter 7, little was said about the quite difficult problem of the amplitude attained by the oscillations, if the gain round the loop is more than sufficient to start oscillation. It is clear however that with a high Q circuit, the amplitude of the oscillations may build up to considerably more than the voltage supply to the transistor. In this case the transistor will be cut off throughout a large part of the cycle, only supplying energy in bursts to the tuned circuit to keep the oscillations going. The output from the oscillator is usually taken from a separate winding coupled to the oscillator inductor to give an output not already connected to any d.c. potential, as mentioned in section 7·9. This output when rectified and smoothed will give a d.c. voltage very much higher than that of the supply. The frequency of the oscillator is usually around 100 kHz, where coils of good Q value can be obtained. The fact that the frequency is so much higher than the 50 (or 60) Hz used in conventional supplies, has the additional advantage that quite small value capacitors give excellent smoothing, as their impedance, $1/C\omega$, is correspondingly reduced. The current available from such an r.f. supply is not large, because loading up the output will reflect back on the tuned circuit, damp the oscillation, and cause the voltage to drop. For running low current devices this is no disadvantage; indeed it has the positive advantage, that if the supply is inadvertently touched by the operator the voltage will drop, and no harm will ensue.

9·4 Electronic regulation of power supplies

Power supplies using filters as described in section 9·2 still exhibit too poor regulation, and too much ripple for use with precision electronic apparatus. Further improvement is based on control devices which exhibit the property of having a voltage drop across them almost independent of the current through them. One of these, mentioned at the end of Chapter 1, is the Zener diode, and Figure 148 is merely Figure 6 redrawn with both the I and the V axes reversed,

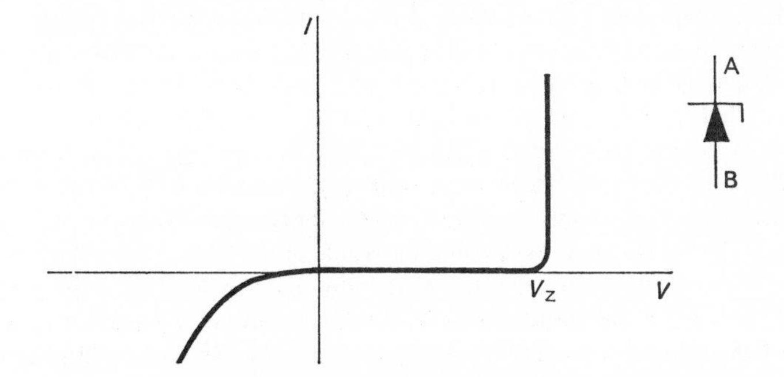

Figure 148. Zener diode characteristic and symbol.

to emphasize the part of the characteristic we are interested in. (Note however that as we have intended to show a smaller value of V_z in Figure 148, the scale of this diagram differs from that of Figure 6.) The conventional symbol for a Zener diode is also shown, the end labelled A being connected to the positive voltage for Zener operation. The Zener voltage, V_z in Figure 148, is a function of the resistivity (that is, of the doping) of the materials from which the diode is constructed. Diodes with values of V_z from a few volts to hundreds of volts are commercially available, and since diodes can be added in series, devices with breakdown voltages anywhere in this range can be constructed. The other device which has a voltage drop across it almost independent of the current through it, is the cold cathode gas discharge. However when transistors supplanted valves, the gas discharge tube similarly gave way to the Zener diode, so we shall not discuss it further. A Zener diode must always have a resistor in series with it to limit the current to the maximum permitted for the particular device; otherwise if the voltage placed across it is greater than the breakdown voltage, its almost vertical I–V characteristic implies that the device will try to draw a huge current, thus destroying itself.

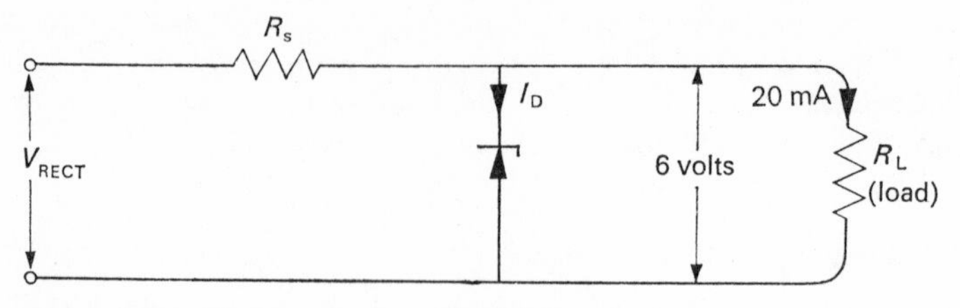

Figure 149. Simple voltage-stabilizing circuit.

Figure 149 shows the simplest way in which a Zener diode may be used to 'regulate' or 'stabilize' a rectified d.c. supply. Suppose we are designing a circuit to provide a regulated supply of 6 volts which is to supply 20 mA, and further assume that this 20 mA may change by ±10 per cent as load conditions vary. We arrange that the current I_D through the diode is greater than these expected fluctuations, say 5 mA in the present case. (We should also check that the selected value of I_D is below the permitted maximum for the diode, which is easily fulfilled in the example we have taken.) The total current normally drawn through the series resistor R_s is 25 mA. The voltage V_{RECT}, from the rectified supply is usually made a few volts higher than the Zener voltage to ensure that even if it fluctuates it is always sufficiently large to keep the diode on. Suppose in the present case it is 9 volts. The value of R_s is now fixed since it must drop $9 - 6 = 3$ volts when carrying a current of 25 mA. This implies that $R_s = 120$ ohms, and the design of the regulator is complete. Its action should also now be clear. If the load current increases to 22 mA, the voltage across the Zener still remains at 6 volts and all that happens is that the current through it drops by 2 mA; the current passing through R_s remains the same as before. Similar remarks apply to a drop in load current. It is also clear that fluctuations in the voltage from the rectified supply are not reflected in the output voltage, provided, as mentioned above, that these are not large enough to render the diode non-conducting at any time. These fluctuations might be due to a change in the alternating mains supply to the transformer, but our remarks apply equally well to any ripple not already removed by filtering.

The above remarks are true only in so far as the Zener diode has an ideal characteristic – that is if the voltage is absolutely independent of the current through it. In a typical real diode, when the current through it changes from 1 to 5 mA, the voltage might change from 5·6 to 6 volts. While this is only a 7 per cent change in voltage for a 400 per cent change in current, it is nonetheless still too great for many applications. A circuit in which the current through the Zener diode is never required to change by more than a small amount will improve matters. Such a circuit is illustrated in Figure 150, which shows a complete rectifier system plus regulator circuit. A regulator of the type shown contains (a) a voltage-reference element (b) a sensing and amplifying element and (c) a control element, either in series, as shown here, or in shunt. In the present circuit the reference element is provided by the Zener diode D_3 (draw-

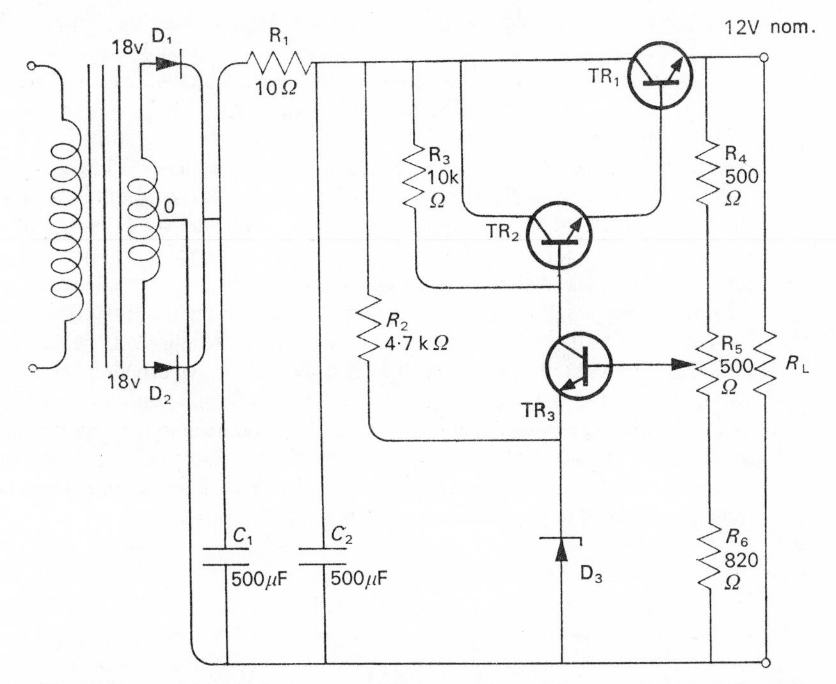

Figure 150. **Regulated power supply using Zener diode as reference element (courtesy Mullard Ltd).**

ing current largely through its associated resistor R_2), although it could equally well, but not so conveniently, be provided by a battery. The sensing element is the transistor TR_3, which compares the fraction of the output voltage which is applied to its base with the fixed Zener voltage on its emitter. If, for example, the output voltage rises (either because the input voltage from the rectified supply rises, or because the load on the regulated supply drops), a positive signal on the base of TR_3 will result. This transistor also acts as an amplifier (collector load R_3), and thus produces a negative signal at the base of TR_2. This transistor and TR_1 are emitter followers in the Darlington connexion (section 5·3), the emitter resistor of TR_1 being the load R_L in parallel with the potential dividing chain R_4, R_5, and R_6. A negative signal on the base of TR_2 thus produces a similar signal on the emitter of TR_1, that is on the output, which tends to correct the rise in output voltage which started the whole sequence of events. (The transistor TR_1 is here the series control element referred to in our general remarks about regulator systems.) We shall investigate shortly how well this arrangement corrects such fluctuations in the output, but first we shall look at some other features of the circuit. As the transistors TR_3, TR_2, and TR_1 are d.c. coupled, with no coupling capacitors anywhere in the loop, slow and d.c. changes can be corrected for, as well as rapid fluctuations like ripple from the rectifiers. Furthermore this d.c. coupling allows the value of the d.c. output voltage to be varied by means of the potentiometer R_5. For example, suppose that the contact were moved so that the base of TR_3 was connected to the point where the 500 ohm potentiometer and 820 ohm resistor meet. If there is to be no correction signal from TR_3 its base must be at approximately the same potential as the Zener voltage, that is 6 volts for the particular diode used here. The junction of the potentiometer and the 820 ohm resistor is thus approximately at 6 volts, so the potential of the output must be $6 \times (500 + 500 + 820)/820 = 13{\cdot}4$ volts approximately. This is in satisfactory agreement with the observed value of 14 volts, considering that we ignored the small base-emitter potential. Similarly if the potentiometer is moved into its other extreme position the output voltage will be approximately $6 \times 1820/1320$ or 8·3 volts (9·5 observed). Thus, we can alter the output voltage through a range of about 5 volts. We should not want to provide an even greater range by increasing the value of R_5, because in moving the point at which the base of TR_3 is connected to the potential chain, we are also affecting the fraction of the voltage fed back when a fluctuation occurs, and thus the ability of the circuit to correct such fluctuations.

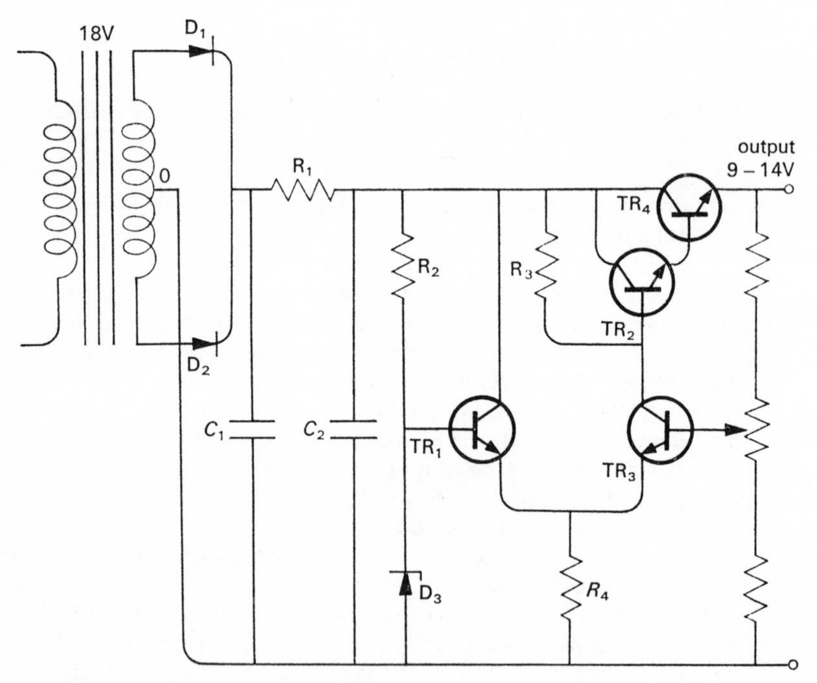

Figure 151. Regulated power supply with more sophisticated sensing amplifier (courtesy Mullard Ltd).

The observed performance of the circuit is as follows:

Output voltage, V_{out}	9·5 to 14 volts
Output current, I_{out}	250 mA max.
$\frac{\delta V_{out}}{\delta V_{in}}$ (under full load conditions)	5 per cent
Regulation (defined here as the fractional change in output in going from no load to full load conditions)	2 per cent
Ripple	<2 mV r.m.s.

Further improvement in performance may be obtained with a circuit of the type shown in Figure 151. Here there is a more sophisticated sensing and amplifying element in the form of a 'long-tailed pair' difference amplifier (TR_1 and TR_3). This compares the voltage on the base of TR_3 (that is a fraction of the output voltage) with the fixed potential from the Zener diode on the base of TR_1 and supplies a correction signal if these are not equal. Furthermore the siting of the Zener diode is more favourable in this case than in Figure 150, as changes in transistor currents will have an almost completely negligible effect on the current through the Zener, and thus on its reference voltage.

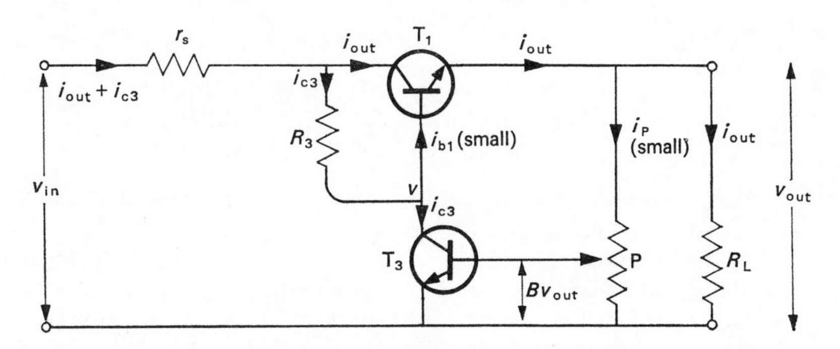

Figure 152. Regulating circuit of Figure 150 redrawn in simplified form.

We finally turn to a simple mathematical analysis of circuits of the type we have been discussing. Figure 152 is Figure 150 in simplified form. TR_1 and TR_2 have been combined into a single equivalent transistor. The Zener diode does not appear at all, as we shall assume in the first instance that it is perfect with zero resistance for signals. v_{in} represents a fluctuation in the rectifier-system output, and v_{out} and i_{out} corresponding changes in the output from the regulated supply. r_s represents the output impedance of the rectified supply, and the current through it, with the approximations noted on the diagram is ($i_{out} + i_{c_3}$).

Let us now follow the sequence of changes round the loop, starting at the base of T_3. The fraction of the output voltage change, Bv_{out}, applied to the base of this transistor produces a collector current, i_{c_3}, given by

$$i_{c_3} = g_{m3} Bv_{out} \qquad 9.2$$

where g_{m3} is the transconductance of T_3 ($=h_{fe_3}/h_{ie_3}$), the suffix $_3$ merely indicating to which transistor the parameters relate. T_1 is really a Darlington pair with high input impedance, so its base current i_{b_1} will be small, as shown, and all the collector current of T_3 will flow through R_3. Hence considering the voltage drop across r_s and R_3 we have

$$v_{in} - (i_{out} + i_{c_3}) r_s - i_{c_3} R_3 = v$$

where v is the voltage change at the collector of T_3. Substituting the value of i_{c3} from equation **9.2** we obtain

$$v = v_{in} - i_{out} r_s - g_{m_3} Bv_{out}(R_3 + r_s) \qquad 9.3$$

which determines v.

Turning now to the transistor T_1, the voltage between its base and emitter is $v - v_{out}$, so its collector current is $g_{m_1}(v - v_{out})$. This, of course, is simply i_{out} so,

$$i_{out} = g_{m_1}(v - v_{out}) \qquad 9.4$$

Now i_{out} is v_{out}/R_L, so we can eliminate v between equations **9.3** and **9.4** to obtain

$$\frac{v_{out}}{v_{in}} = \left\{1 + \frac{1}{g_{m_1} R_L} + \frac{r_s}{R_L} + Bg_{m_3}(R_3 + r_s)\right\}^{-1} \qquad 9.5$$

Looking at the relative magnitudes of the terms on the right-hand side of this equation, and noting that R_L cannot be less than 50 ohms or the maximum allowable load current of 250 mA will be exceeded, and that r_s will probably be of the order of some tens of ohms, it is clear that with the usual sort of values for g_{m_1} and g_{m_3}, the last term is by far the largest. Furthermore in this last term we can neglect r_s in comparison with R_3, to obtain finally a 'stability factor' v_{out}/v_{in} given by

$$\frac{v_{out}}{v_{in}} = \frac{1}{Bg_{m_3} R_3} \qquad 9.6$$

(Sometimes the inverse of this quantity, the stabilization factor S, is used to describe the situation, and sometimes S is defined slightly differently as the fractional change in the input divided by the fractional change in the output, or $(v_{in}/V_{in})/(v_{out}/V_{out})$. As V_{in} and V_{out} usually differ by only about 25 per cent due to the drop across the emitter follower, these two definitions of S are not very different.)

The quantity $g_{m_3} R_3$ in equation **9.6** is the voltage gain A of the amplifier stage T_3, so we can also write the result as

$$\frac{v_{out}}{v_{in}} = \frac{1}{AB} \quad \textbf{9.7}$$

The familiar quantity AB underlines the fact that we are working with a feedback loop. In practice the gain of T_3 will not be $g_{m_3} R_3$ because of the non-zero dynamic resistance of the Zener diode in its emitter circuit, which in turn follows from the fact that the slope of the Zener characteristic is not infinite. Calling this resistance r_z, we remember from our discussion of negative current feedback in section 6·4, that the signal current in T_3 is the input voltage divided by the emitter resistance r_z, and the voltage gain A is R_3/r_z. This is the value we must now use in equation **9.7.** For $R_3 = 10$ kohm, $B = \frac{1}{2}$, and $R_z = 150$ ohms (a quite possible value for a Zener current of 2 or 3 mA), we obtain $AB = 33$, and the stability factor is thus 3 per cent. Considering the simplicity of the model we are using, this is in quite good agreement with the previously quoted experimental value of 5 per cent.

The output impedance of the regulated supply can be obtained as usual from the quotient of the open-circuit voltage and the short-circuit current. The former we have already obtained, since in deducing equation **9.7** we neglected the terms in $1/R_L$, thus effectively putting $R_L = \infty$. Thus the open-circuit voltage is v_{in}/AB. The short-circuit current is found by putting $v_{out} = 0$ in equations **9.3** and **9.4** and solving for i_{out} giving

$$i_{out} = \frac{v_{in}}{r_s + \dfrac{1}{g_{m_1}}} \approx \frac{v_{in}}{r_s} \quad \textbf{9.9}$$

(and this is true whether r_z is zero or not). The output impedance is thus approximately r_s/AB, that is, the output impedance of the rectifier system has been reduced by the same factor as that in equation **9.7.** (This result might have been obtained even more easily by considering an increase in load current i_{out} with v_{in} constant – due say to a change in load. The drop at the input of the regulating circuit proper would be $i_{out} r_s$ approximately (as i_{c_3} is small compared with i_{out}). This voltage change will be reduced by a factor AB by the regulating system, thus producing a change in output voltage of $i_{out} r_s/AB$. Expressing this as $i_{out}(r_s/AB)$ we see that a change in current i_{out} means a drop in voltage of $i_{out} r_s^*$, where r_s^* is the output impedance of the whole system, and equals r_s/AB as before.) It is impossible to check exactly how well these predictions agree with practice for the circuit of Figure 150, as we have no information on the output impedance of the rectifier system itself. For d.c. and slow changes the output impedance must be at least 10 ohms, as there is a physical resistor of this size in the π-section filter. Using the previously calculated value of $AB = 33$ (which we already know to be a bit optimistic), this would imply an overall output impedance of $\frac{1}{3}$ ohm. The observed performance in this connexion is that there is a change of 2 per cent in output voltage in going from no load to full load – the latter corresponding to 250 mA, or $R_L = 50$ ohms. This implies that the output impedance is 1 ohm, which is in reasonable agreement with the minimum value of $\frac{1}{3}$ ohm deduced above.

Reduction of both the stability factor and the output impedance by at least an order of magnitude, can be achieved with circuits not very much more sophisticated than the one we have been discussing.

Reference and further reading

1. J. MILLMAN and C. C. HALKIAS, *Integrated Electronics,* McGraw-Hill, 1972.

CHAPTER 10

Some applications in nuclear electronics

10·1 Systems of interest

In this chapter we discuss the application, in various electronic instruments, of some of the circuits we have discussed. It would be quite impossible, within the scope of this book, to attempt to give even a cross-section of electronic instrumentation, in general, so we shall restrict ourselves to the special area of nuclear electronics. Even here the growth has been so rapid in recent years that we cannot hope to give anything like complete coverage.

We shall first give a general description of the mode of action, and application of a number of these instruments, and then go on to see how this action is obtained. Figure 153 shows the arrangement for counting and processing pulses from a nuclear radiation detector with a number of alternative possibilities, of increasing complexity from top to bottom, in the later stages. The detector could be an ionization chamber, a geiger, proportional or scintillation counter, or a gas or solid-state detector. The bias supply provides the voltage necessary for the functioning of the detector itself, and this, depending on the particular detector in use, would be in the hundreds to thousands of volts range. The charge arriving at the collector of the detector on the passage of a nuclear

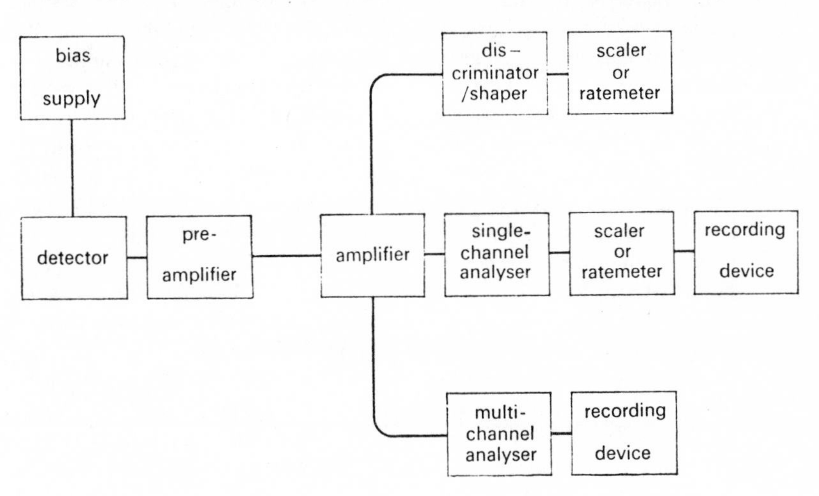

Figure 153. Counting systems (schematic).

particle produces a step of voltage there if the pre-amplifier is of the voltage-sensitive type, or is converted into a voltage step by the pre-amplifier, if this is of the charge-sensitive type. The signal then passes to the main amplifier, which may be some distance away, where further amplification and shaping of the pulse takes place. It may then be routed in one of three ways as shown. In the simplest system, the pulses go to a discriminator and shaper, which accepts all pulses over a certain size, and shapes them into a form suitable for registering on the scaler. This is a device which records the total number of pulses arriving in a given time. Alternatively a ratemeter may be substituted for the scaler: this device indicates the *rate* at which pulses arrive. Because the scaler and ratemeter to a large extent perform the same function, we shall discuss only the former: the principle of operation of the ratemeter may be found, for example, in reference 1. (In the very simplest system using a geiger counter, it may be possible to eliminate the amplifier, and even the pre-amplifier, if the scaler is close to the detector, and the length of the connecting cable, and therefore its stray capacity, is small.)

We noted in Chapter 1 that it is a property of most detectors (the Geiger counter being an exception) that the size of the charge produced at its collector is proportional to the energy of the particle producing it, and we may wish to take account of this in the recording equipment. It is possible to do this using the apparatus we have described, provided the threshold of the discriminator is variable: it is more convenient and accurate, however, to use the apparatus indicated in the middle line of Figure 153. Here the pulses, after amplification, go to a single-channel analyser. To understand its action, let us first suppose that the gain of the preceding amplifier has been adjusted so that the maximum pulse size at the analyser input (corresponding to the most energetic particle being detected) is 10 volts. The analyser has the property of providing an output signal to the scaler only when the input pulse height to it is in a predetermined size range. Suppose, for example, we have selected this range to be from 4 to 4·5 volts. The lower limit, 4 volts in this case, is known as the 'channel level', while the band in which the pulses are counted, 0·5 volts in this case, is known as the 'channel width'. Both channel level and channel width can be varied separately. We begin by setting the channel level at zero and the channel width at say 0·5 volts, and noting the count on the scaler for a certain time. We then move the channel level to 0·5 volts, leaving the channel width as before. We can thus count the number of pulses arriving in the same time, in the range 0·5 to 1·0 volts. The channel level is then moved up in further steps of 0·5 volts, until the whole range from zero to 10 volts in which pulses are appearing has been covered. The results may then be presented in the form of a histogram (Figure 154), where a smooth curve has also been drawn through the tops of the histogram elements to represent the energy spectrum (more precisely the 'differential' spectrum) of the particles or radiation incident on the detector. The spectrum shown is what would be obtained from a gamma-ray-emitting isotope using a scintillation counter. In the present example we have taken a spectrum with twenty channels: if more

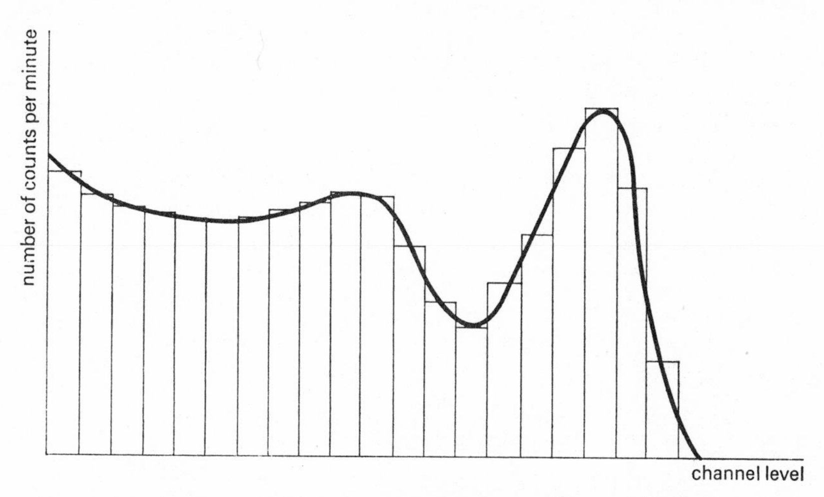

Figure 154. Gamma-ray spectrum obtained with single-channel analyser.

detail is required the number of channels will need to be increased by reducing the channel width. With such large numbers of channels it is possible and convenient to have the data accumulated and recorded automatically.

The single-channel analyser is very uneconomical as a data gatherer as it collects information one channel at a time, and for the twenty channel case fails to record $\frac{19}{20}$ of the data arriving at any particular moment. If we had twenty single-channel analysers suitably coupled together, and twenty scalers, every arriving pulse could be processed and routed to the appropriate scaler, with a corresponding reduction in the time required to produce a spectrum. An arrangement similar to this could form the multi-channel analyser shown in the bottom line of Figure 153, although as we shall see later, there are better and more economical ways of achieving the same result. Multi-channel analysers with over 1000 channels are commonly available, as are a wide variety of recording devices, which print, type, or punch the results on tape, the latter for direct computer processing.

Figure 155(a) shows two detectors operating 'in coincidence'. If a nuclear particle passes through both detectors, as shown by the arrow in the figure, pulses will be produced simultaneously (or almost simultaneously) in both detectors. These pulses are fed through their respective pre-amplifiers to a 'coincidence circuit' which recognizes the simultaneous arrival of the pulses and gives an output pulse to the scaler under these circumstances. We can thus count only those pulses which trigger both detectors. In the present case this could give, for example, the number of cosmic ray particles passing per minute through the detectors in a vertical direction. On the other hand, we may wish, as shown, in Figure 155(b) to count particles from a radioactive source which enter the lower detector from beneath and are completely absorbed in it, while at the same time rejecting counts from cosmic ray particles passing through this

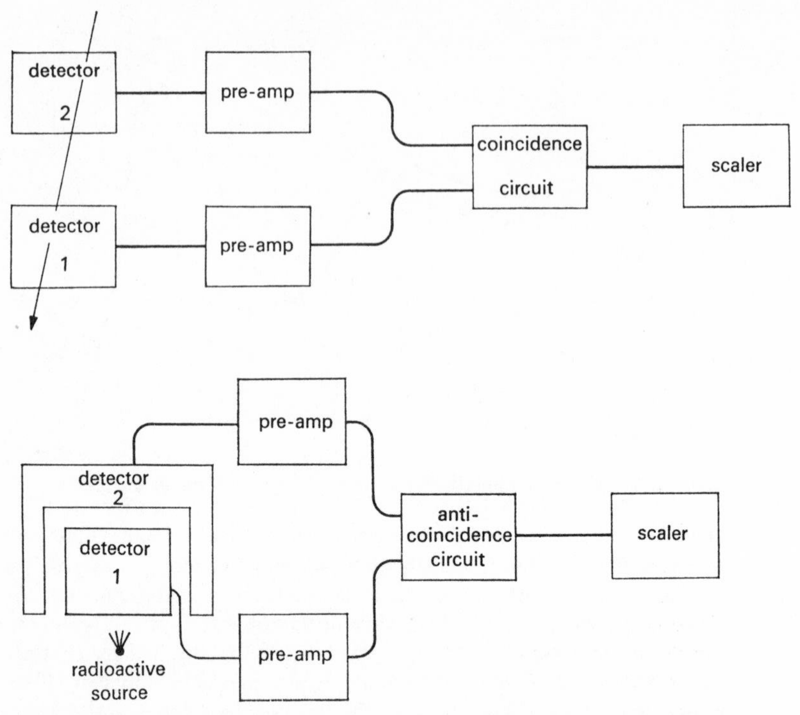

Figure 155. (a) Detectors feeding coincidence circuit. (b) Detectors feeding anti-coincidence circuit.

detector. In this case it is usual to place a second 'guard' detector almost completely surrounding the first, and to connect both to an 'anti-coincidence circuit'. This, as its name implies, sends a pulse to the scaler when it receives a signal from the lower detector only, but gives no output if signals are received from both detectors. In fact, it arranges for a signal from the guard detector to 'veto' that from the other one. The example given is just one of the applications of anti-coincidence circuits. We shall find another connected with the processing of pulses, when we discuss the single-channel analyser in detail.

10·2 Scalers

We return now to a more detailed consideration of the various pieces of apparatus which we have described in a general way. We start with the components for the simple counting system. We have already discussed bias (that is, power) supplies, pre-amplifiers, and amplifiers. We have also met the discriminator/shaper unit as it is simply a monostable multivibrator. Any pulse greater than the discriminator threshold will produce a standardized output pulse to be recorded on the scaler. The threshold is useful in that small spurious pulses, due for example, to electrical interference from near-by equipment, will be rejected. Although in this case the threshold is not easily variable, we can effectively set it where we desire, by altering the gain of the amplifier, and thus increasing or decreasing the size of the complete spectrum. Thus, the only new device in this counting system is the scaler and we now proceed to discuss it in some detail.

The scaler records and stores (usually on its own, but sometimes in conjunction with a mechanical register) the number of pulses which arrive at its input during the time it is connected to the source of pulses. Its basic component is the bistable multivibrator of Chapter 7, or 'binary' as it is often called in counting applications. The binary of Figure 93, redrawn in Figure 156(a), is returned to its initial state after two identical pulses have been applied to its input. Figure 156(b) shows the input and output waveforms for this operation. The input consists of two short rectangular pulses (which would come in practice from the shaper of Figure 153), and the output, taken from the collector of T_2, merely reflects the fact that T_2 (assumed to be initially conducting in this case) has gone off and come on again. So an input of two rectangular pulses produces an output of one rectangular pulse: we have a 'divide by two' action, and hence yet another name for the binary – 'scale of two'. We may continue the process by feeding the output from the binary of Figure 156(a) into a similar binary: the resulting waveforms are shown in Figure 157. The second binary will be changed over by the pulses from the output of the first binary, these change-overs occurring on the positive going edge. On the final output there is now one rectangular pulse for four input pulses, thus achieving a 'divide by four' action.

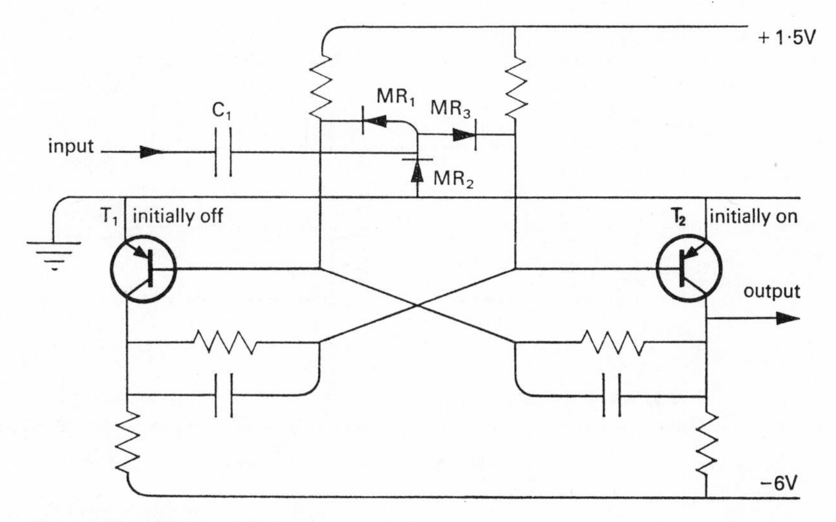

Figure 156. Binary (a) circuit.

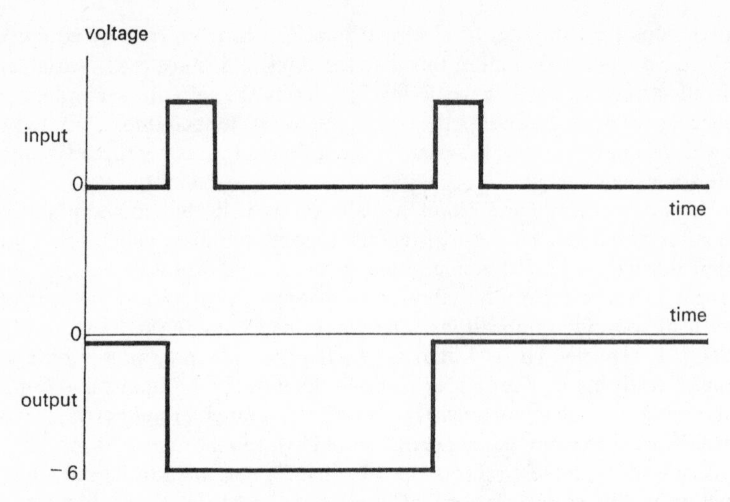

Figure 156. Binary (b) waveforms.

The 'scale of four' we have just constructed is in fact a memory which tells us by the state of the two binaries at any moment, how many pulses have arrived. Suppose that, in order to check which of the transistors in the binary is conducting at any time, we connect across the collector load of the left-hand transistor in each binary a voltmeter which reads –6 volts full scale, and watch the voltmeter readings as the pulses arrive. Figure 158 shows the result. (The numbers above the voltmeters will be explained later.) Before any pulses arrive

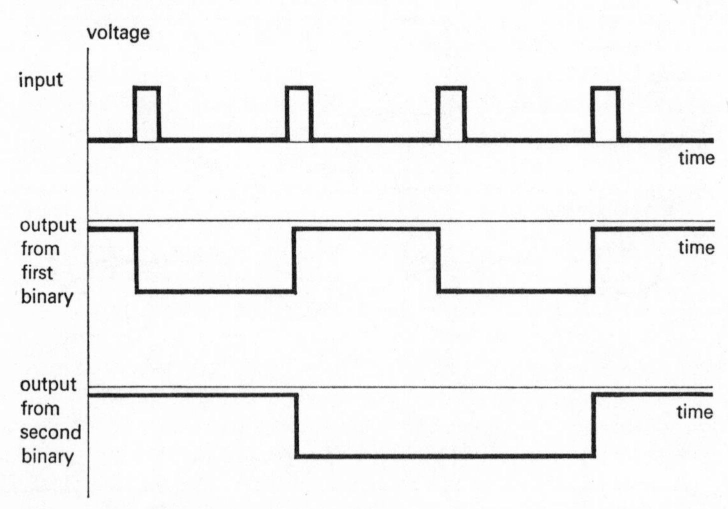

Figure 157. Waveforms in a scale of four.

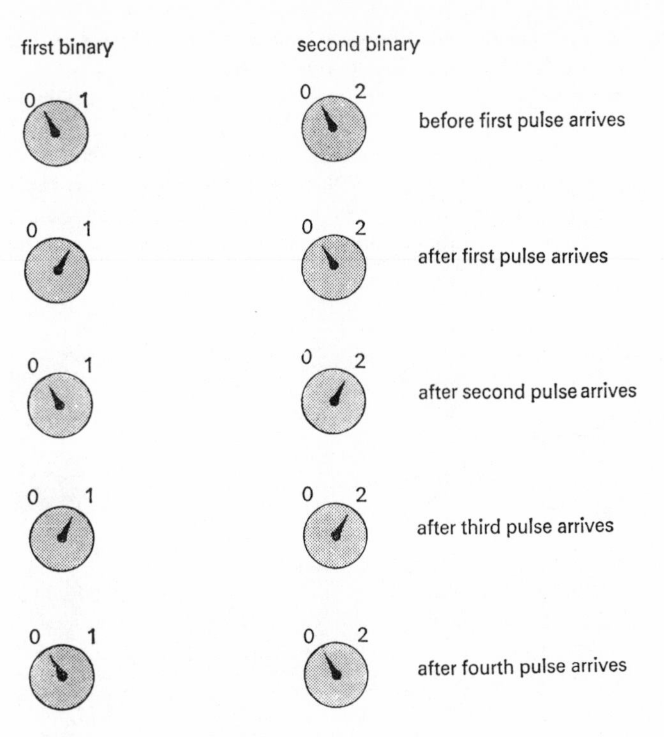

Figure 158. Indicators for a scale of four.

we shall assume that the left-hand transistor in each binary is non-conducting, so there is no current flowing in their collector loads, and thus no voltage indicated by the voltmeter connected across these resistors. When the first pulse arrives, the first binary makes a transition, and its left-hand transistor conducts; binary two is unaffected – see Figure 157. The left-hand voltmeter thus reads full scale, while the right-hand one still reads zero. After the second pulse, Figure 157 shows that the first binary returns to its initial state while the second makes a transition: the third line of Figure 158 shows the corresponding state of the voltmeters. After the third pulse both voltmeters read, and after the fourth, the binaries and voltmeters return to their initial condition. There are three different arrangements of the meter needles corresponding to the arrival of 0, 1, 2, or 3 pulses, and simply by looking at the voltmeters, we can tell how many pulses have arrived. By labelling the two positions of the needle in the left-hand meter as 0 and 1, and those of the right-hand meter as 0 and 2, we can read off directly, by adding the indications of both meters, just how many pulses have arrived. It is not even necessary to have voltmeters: in principle small bulbs which would go off and on would do. In practice this would throw too much load on the circuit, and interfere with its operation, and it is usual to interpose another transistor to drive each indicator lamp. More sophisticated forms of readout will be discussed later in this chapter.

A scaler which only counts up to three before resetting to its original condition, is not of much practical use, but clearly scales of 8, 16, 32, etc., can be built simply by adding more binaries, and indication of the count stored can be given in exactly the same way. Figure 159 shows the four voltmeters in a scale of 16, indicating (a) a count of 8, (b) a count of 9, (c) a count of 14, and (d) a count of 15. At a count of 16 all the voltmeters will return to their zero position.

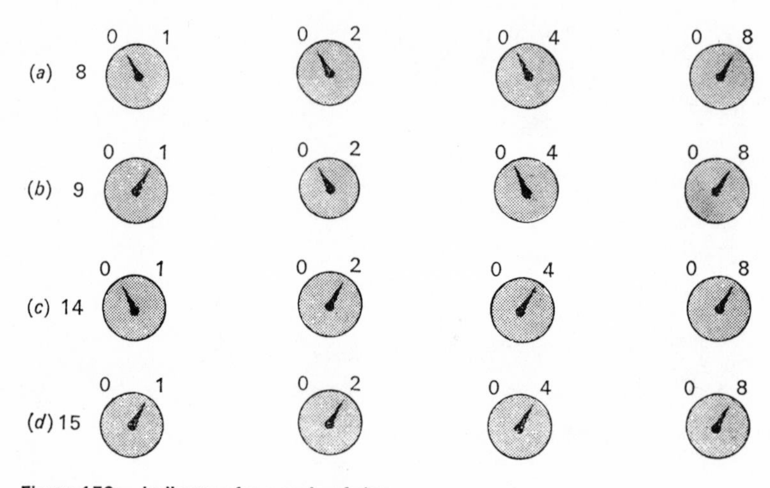

Figure 159. Indicators for a scale of sixteen.

Indeed we can make the system simpler by expressing the count in binary notation. If we label any meter which shows a deflection as '1', and one which does not as '0', then reading line (a) of Figure 159 from right to left we obtain 1000, which is binary notation for 8. Line (b) is 1001 or 9, and similarly for the others.

Integrated circuit technology allows us to have a complete scale of 16 (and indeed a great deal more) on a single chip. Figure 160 shows schematically four binaries or 'flip-flops' so arranged, each flip-flop doing the work of our circuit of Figure 156(a).

The input in each case is to the terminal C, whence, as previously, it is 'steered' to the correct half of the flip-flop. The terminals Q and $\overline{Q}$ correspond to the two collectors of Figure 156, the terminology Q and $\overline{Q}$ underlining the fact that these are complementary outputs, that is when one is at a high voltage the other is low and vice versa. The J and K terminals are further inputs, one to each side of the flip-flop for use in other applications of the circuit, and are not used here.

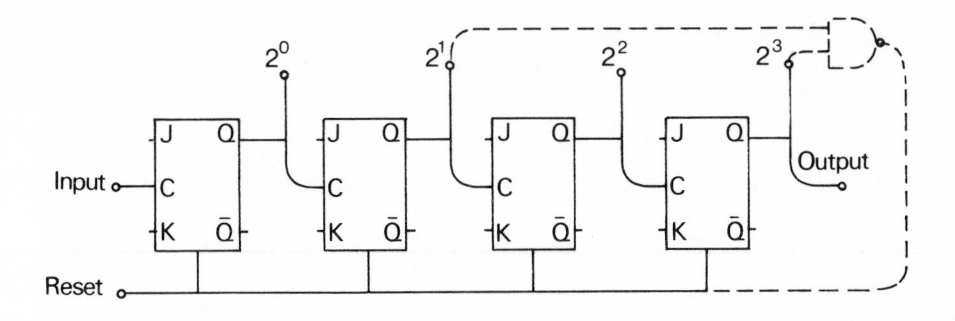

Figure 160. Counting circuit using flip-flops.

(They give however, their name to the circuit, the J-K flip-flop.) The circuit recycles after a count of 16: intermediate counts, from 0 to 15, can be read out from the output of each flip-flop as indicated. The reset line allows us, after finishing a counting run, to momentarily apply a suitable d.c. voltage to one side of all the flip-flops, thus ensuring that any flip-flop not in the zero count position is reset to this state. (The action of the part of the circuit shown dashed will be discussed later.) The most commonly available integrated flip-flops use a positive 5 volt supply voltage (not shown in the previous diagram) and positive input pulses. The wave forms from such a counter will not differ in principle from those of Figure 157, but of course the voltage changes in the second and third lines of the diagram will be between zero and +5 volts, rather than between zero and −6 volts. Note that the transitions of the binaries will now occur on the negative going edges of the pulses, and this will be true also for the transitions caused by the original input pulses. Counting rates of the order of 10^7 per second are obtainable.

It would obviously be much more convenient if, instead of counting pulses in a 'scale of 16' we could do so in a 'scale of 10'. There are a number of ways in which we can convert the circuit of Figure 160 to a scale of ten, some employing the previously unused JK inputs. However a comparatively simple way to do so is shown by the dashed part of the circuit we previously ignored. The idea is to allow the original circuit to count as before up to a count of nine. When the tenth input pulse arrives we use the additional circuitry to reset all the flip-flops to zero and thus complete the cycle. We do this by using the 'gate' whose symbol is shown in the top right-hand corner of the diagram. We shall be discussing gates in more detail later in this chapter: for the moment all we need to know is that this particular gate produces zero voltage at its output (which is what is required to operate the reset line) only when both inputs are at a high voltage. We note that since the inputs of the gate are connected to the second and fourth flip-flops, this will only happen when a count of $2^1 + 2^3 = 10$ has been reached, and our aim has been achieved. Since the intermediate counts (up to 10) are still read out in binary notation, we refer to the circuit as a binary coded decimal

(BCD) counter. We will later see how this type of read out can be converted to direct decimal form. A typical scaler might consist of six scales of ten in cascade, giving a total count capability of 10^6.

One concluding remark on terminology should be made. The counter we have been discussing (whether in binary or BCD form) is known as an 'asynchronous' or 'ripple through' counter. This refers to the fact that the input signal is applied to the left-hand flip-flop, and any subsequent changes in the others occur in sequence. Since there are slight, but significant, delays in each flip-flop, it is clear, if we are pushing the counter to the limit of its capabilities, that a fresh pulse may be appearing at the input before the later stages have completed the transitions initiated by the previous pulse. It may thus not be possible to read the state of the counter unambiguously between input pulses. In such cases we can effect a considerable improvement by using 'synchronous' counting. Here the input pulses are applied simultaneously to all the flip-flops and a series of gates allows only the appropriate ones to change over. Obviously since all flip-flops change together we can read out the state of the circuit between pulses at rates where it is not possible with the asynchronous method.

10·3 Other forms of decade counter

A number of devices have been produced which perform the function of decade scaling in a single valve envelope. One such device, the 'dekatron'*, consists of a gas filled tube with an anode surrounded by ten cathodes. The discharge is originally formed between the anode and a particular cathode, and progresses round the other cathodes in turn as pulses are applied. The 'trochotron' is a form of magnetron, this time with a single heated cathode and ten anodes, to which the stream of electrons is directed in turn as the pulses arrive. Finally there is a counting tube devised by Philips consisting of a miniature cathode ray tube with the beam passing through ten slots placed between gun and screen, thus providing ten stable positions for the beam, through which it is driven sequentially by the arriving pulses. The advantages associated with such counters are not as great as would first appear. A fair amount of external circuitry is still needed to provide pulses suitable for driving them, and their counting speed is less, in some cases very much less, than conventional scalers. Furthermore, integrated circuit techniques, by which a complete decade scaler, together with count indicator, can be obtained in a package of centimetre dimensions, have completely removed the size advantage which dekatrons and similar devices held so markedly over thermionic valve scalers. We shall not therefore discuss these types of decade scaler further.

10·4 Gates and coincidence circuits

We start with a useful preliminary analogy in Figure 161(a). With one or both switches closed the point X is at zero potential. With both switches open the output at X rises to the full voltage V of the supply. This is our basic coincidence

* 'Dekatron' is a registered trade mark of Ericsson Telephones Limited.

circuit which, by a rise in voltage at the output, indicates the simultaneous opening of SW_1 and SW_2. In electronic circuits we use diodes as switches: they conduct or fail to conduct depending on the voltage placed across them. Figure 161(b) shows a simple circuit of this type. If either or both diode inputs are held at zero potential then the relevant diode or diodes conduct and the output of the circuit is close to ground: however if both inputs are caused to rise to the supply voltage level by applying simultaneously to both a d.c. voltage or (as shown) pulses of height V_{YY}, then both diodes cease to conduct and the output rises to the supply voltage level for as long as the inputs are present. Again we have an indication of a coincidence and we could thus call this circuit a diode coincidence circuit. Alternatively, in the terminology of logic circuitry we could refer to it as an AND circuit. If we label zero potential as logical 0, and the supply voltage as logical 1, then if we apply a 1 to one input AND a 1 to the other we get a 1 on the output. The number of diodes in an AND gate need not of course be restricted to two, and we can have a three, four, etc., input AND gate.

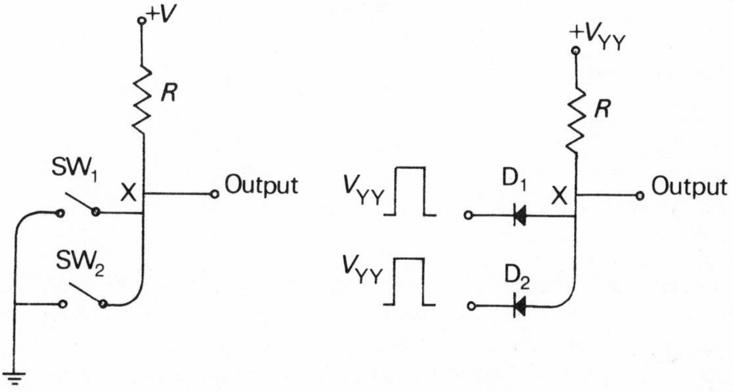

Figure 161. Diode coincidence circuit (a) switch analogy (b) actual circuit.

We can improve the simple gate of Figure 161(b) in a number of ways. Firstly, we can add an output stage to improve its capabilities in driving further circuitry (see Figure 162, where for a change we have shown a 3-input gate). The transistor output is at V_{CC} (logical 1) when zero voltage is applied to its base, and switches to almost zero volts when a high voltage is applied to its base and the transistor saturates. (So we are here using the transistor as a switching device rather than a linear amplifier.) Secondly we must take account of the imperfections in the diodes forming the gate (D_1, D_2, and D_3). We had previously assumed them to be ideal, but in practice of course, there exists a finite impedance even for a conducting diode. As a consequence, in our previous Figure 161(b) we do not get quite zero voltage at X when one diode alone is conducting, but the fraction of a volt associated with such a conducting diode. We do not wish to take the chance that this small voltage (plus whatever transients or electrical noise that might accidentally be added to it) would start turning on the transistor, so we add a diode D_4 (Figure 162) which must be turned on before the transistor can conduct, and thus adds a barrier which prevents the spurious turn-on of this transistor. Just be on the safe side (and to take account of any additional voltage at X due to the impedance of whatever sources may be driving the gate diode) we add a second diode D_5 to D_4. (Note incidentally that we have dealt with the most likely case for false turn ons - when only one gate diode conducts. When two diodes in Figure 162 are conducting then clearly we have the two diode impedances in parallel, and the unwanted voltage is even smaller.)

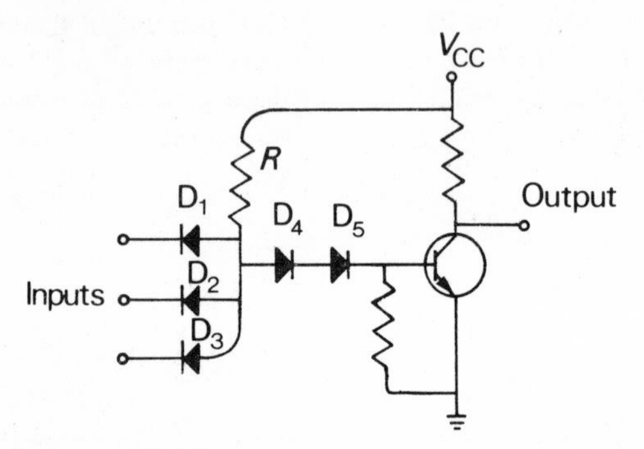

Figure 162. Improved diode gate.

Apart from improving the preceeding circuit we have also made an important change in the logic. Because the transistor we have added inverts the state of the logic signal applied to it, it is clear that when a coincidence occurs (that is all inputs of Figure 162 high at logical 1) the circuit indicates this by producing an output at logical 0 (and not at logical 1 as in the case of the simple circuit of Figure 161(b)). This is of course a perfectly satisfactory coincidence circuit, but from the point of view of logic we must classify it as a NAND circuit (that is an AND circuit followed by a NOT circuit or inverter). A NAND circuit thus produces a logical 0 output only if all inputs are at logical 1. If we really hanker after an AND circuit we can reconvert our NAND gate to an AND gate by adding a separate NOT gate on the output. Indeed it can be shown that any circuit performing a complex logical function can be implemented with NAND and NOT gates only - and a NOT gate is really only a NAND gate with a single input, or with all its inputs joined, if there are more than one. It is not surprising therefore, that the NAND gate is by far the most commonly available logical element. Figure 163 shows the symbols for the logic elements we have already met: an AND gate, a NAND gate (both in 3-input form) and an inverter or NOT gate. The small circle at the output is used to indicate the inversion occurring in the NAND and NOT gates.

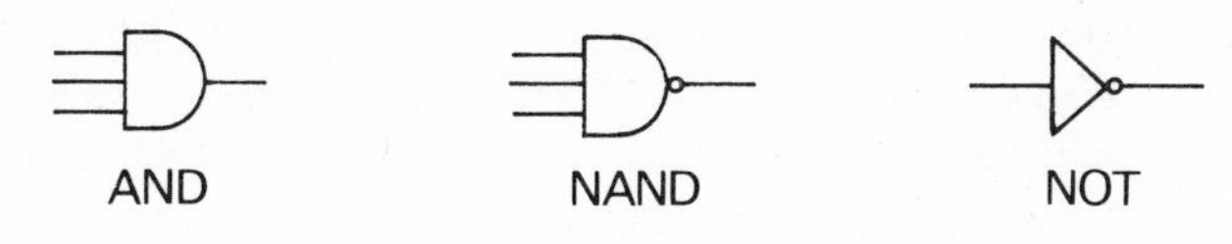

Figure 163. Symbols for logic elements.

It would not be appropriate in the present text to develop in a major way further topics in logic. For a discussion, among other things, of OR gates (which give an output if one OR more of a number of inputs are present), Boolean algebra, truth tables and logic circuits using MOSFETs, reference 2 should be consulted, or any of the texts devoted exclusively to this important field. Here we will merely mention a further development of the NAND gate of Figure 162, and add some examples of its use particularly appropriate to the aims of the present chapter.

We return then to Figure 162. The NAND gate there belongs to a logic family known as positive diode transistor logic (DTL). The 'diode transistor' part of the terminology occurs for the obvious reason that both diodes and a transistor are used in the circuit: 'positive' indicates the polarity of the power supply employed. When such a circuit is manufactured in integrated form it turns out to be more convenient to fabricate the gate diodes D_1, D_2, D_3, ... plus the diode D_4 as a single transistor with mutiple emitter (see Figure 164). The base-emitter diodes play the role of D_1, D_2, and D_3, while the base-collector diode acts as D_4, and is of course oriented in the correct sense. Usually a more sophisticated output stage is also employed, but this makes no difference in principle. Such a NAND gate is said to employ positive transistor-transistor logic (TTL) and the terminology should be self explanatory.

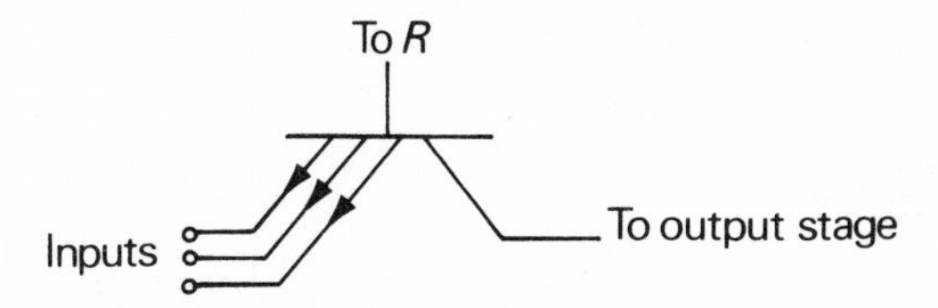

Figure 164. Multi-emitter transistor.

We have already had two examples of the use of NAND gates in our context - as a coincidence circuit and as a device for converting the binary circuit of Figure 160 to BCD form. We add two further examples, also connected with Figure 160. Although that circuit has, with the aid of a NAND gate, been made to recycle after ten input pulses, the intermediate read out is still, as was pointed out, in the binary code. It would clearly be preferable if this also could be read out in decimal notation. Figure 165 shows how this is done with a decoder using ten NAND gates, the whole lot available in a single integrated package. The inputs (at the top) are from the output and complementary output (Q and $\bar{Q}$) of each flip-flop of Figure 160, Q_0 and $\bar{Q}_0$ referring to the flip-flop at the extreme left-hand side. The decimal outputs appear down the right hand side of Figure 165. We need only examine a couple of these outputs to understand the operation of the circuit. Take for example the decimal 1 output, and its corresponding NAND gate. Remembering that when a flip-flop is indicating a count its Q output is high (logical 1) and its $\bar{Q}$ output is low (logical 0), we expect the decimal 1 NAND gate to have an input connected to Q_0 (which it has) and this indicates when a count of one has occurred at the BCD input. However we must avoid the decimal 1 gate operating when Q_0 is also high at, for example, counts of $1 + 2$ (=3), $1 + 4$ (=5), and $1 + 8$ (=9). Thus this gate has also inputs connected to $\bar{Q}_1$, $\bar{Q}_2$, and $\bar{Q}_3$. An output from this gate occurs then only at a BCD count of one. Similarly looking at the gate connected to the decimal 6 output, we expect it to have inputs to Q_1 and Q_2 to indicate a count of $2^1 + 2^2 = 6$. In principle it should also have inputs to $\bar{Q}_0$ (to avoid giving an indication at a count of $2^0 + 2^1 + 2^2 = 7$) and to $\bar{Q}_3$ (to avoid giving an indication at a count of $2^1 + 2^2 + 2^3 = 14$) but of course the last connection is unnecessary because the whole system recycles at a count of ten. Thus there are only three inputs to this particular NAND gate, as shown, and the NAND gates for the 8 and 9 outputs require only two inputs. (Note also that since we are using NAND gates, a decimal output is indicated by the output in question going low. This is no disadvantage since we can connect this output to one side of our indicator device - whether small bulb or light emitting diode - the other side of which goes to the positive supply. Our indicator then operates when the output from the gate goes low.)

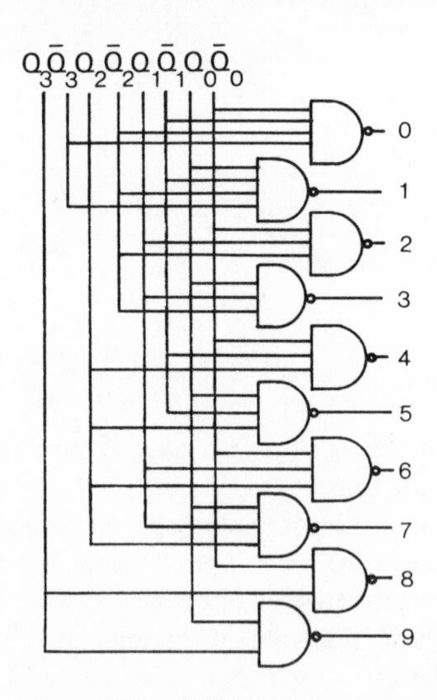

Figure 165. BCD to decimal converter.

If we are interested in having our decimal output in the form of the more compact seven segment display (Figure 166) the problem, while the same in principle, becomes more complex in practice, since for each particular decimal output we have to select a different arrangement of segments. In fact we need a decoder with 16 gates to do this, but such is the state of integrated circuit technology that these, together with a BCD counter and some additional circuitry can be accomodated on a single silicon chip, and the whole device including the seven segment light emitting diode display contained in a package of centimetre dimensions.

Figure 166. Seven segment display.

As a last example of logical circuitry we discuss the anti-coincidence circuit, mentioned in an earlier section. In this case we wish to indicate the arrival of a pulse (or of two, three, four, ... pulses in coincidence) provided another specified pulse is absent. We can do this by putting a NOT gate in one of the inputs to a NAND gate. (Figure 167 shows the arrangement.) If no pulse is applied to the lower of the three inputs (that is if it is at logical 0) then on the other side of the NOT gate (that is at the actual input of the NAND gate) we have a logical 1. Consequently if two simultaneous pulses are applied to the other inputs, an output appears as shown. If however a pulse is applied at the lower input simultaneously with the other two, then a logical 0 is produced at the lower input of the NAND gate, and the output pulse is inhibited (that is, does not occur). We may thus call this arrangement either an anti-coincidence circuit or an INHIBIT gate.

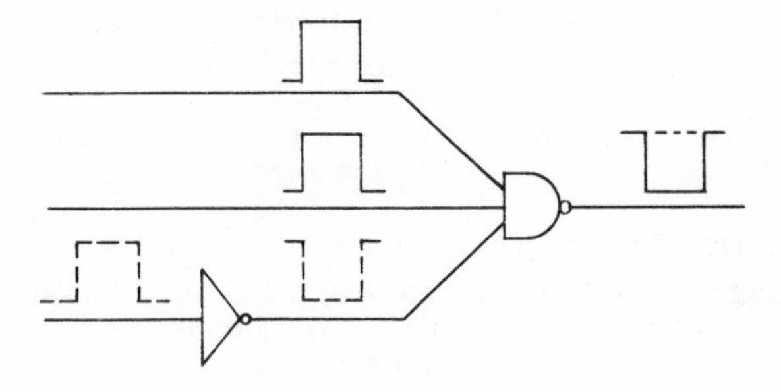

Figure 167. Anti-coincidence circuit.

We close this section with an explanation of the 'resolving time' of a coincident circuit. Compared with idealized circuits, real coincidence circuits record not only true coincidences, but also the arrival of pulses on the two channels which although not coincident overlap to some extent in time. The 'resolving time' of a coincidence circuit is defined as the time before and after the leading edge of a given pulse, during which if a pulse arrives on the other channel, a recordable output will be produced. The resolving time is usually determined empirically using two pulse generators, and once known can be used to correct for spurious coincidences due to uncorrelated pulses arriving, one on each channel, at a small time interval apart.

10·5 Single-channel analysers

Single-channel analysers are devices which produce an output only when the input pulse lies in a pre-determined size range. Figure 168 shows a block diagram of such an analyser, made up from familiar components. The input pulse is applied simultaneously to two discriminators of the Schmitt trigger type,

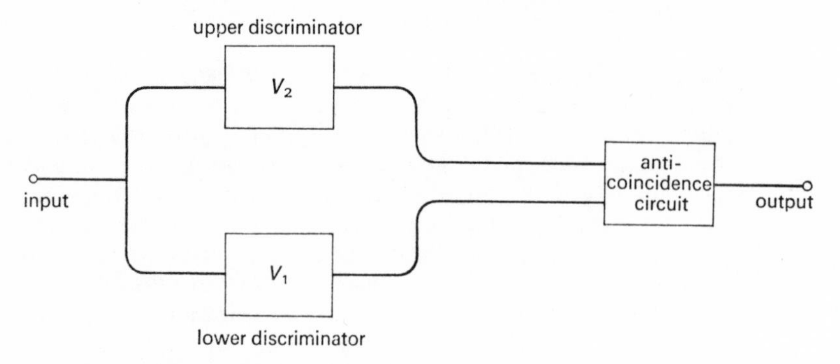

Figure 168. Block diagram of single-channel analyser.

with thresholds set at voltages of V_1 and V_2 respectively ($V_1 < V_2$). The outputs from these are fed to an anti-coincidence circuit. An input pulse less than V_1 in size fails to trigger either discriminator, and no output is obtained. An input pulse greater than V_2 triggers both discriminators, but again no output is obtained, because the output from the upper discriminator is used to veto that from the lower, in the anti-coincidence circuit. An output is obtained only when the input lies in the range between V_1 and V_2 because here the lower discriminator is triggered, but no veto is applied, as the upper one is not. The required action is thus achieved. The 'channel level' is defined by V_1, and the 'channel width' (also known as the 'window width') is given by $\Delta V = V_2 - V_1$. However it is preferable to have an independent control of ΔV, rather than to have to set it as the difference of V_1 and V_2. For details of how this is achieved, and also on the use of a 'window amplifier' to produce greater accuracy in the 'window-width' control, reference 3 should be consulted.

Two points about time relationships between signals in discriminators and analysers are worth making. Figure 169 shows a slowly rising pulse, and the corresponding outputs from the upper- and lower-level discriminators of a single-channel analyser. (We note, as discussed in section 7·4, that the output pulse from each discriminator ends slightly later than one would expect, due to hysteresis in the Schmitt circuits. This however is not relevant to the matter in hand.) What is important is that the pulse from the upper discriminator, which is to be used as the veto pulse in the anti-coincidence circuit, starts appreciably later than that from the lower discriminator. It is conceivable therefore that before the veto pulse has time to arrive, part of the lower discriminator pulse may have passed through the anti-coincidence circuit and been recorded, wrongly of course as there was a veto pulse present. Furthermore the delay between the start of the pulses from the two discriminators is a function of the level settings and input pulse height, so a simple delay on the lower discriminator will not be satisfactory. Without going into the details of the solution, it is clear that if we use signals obtained from the trailing edge of both pulses, the veto signal, if present, precedes that from the lower discriminator, and the problem is solved.

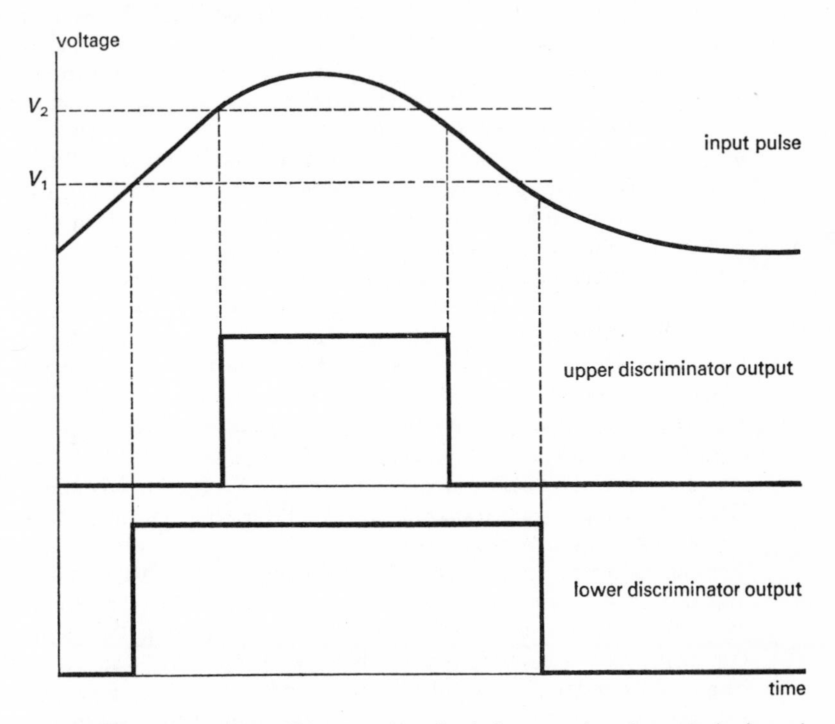

Figure 169. Time relationship between discriminator pulses in a single-channel analyser.

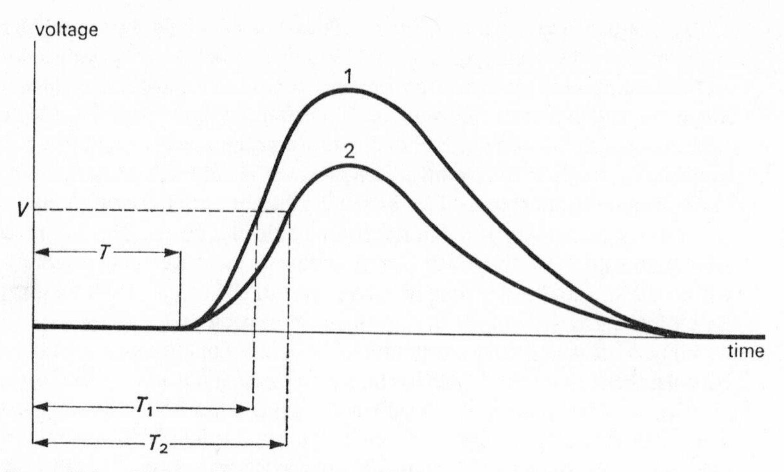

Figure 170. Timing errors in a discriminator.

A second and related problem arises even in the case of a single discriminator, if we are making accurate time measurements. In Figure 170 a pulse from a nuclear radiation detector (labelled 1) is shown crossing a discriminator level V, which is set at some non-zero level, to reject, say, other unwanted nuclear events or noise from electronic equipment. While the pulse actually commences at some time T after a selected time origin, no indication of the event is received until the discriminator has triggered at a later time T_1 as shown. Further if the pulse had been of smaller amplitude (curve 2), then we would have had a different indication, T_2, of the time of arrival. This time shift as a function of amplitude – the so-called 'walk' – makes it difficult to establish coincidence, or any other form of exact time relationship, between this pulse and that say from another detector. In many cases we should not object to the time difference $T_1 - T$ between the actual start of the pulse and the time when it triggered the discriminator, provided that this was the same for all pulses regardless of amplitude. Figure 171 shows a way of achieving this.

In this case we have doubly differentiated the pulses so that they cross the time axis again in a manner described in Chapter 3 (equation **3.22** and Figure 37). Both pulses are shown crossing the axis at the same time, independent of their amplitude, and we must now justify this. Let us look first at equation **3.22**, even though it represents an ideal pulse with zero rise time, while those we are dealing with at present in Figure 171 have a finite rise time. Putting $v_{out} = 0$ there, to find the time of crossing of the axis, we note that V_0 disappears, and the answer depends only on T_F and T_F^*, but not on V_0, which characterizes the height of the pulse. (On the other hand to find where such a pulse crosses a *non-zero* discriminator level, one puts v_{in} equal to the level in question V, and the solution clearly depends on V_0.) A pulse like those shown in Figure 171, with non-zero rise time will contain three exponentials, and three constants, T_F, T_F^* and another defining the rise time, in the mathematical expression for it. However the general conclusion about the invariance of the crossing point with respect to amplitude remains unaltered. If we now set a discriminator with its triggering level for pulses equal to V_H, the hysteresis of the Schmitt trigger circuit, the output will terminate when the input pulse returns to zero (see Figure 171) as we saw in Section 7·4. Thus, the end of the rectangular pulse from the discriminator provides a timing signal which is independent of the amplitude of the incoming pulses. Circuits of this kind are known as 'zero-crossing' or 'cross-over-pickoff' discriminators. The same result can be obtained by the use of a 'constant fraction discriminator' (reference 4), in which the pulse is inverted, delayed, and added to an attenuated version of itself, again producing a walk-free zero crossing point. A further development of this idea allows us to eliminate timing errors due not only to different pulse heights but also to different pulse rise times.

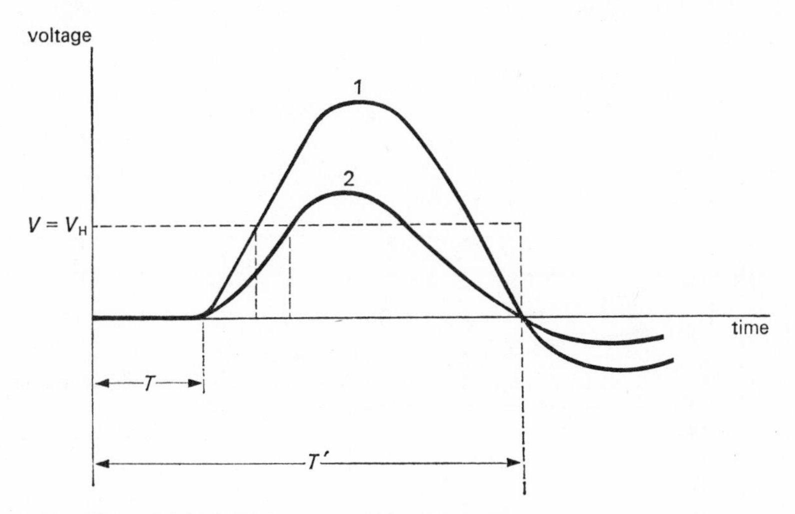

Figure 171. Principle of the zero-crossing discriminator.

10·6 Multi-channel analysers

We mentioned in the first section of this chapter that one could construct, say, a twenty-channel analyser, to analyse pulses in the zero to twenty-volt range, by stacking twenty single-channel analysers each with a channel width of one volt. Twenty complete analysers are, in fact, unnecessary because the upper discriminator of one analyser is doing the same job as the lower discriminator of the next highest analyser. Instead, we can use twenty-one discriminators, suitably interconnected with anti-coincidence circuits, and such multi-channel analysers have actually been built. But, because of the complexity of such devices for large numbers of channels, and the difficulty in eliminating drift in the channel positions, modern analysers almost invariably use a method which depends essentially on more easily made time measurements.

The basic idea (originally due to D. H. Wilkinson) lies in the conversion of the input pulse into a rectangular pulse lasting for a time proportional to the height of the incoming signal. During this time, the number of pulses produced by an accurate oscillator is recorded: we thus obtain a number which is proportional to the height of the input pulse. This attaches to the pulse a 'code number' which allows us to route a signal to a scaler appropriate to the pulse height in question, or to a part in some other type of memory in which the event should be tallied. The device which converts the pulse height into number form is known as an 'analogue to digital converter' (ADC), because the information on the energy of the particle which entered the detector, originally in analogue form (that is, expressed as the magnitude of the corresponding voltage pulse) is converted by it into digital, that is, number, form. Figure 172 shows simplified waveforms at various points in the analyser, for

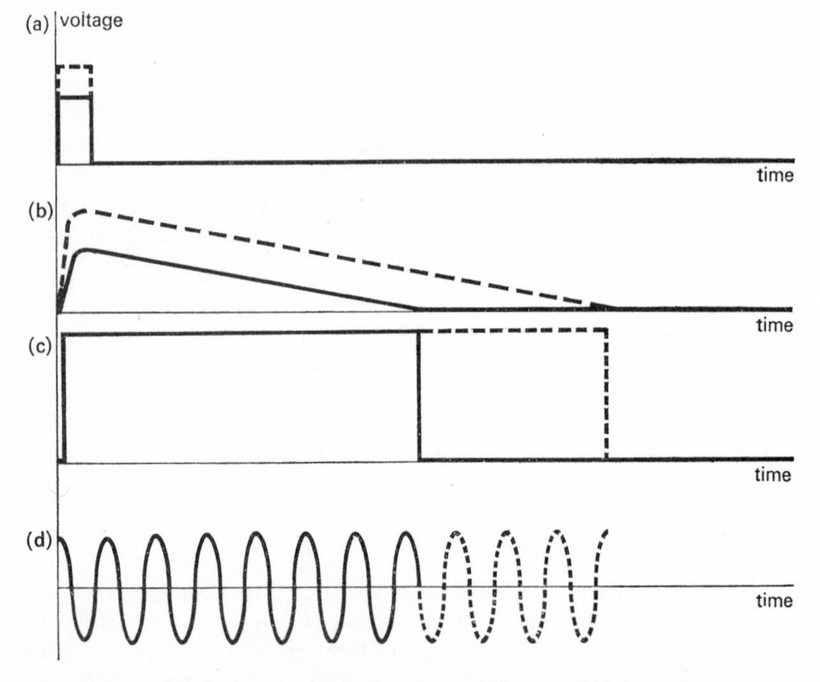

Figure 172. Waveforms at various points in an analogue to digital converter.

two successive input pulses of different amplitudes (one shown with a solid, and the other with a broken line). They are assumed to be short delay-line-shaped pulses, which because of their flat tops are better suited to this type of analyser than those with RC shaping (line (a) in diagram). These are converted by a 'pulse stretcher' in a manner described below into pulses of corresponding height, but with long linearly sloping tails (Figure 172(b)). A Schmitt discriminator, set to trigger at a low level, then produces rectangular pulses from the previous waveforms. The length of these is thus proportional to the original pulse heights (line (c)). Finally a train of pulses is produced by an oscillator which is turned on and off by the leading and trailing edges respectively of the discriminator pulse. These can be counted on a scaler (the 'address' scaler) to record a number proportional to the initial pulse height, and we can therefore select a scaler or a particular part of a memory, where a count should be added. Incidentally, it takes an appreciable time to analyse a pulse, particularly a large one, because we must wait until the oscillator produces the pulse train. For example with an oscillator of 10 MHz frequency it would take 100 μsec to digitize a pulse producing 1000 waves in the pulse train, that is one which would eventually be recorded in the thousandth channel of the analyser. Before proceeding to see how, once digitized, the pulse is stored in the memory, we must return to indicate how the pulse shape of Figure 172(b) is obtained.

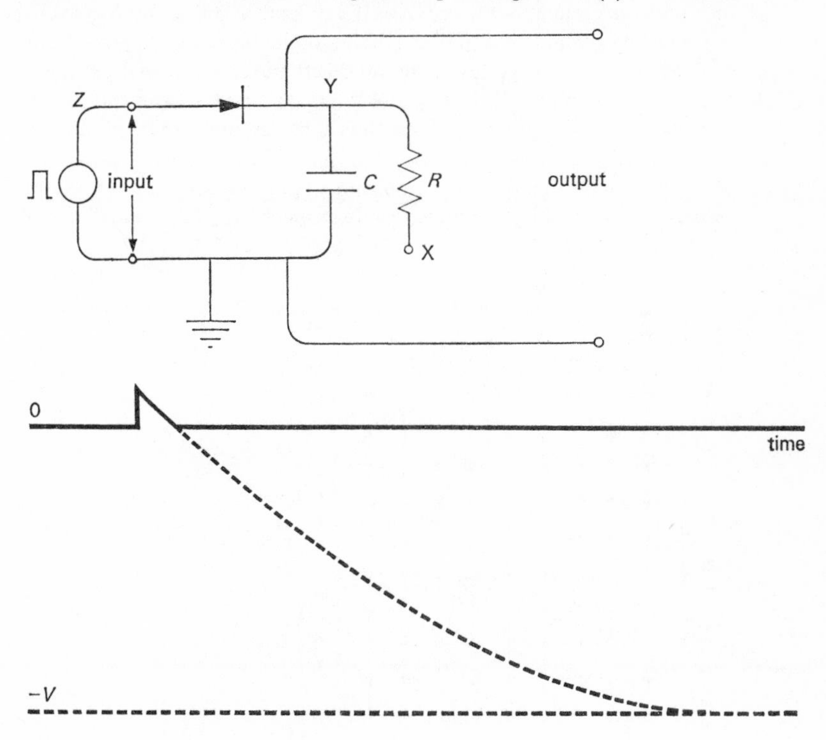

Figure 173. Production of pulse with linear tail
(a) circuit (b) waveform.

In Figure 173(a) let us assume, in the first instance, that the point X is connected to ground. In that case when a short pulse is applied to the input from the generator, as shown, the diode conducts, and the capacitor charges quickly to the peak voltage of the pulse. When the pulse disappears, the point Y is at this positive peak voltage, while the point Z has returned to ground, and so the diode ceases to conduct. The capacitor then discharges through the resistor R with a time constant RC, which in principle, can be as large as we please, thus producing a long tail on the pulse. However, it is an exponential tail, not a linear one. We can overcome this by using integrating circuits such as we have previously discussed, but a simple and effective way is to return the point X, not to ground, but to a large negative potential – say several hundred volts. Let us suppose that the input pulse height is six volts. Before the arrival of the pulse the point Y is at ground potential, assuming the diode to be ideal and the internal impedance of the generator zero. Some current is flowing through the large resistor R and the diode, but this is of no significance. When the pulse arrives, Y rises to six volts, and when the pulse disappears, the diode becomes non-conducting as before. Since X is now not at ground, but at a high negative potential $-V$, say, the voltage of the charged capacitor will decay exponentially, not towards ground, but towards $-V$ (see Figure 173(b)). When the voltage at Y has reached ground potential, the diode conducts once more, and the operation is finished. As V is very large compared with the six-volt height of the pulse, we have travelled only a little way along the exponential, and this approximates closely to a straight line. We have thus achieved our aim of producing a pulse whose height is the same as the input pulse, but which possesses a long linearly-falling tail, whose slope is also independent of pulse height, provided this height is very much less than V. In practice the imperfections of a real diode (non-zero

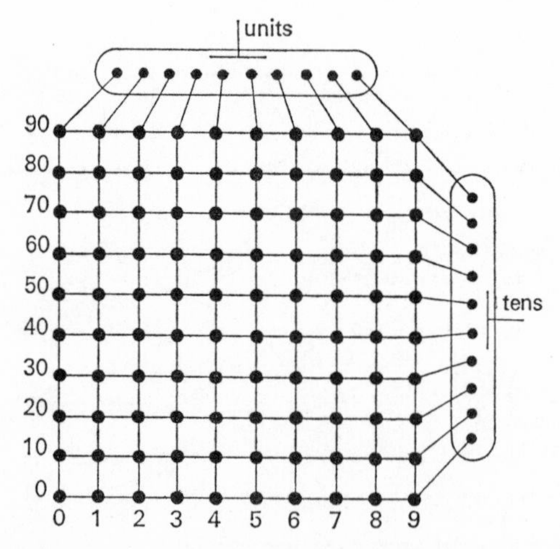

Figure 174. Address scaler and recording system (schematic).

forward resistance and non-infinite back resistance) as well as the non-zero impedance of the generator, will cause some departure from the ideal output pulse shape described.

We now return to our main discussion. Having recorded a number in the address scaler, which identifies numerically the height of the pulse, we can use this to record the event by adding a count in an appropriate recording scaler, that is, at the correct 'address'. Let us suppose for simplicity that the analyser has only 100 channels, and that the address scaler consists of two dekatrons, one for units and the other for tens. Figure 174 shows schematically a mesh of wires, the extremities of the rows connected to the corresponding ten output electrodes (cathodes) of the 'tens' dekatron, and the extremities of the columns to the ten output electrodes of the 'units' dekatron. A coincidence circuit is placed at each intersection between wires, each wire feeding an input, and with the output fed to a scaler. We thus have in this case 100 coincidence circuits, and 100 scalers.

Suppose that the address scaler has just recorded a count of 47. When the glow settles on the corresponding electrodes, we have signals passed along the wire labelled 40 and that labelled 7. The coincidence circuit at the intersection of these two wires (and only this coincidence circuit) produces an output, which is recorded on the scaler appropriate to channel 47.

Even with the simplest recording scalers it is clear that an analyser with more than say 100 channels is going to be enormously expensive and complicated. A big advance was made by the use of a 'ferrite-core memory' to replace the assembly of scalers. Ferrite is a ceramic material that can be made to have an almost square magnetic hysteresis loop (Figure 175) and the memory is made up of a three-dimensional array of ferrite toroids each of about a millimetre in

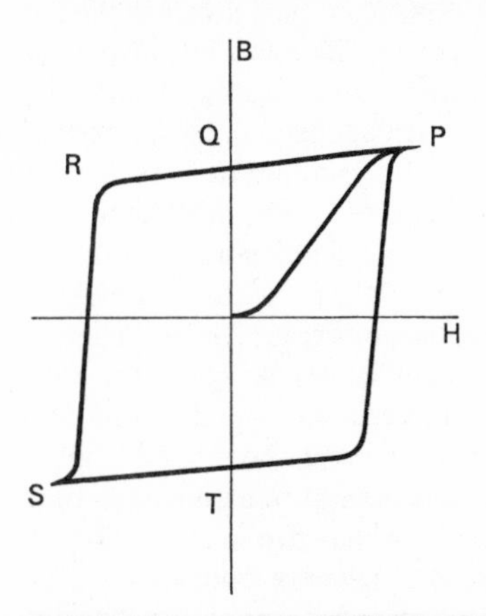

Figure 175. Ferrite-core hysteresis loop.

diameter. A toroid may be magnetized in either direction by a current in the appropriate sense in one or more turns of wire wound on it. Such a 'core' is in fact a form of bistable. If we apply a magnetic field so as to reach the point P (Figure 175), and then remove the field, the core will move to the point Q, and rest stably there. Under the application of a small field it will move a little from Q, but will return there after the field is removed. If however, a field in the opposite sense to the first one, and sufficiently large to bring it 'round the corner' at R is applied, the operating point will move to S, and after the field is removed it will go to the other stable position at T. The more square the hysteresis loop, the closer will the core approximate to a true bistable 'trigger'. With suitable coupling between them a string of these devices could function as a scaler; however it is not as scalers that we propose to use them, but rather as indicators in the way we used voltmeters or indicator lamps with conventional scalers. The difference, however, is that if a voltmeter displaying a '1' (that is reading full scale) were disconnected from the scaler, the voltmeter would return to zero, and the information would be lost: with the ferrite core it would still retain its magnetization and hence the information. We can thus use one scaler to store information in a number of places.

Let us imagine that we have a vertical stack of cores, and that we use the lowest core to indicate 0 or 1, the next 0 or 2, the next 0 or 4, and so on, in the usual manner. We can thus store in binary notation any number by the state of magnetization of the cores. By placing beside the original stack, other vertical stacks for the other channels in the analyser, we provide an array which serves as a very compact memory for storing the complete information acquired by the analyser. Furthermore the necessary information can be inserted into the memory in a surprisingly economical way, which we shall describe only in general terms. A detailed account is given in reference 3.

The memory consists of a three-dimensional array of cores, with the number in the x and y directions depending on the number of channels in the analyser. A hundred channels (10×10), 400 channels (20×20) and 1024 channels (32×32) are popular commercial arrangements. A string of cores in the z direction belongs, as we have seen, to a particular channel, the number of cores depending on how many counts the channel is required to store. Twenty and twenty-four, storing $2^{20} - 1$, and $2^{24} - 1$, respectively, are commonly used. Any xy plane in the array will look something like Figure 174, although the address scaler will, in modern equipment, be of the transistorized BCD bistable type; where pairs of wires cross, there will be, not a scaler, but a single core. The coincidence circuit at each crossing point will also be gone, the coincident property now being provided by the 'trigger' action of the core itself. For example if, as before, the address scaler has recorded a count of 47, currents are sent down the 40 and the 7 wires. These currents can be arranged to be of such magnitudes that only the core where the two wires cross is changed over to its other state; the single current pulse applied to the other cores along the 40 and 7 wires being too small to cause a permanent change. Each core in the xy plane in question belongs to a different channel, so the address scaler effectively locates a core in a particular channel only. Furthermore, all the cores in a particular xy plane perform the same function for their particular channel; if for example, we are looking at the fourth plane from the bottom, all the cores will be used to store either zero or $2^3 = 8$.

To add a count to a particular channel, corresponding to an input pulse which has just been analysed, we need one further scaler-register, variously known as the 'add-one' scaler (or 'add–subtract' scaler, to take account of a more sophisticated function which we shall not discuss), count scaler, data scaler, temporary memory register, or word register. The applicability of these names should be clearer as we describe its functions. This scaler-register will have the same number of bistables as there are cores in the z direction (20 or 24, say) because its function will be to store temporarily the information in a particular channel. Each bistable will be connected to *every* core in the corresponding plane, by a winding threading them all (not shown in Figure 174). For example the bistable which stores 0 or 8 will be connected to all the cores in the fourth plane from the bottom, which also store 0 or 8. We are now in a position to understand how a count can be recorded in a particular channel.

Let us imagine that the analyser has been previously running, and that some counts are already stored in each channel. Now suppose that a further pulse arrives at the input, and, as in the example we have been taking, the address scaler produces a count of 47. Let us first look at the cores in a particular xy plane, the fourth from the bottom, say, again. The only core here to receive a large pulse will be, as we have seen, the one where the wires corresponding to 40 and 7 cross. We make the polarity of the signals in these wires such as to turn the core from its '1' state (that is where it is recording 8) back to its zero state. If an 8 is stored in the core, when it flips back to zero, a pulse will be induced in the additional winding (the 'sense' winding) which passes through this core (and indeed all the cores in this plane) to the corresponding bistable of the temporary-message register. We assume that all these bistables have been previously reset to zero, and the pulse in the sense winding is arranged to turn the bistable to its other state. The '1' (=8) which was stored in the core, has now been transferred to the corresponding bistable of the temporary memory register. If there were originally a '0' in the core, pulsing it towards it '0' state would have no effect, produce no signal in the sense winding, and so leave the bistable in its '0' state also. Since the address scaler is connected to all the xy planes in the same way, this process of data transfer from all the other cores belonging to channel 47 happens simultaneously. We thus move all the information in channel 47 into the temporary-memory register at the same time. The reason for this name for it should now be clear. As the various bistables of the register are connected together to form a scaler, we can add one to the number stored in it: we must do this because an input pulse of size corresponding to channel 47 has been indicated. (This function of the scaler-register is responsible for its name 'add-one' scaler.) The increased number in the register is then transferred back to the ferrite cores by a process analogous to the previous one, and the action is complete. Although for simplicity we have talked in terms of binary numbers, the temporary-memory register (alias the 'add-one' scaler), is usually in BCD form as is the information in the cores. The 20 bistables mentioned above will now give a total storage of $10^5 - 1$, as we can turn 5 scales of 16 into 5 scales of 10. Twenty-four bistables will allow us a maximum capacity of $10^6 - 1$ counts.

At the end of the complete run, in order to record the final count accumulated, the contents of the channels can be transferred, one at a time, to the temporary-memory register, from where they can be printed out channel by channel, plotted automatically by a chart recorder, or displayed on the screen of an oscilloscope as a spectrum. This sampling of channels and oscilloscope display can take place even when counts are being accumulated, allowing the operator to watch the spectrum of interest develop.

A ferrite core memory, as we pointed out, has the advantage that it is 'non-volatile' that is, if we switch off the analyser, or the power fails, the memory retains the information already in it in magnetic form. In circumstances where this is not essential, it is now more economical to use a semiconductor memory, that is one in which each ferrite core is replaced by a 'flip-flop' either of the bipolar or MOS variety. A typical arrangement might, as before, consist of a 1024 channel memory with each channel capable of storing 10^6-1 counts. In such a memory we would have, corresponding to a particular xy plane in the ferrite core case, just one integrated circuit (in a compact 18 pin package). It would contain 1024 flip-flops arranged in a 32 x 32 matrix and would be addressed in binary code from the ADC. It would thus need only five input wires for the x address and five more for the y address (since $32 = 2^5$) and these would be decoded in the circuit itself to the 32 x 32 form. (We thus refer to such a circuit as a 'fully decoded 1024-bit RAM', where 'RAM' means 'random access memory', and '1024-bit' means that we can store one binary digit or 'bit' at each of 1024 addresses.)

A 'write' wire goes to one side of all the flip-flops in the circuit. It enters each flip-flop through an AND gate, the other two inputs to the gate being the x and y addresses for that gate. Thus when one puts a logical 1 on the write line, it is only written into the flip-flop whose address has been activated. A similar wire to the other side of all the flip-flops allows us to write a logical 0 in a particular flip-flop. Reading of the stored data can also be done from the 'write' lines, again in conjunction with a particular address. (Indeed the semiconductor memory has some advantage here, inasmuch as if we are only reading out data from the memory we can do so without, as in the ferrite core case, having to destroy and then rewrite the information there, that is, we have 'non-destructive' read out.)

In the ferrite core case, the number of cores in the z direction corresponded to the number of counts we required to store per channel. Here the total number of integrated circuits will be similarly defined. For example, if we wish to count up to 10^6-1 in BCD before overflow we must have six decades available, and since each scale of ten requires four flip-flops, we will need just twenty-four integrated circuits in all. Our memory is thus compact as well as economical. The role and operation of the 'add one' scaler will be similar to our previous description of it.

Finally, we deal with a few additional items associated with a multi-channel analyser. There must first be some sort of blocking device on the input which

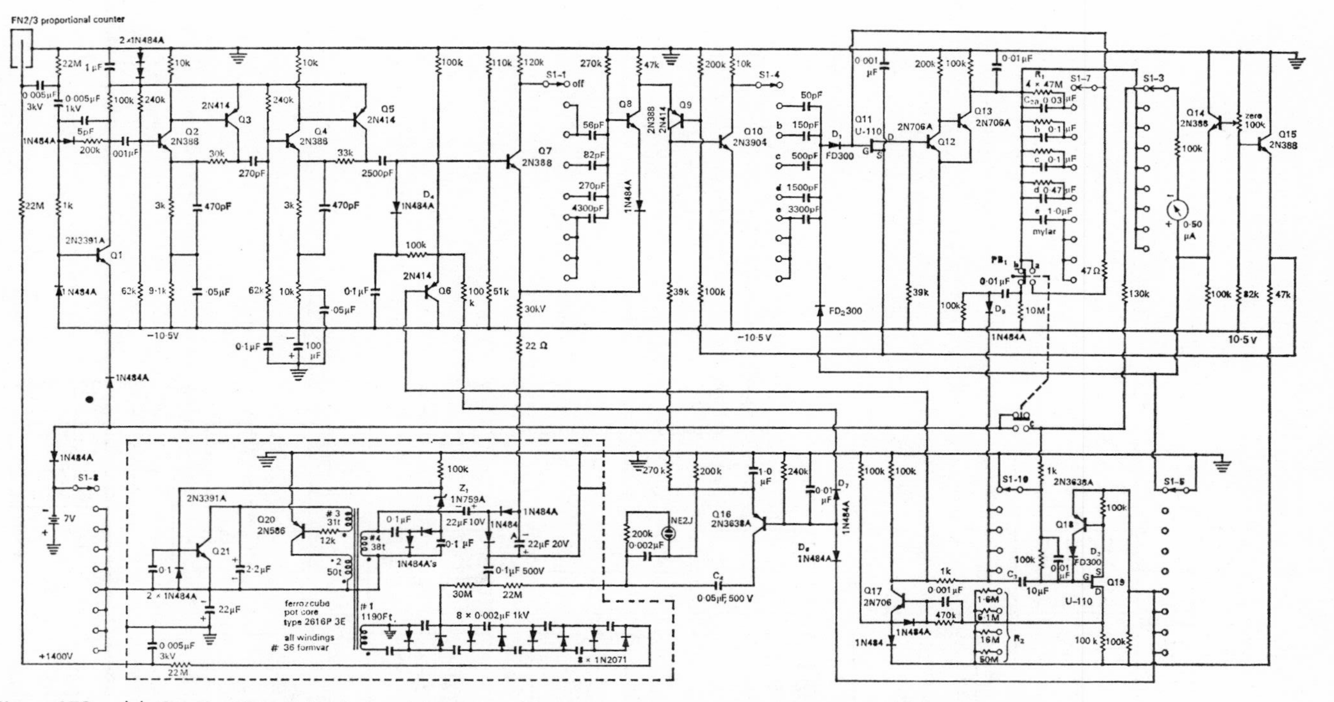

Figure 176. (a) Small self-contained counting system. (Courtesy of the Institute of Electrical and Electronic Engineers, Inc.)

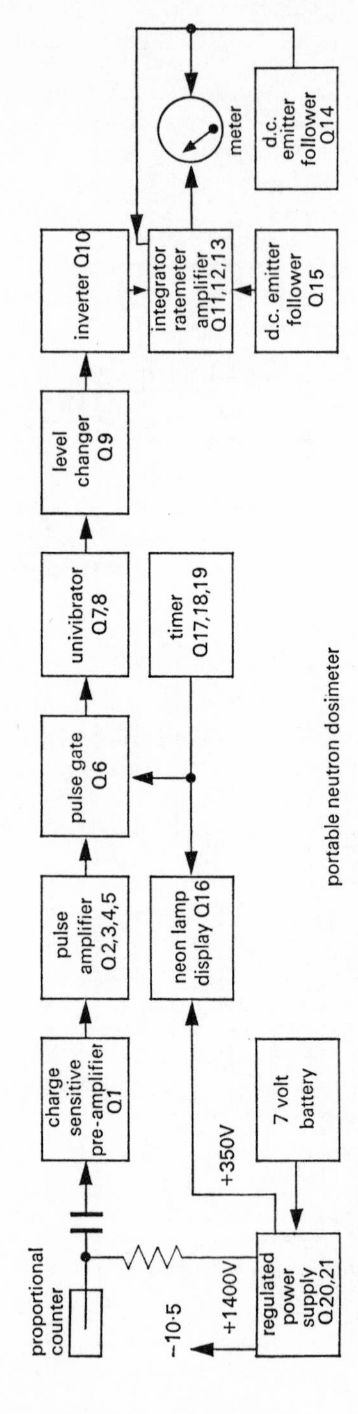

Figure 176. (b) Block diagram for circuit of Figure 176(a).

comes into action as soon as the analysis of a pulse has begun. This prevents a further pulse from entering the analyser and interfering with the operations, until the processing of the first is complete. If we are concerned about the actual numbers arriving at the input of the analyser (and not just the relative numbers with different heights), we must take account of those pulses which arrive to find the analyser blocked and dealing with another pulse. In fact we must have a device for determining the 'dead time' of the analyser, or its complement, the 'live time'. Usually also included in an analyser are an upper and lower discriminator, which operate on the input pulses, and prevent the analyser wasting time on pulses of a size in which the experimenter is not interested. To make the analyser a self-contained device for more complicated experiments, coincidence and anti-coincidence gates are often provided at the input, so that the analyser is permitted (using the coincidence facility) to analyse a pulse from one detector, provided a pulse from a second detector is also present; or alternatively (using the anti-coincidence facility) is inhibited from doing so. If we demand information about the sizes of the pulses arriving in coincidence, as well as about our primary pulses, we move into the field of multi-parameter analysis, which is outside the scope of this book.

10·7 A complete counting system

In this chapter we have discussed basic 'building blocks' which can be assembled in various ways to produce systems which will count, analyse and process pulses from nuclear radiation detectors. Because of the increasing sophistication of such systems, the experimenter in nuclear physics also depends more and more on commercially available modules – amplifiers, single-channel analysers, coincidence circuits, etc., which he assembles into the required configuration. It will be appreciated then that, not only is there no such thing as a 'typical' assembly of this kind, but also that a single diagram giving the complete circuit of a particular assembly would be of immense complexity.

Figure 176(a) shows instead the circuit of a small, self-contained instrument, a portable neutron dosimeter, which contains at least some of the elements we have discussed. The block diagram (Figure 176(b)) will help the reader to identify a pre-amplifier, amplifier, univibrator discriminator, an integrator amplifier whose output voltage indicates the accumulated number of input pulses for a period determined by the timer, and a power supply of the oscillator type driven from a seven-volt battery, which provides both the voltage for the transistors and the high voltage for the proportional counter.

References and further reading

1. W. J. PRICE, *Nuclear Radiation Detection,* McGraw-Hill, 1964.
2. J. MILLMAN and C. C. HALKIAS, *Integrated Electronics,* McGraw-Hill, 1972.
3. R. L. CHASE, *Nuclear Pulse Spectrometry,* McGraw-Hill, 1961.
4. D. A. GEDCKE and W. J. MCDONALD, 'Design of the constant fraction of height to trigger for optimum time resolution', *Nuclear Instruments and Methods,* vol. 58, 1968, p. 253.

5. P. W. NICHOLSON, *Nuclear Electronics,* Wiley, 1974.
6. *IEEE Transactions on Nuclear Science,* (Tutorial issue on high speed pulse instrumentation techniques) NS–**20**, no. 5, 1973.

CHAPTER 11

Noise

11·1 Introduction

If the gain of an amplifier is sufficiently great, an output can be observed with no applied input. (We are not referring to the possibility of oscillation due to positive feedback – we assume that the amplifier is inherently stable. Nor are we referring to the case where although there is no apparent input, there is in fact an input signal, picked up, for example, from a nearby electrical machine, or introduced into the amplifier because of insufficient smoothing of the d.c. power supply.) The amplifier output under these conditions, when viewed on an oscilloscope with a fast sweep, is seen to consist of a sequence of sharp pulses of random height, following directly on one another. This spurious output is called 'noise', and because of the progressive amplification by the successive stages of the amplifier, it clearly comes, in the main, from the first stage. We shall see that it arises both in the transistor or tube itself, and in the resistor at the input of the amplifier. We are interested in noise because, as we shall see, its presence places a limit on the smallest genuine signal we can detect.

11·2 Thermal noise

Thermal noise, or Johnson noise (after its first investigator) refers to the noise generated thermally in a resistor (or other electrical circuit). We can think of the random Brownian motion of the electrons as resulting in the appearance of small fluctuating voltages across the resistor. Clearly there can be no steady voltage in any one direction (that is the average voltage $\bar{v} = 0$), but the mean square voltage, $\overline{v^2}$, need not be zero, and its value was in fact deduced by Nyquist in 1928 (reference 1). He found that the mean square voltage developed by a resistor in any frequency interval df is given by

$$\overline{v^2} = 4kTRdf \qquad \textbf{11.1}$$

where T is the absolute temperature, k is Boltzmann's constant, and R is the size of the resistor. The noisy resistor R can be considered as made up of a noiseless resistor R, in series with a noise voltage source of mean square value as given by equation **11.1**. In terms of an equivalent current source:

$$\overline{i^2} = \frac{4kTdf}{R} \qquad \textbf{11.2}$$

where $\overline{i^2}$ is the mean square current. In terms of the available power, P (that is, the power which will be delivered to a load resistor of similar size), this becomes

$$P = kTdf \qquad \textbf{11.3}$$

which is, not unexpectedly, independent of R.

Nyquist's proof of the relation **11.3** (from which the others can be deduced) was as follows. He considered two identical resistors at a temperature T, connected together by an idealized delay line, of characteristic impedance R, the size of the resistors at each end. Each resistor can be thought of as sending power along the delay line, this being fully absorbed in the correct terminating resistor at the far end. At some instant, after equilibrium has been established, let the line be isolated from the resistors, and the ends short-circuited. Under these conditions energy in transit remains trapped in the line, with perfect reflections taking place at both ends. Alternatively we can consider the line as vibrating with its natural frequencies: the fundamental oscillation will be with a voltage node at each end, and of wavelength $2L$, where L is the length of the line. If U is the velocity of propagation of waves in the line, then the corresponding frequency of oscillation is given by $f = U/2L$. The next highest frequency (that with a node at each end and one in the centre) will be given by $f = 2(U/2L)$, the next by $3(U/2L)$, and so on. In fact, the natural frequencies are spaced apart by the amount $U/2L$. Consequently if L is assumed to be large, so that this spacing is small, in any frequency interval df there will be $df/(U/2L)$ natural frequencies of vibration. Each of these standing waves can be considered as a degree of freedom of the system, and on the basis of the equipartition law have associated with it an energy of kT ($kT/2$ for magnetic energy, and a further $kT/2$ for electrostatic energy). The total energy of the vibrations in the frequency interval df is thus $(2LkTdf)/U$. This represents energy in transit from the first resistor to the second and vice versa, when they were connected to the line, so the energy delivered to the line by a single resistor is $(LkTdf)/U$. This occurs in a transit time L/U, so the power delivered to the line by a resistor is $(LkTdf)/U$ divided by L/U, or $kTdf$, as given in equation **11.3**.

Equations **11.1**, **11.2** and **11.3** are quite general and are not related to any particular type of resistor, nor do they depend on any assumptions about the fundamental nature of electricity. It was convenient to think of Johnson noise as due to random motion of electrons, but in fact the electronic charge e, does not appear in the equations we deduced. The corpuscular nature of electricity, however, is obviously important in noise theory, and it is discussed in the next section. Before leaving thermal noise, it should be pointed out that the noise received from a resistor depends also on the input capacity and frequency pass band of the amplifier to which it is connected; this will be discussed in more detail later.

11·3 Shot noise

Shot noise was first investigated by Schottky in connexion with the fluctuations in the current in thermionic diodes and triodes, and he attributed these to the fact that the current consisted of a stream of individual charged particles. The effect is not of course confined to currents in thermionic tubes, but can occur, for example, in the gate current of an FET, or indeed in principle anywhere currents flow - although we shall have a little more to say about when we expect shot noise to be significant when we have derived an expression for it. This we now do as follows.

Suppose we are attempting to measure a current by counting how many electrons pass in succesive intervals of time t_0. If $\bar{N}$ is the average number passing in this time, the average current $\bar{I} = \bar{N}e/t_0$ where e is the electronic charge. The number N found in successive measurements will fluctuate statistically around $\bar{N}$, with a deviation given by $(\bar{N})^{1/2}$. Expressing this in terms of $\bar{I}$, the fluctuation in number is given by $(t_0\bar{I}/e)^{1/2}$. The current fluctuation is obtained by multiplying this fluctuation in numbers by e and dividing by t_0, to give $(e\bar{I}/t_0)^{1/2}$. Squaring this up we obtain for the mean square fluctuation:

$$\overline{i_1^2} = \frac{eI}{t_0} \qquad 11.4$$

To obtain the corresponding expression in terms of frequency is not easy, but the following plausible argument may be used. To be precise let us imagine that the time t_0 over which we make the measurement is one millisecond. Thus only fluctuations with time constants greater than about a millisecond will be discerned, as those with time constants much less than this will be smoothed out over the time t_0 of the measurement. In terms of frequency the previous expression for the noise can be written as $\overline{i_1^2} \approx e\bar{I}f$, where $f = 1/t_0$, and from what we have said, this represents the noise contribution from a frequency f down to zero. In the small frequency interval df, the fluctuation can therefore be written

$$\overline{i^2} \approx eIdf$$

Detailed analysis (see references 2 and 3) shows that the exact result should be twice what we have obtained, that is,

$$\overline{i^2} = 2e\bar{I}df \qquad 11.5$$

which is our final expression for the shot noise.

In the original investigation into shot noise in a thermionic diode, the formula we have just derived was found to be applicable only when the diode was operating in its 'emission limited' mode. Without going into too much detail this meant, in simple terms, that the diode voltages and currents were such that as soon as the electrons were emitted by the cathode, they were immediately pulled across to the anode. On the other hand the observed noise dropped very markedly when the diode was operated in its 'space charge limited' mode. Again in very simple terms, this meant that in this case there was a copious supply of electrons in a space charge near the cathode 'waiting' to move across to the anode. It is not hard to believe then that the effect of this space charge reservoir of electrons would be to markedly reduce the shot noise, since its smoothing action will make our assumption about the statistical nature of the electron stream no longer valid.

Correspondingly, in terms of transistor operation we may reasonably expect for the case, for example, of FET gate currents, or bipolar transistor base currents (that is for minority carrier currents) that the usual shot noise theory will apply. However for, say, the drain current in an FET (that is where we have large numbers of majority carriers present) we can expect (from analogy with the case of the space charge limited diode) that the shot noise due to the current will be small, and consequently that the thermal noise in the channel (that is, the noise generated by the random movement of the carriers, as in a resistor) will be dominant.

Returning again to our main discussion on shot noise, we should finally add that, just as for thermal noise, the frequency characteristics of the following amplifier stages will affect the amount of shot noise actually appearing at the amplifier output.

11·4 1/*f* noise

Thermal and shot noise are both referred to as 'white' noise, as they extend uniformly over the frequency spectrum. In valves and transistors another noise source is usually present, with a frequency dependence varying approximately as $1/f$ (hence its name). It arises from different physical mechanisms in valves and transistors, although neither is well understood. In valves where the term 'flicker noise' is also used, it is believed to arise from relatively slow changes in the electron emitting properties of the cathode: in transistors, where it is also known as modulation noise, it is thought to be due to crystal imperfections and surface effects. Without prejudice as to how it arises, we can write it as an equivalent noise source at the input of our device given by:

$$\overline{v^2} = A_f \frac{df}{f} \qquad 11.6$$

where A_f is the $1/f$ noise constant. Again, the amount of $1/f$ noise at the output will depend on the frequency response of the amplifier – in this case quite drastically, because of the large part of the noise appearing in the low-frequency region of the spectrum.

11·5 Choice of amplifier time constants

We can now discuss the total noise produced by an amplifier input stage, whether thermionic valve, bipolar transistor, or field effect transistor, although we shall restrict the discussion to the FET case, for the following reasons. The bipolar transistor is markedly inferior to valves and FETs for most low noise applications, because of the noise associated with the appreciable input (that is, base) current. The valve and the FET are almost equally good at normal working temperatures. However, the FET can easily be cooled to dry ice or liquid nitrogen temperatures, with a big reduction in thermal noise, and under these conditions it is superior to the thermionic valve as a low-noise device. This, together with its small size, makes it the preferred input device for low-noise pre-amplifiers.

The most favourable frequency pass band for the amplifier – that is, the values of the amplifier rise and fall times, T_R and T_F – must first be chosen, since this determines which part of the noise, generated in the first stage, reaches the output. We saw in section 3·7 that there was no point in having the fall time, T_F, unnecessarily large, once the information in the leading edge had been obtained; but on the other hand if T_F were decreased to the same sort of size as the rise time T_R, the pulse height is reduced considerably. In the simple situation discussed in Chapter 3, we settled for $T_F \approx 10T_R$. The present situation is more critical. The smaller the frequency pass band, the more noise will be cut out, so it is possible that with quite a narrow pass band, that is, with T_R and T_F close together, we may be in a better position regarding signal to noise ratio, even though the height of the signal itself has been reduced. It is not difficult to believe that the most favourable situation for good signal to noise ratio is when $T_R = T_F$ and we shall accept this here without proof (and return to discuss the question more rigorously in Chapter 13). For simplicity we put $T_F = T_R = \tau$.

Now τ must be chosen (that is, the frequency pass band positioned) in such a way that noise from the various sources is minimized. However, the rise time T_0 of the actual signal from the detector may limit the choice. There is certainly no harm (unless we are concerned with the fast timing of events) in having τ greater than T_0; it merely means that the information is there waiting, until the slower rise time of the amplifier presents it. Can τ be less than T_0? The answer is no, because τ is also the fall time, so if it is very much less than T_0, very serious attenuation of the signal will result, with a corresponding rapid degradation of the signal to noise ratio. Thus, τ should be made of the order of, or larger than, the detector rise time. For the important case of pre-amplifiers for semi-conductor detectors, it is fortunate that the values of τ of a few microseconds, which are required for optimum noise performance, fulfil this condition.

11·6 Noise calculations for the field effect transistor

The four noise sources we shall discuss in connexion with the FET are (not necessarily in order of importance):

1. Thermal noise in the gate resistor R_g, which acts also as detector load (see Figure 177). We have shown here a semiconductor detector – currently the most interesting case.

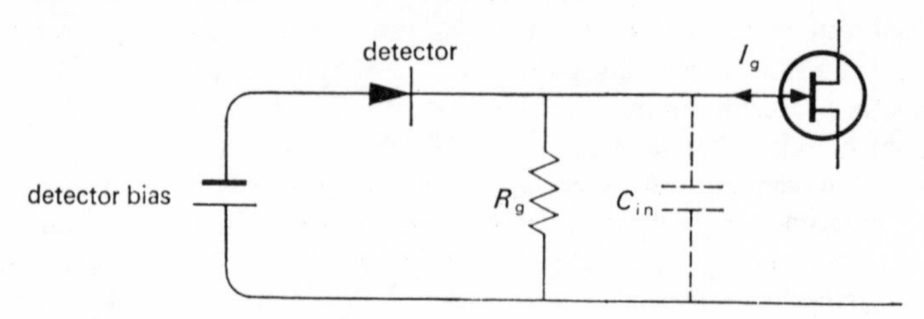

Figure 177. Detector and input circuit of FET.

2. Shot noise from the gate current I_g. This is the usual small current in a reverse biased junction. (We should take account in addition of the leakage current in the semiconductor detector, which is also a reverse biased diode.)

3. Thermal noise in the FET channel. This takes the place of the shot noise we might have expected to find there: we discussed the reason for this in Section 11·3.

4. $1/f$ noise in the FET channel.

We now discuss each of these in detail (references 5, 6, and 7).

11·6·1 *Thermal noise in the gate resistor*

The noise R_g can be thought of, as we have seen, as being generated by a separate noise source of size $4kTR_g df$ in series with a noiseless resistor R_g. Alternatively it could be considered as being generated by a separate current source of size $\overline{i^2} = 4kTdf/R_g$ in parallel with a noiseless resistor R_g. The latter representation is more appropriate here (see Figure 178) because R_g is large. Indeed R_g will usually be large enough so that the noise current will flow almost entirely through C_{in}, the input capacity of the FET plus that of the detector. (Although by implication we are talking about a voltage sensitive stage, we shall see later that our remarks will apply equally well to a charge sensitive stage.) The mean square thermal noise voltage appearing at the FET input is thus given by the value of $\overline{i^2}$ multiplied by the square of the impedance of C_{in}, that is

$$\overline{v_R^2} = (4kTdf/R_g)\,(1/C\omega)^2$$

or putting $df = d\omega/2\pi$

$$\overline{v_R^2} = \frac{2kT}{\pi\omega^2 R_g C_{in}^2} d\omega \qquad 11.7$$

Note incidentally that the larger R_g is, the less its noise contribution, a fact we have alluded to, without proof, on a number of occasions.

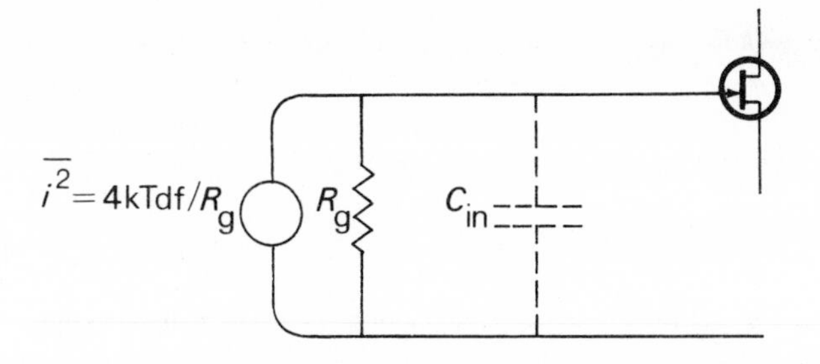

Figure 178. Circuit for gate resistor noise calculations.

11·6·2 *Shot noise from the gate current*

From the discussion of the shot effect this is given by

$$\overline{i^2} = 2eI_g\,df$$

This current flows almost entirely through the capacitor C_{in} (see Figure 177), as R_g has been chosen to be large. The magnitude of the mean square voltage at the f.e.t. input due to this source is

$$v_G^2 = \frac{2eIg}{\omega^2 C_{in}^2}\,df = \frac{eI_g}{\pi\omega^2 C_{in}^2}\,d\omega \qquad 11.8$$

Any leakage current due to the detector can be included with I_g.

11·6·3 *Thermal noise in the FET channel*

In Chapter 2 it was shown that as the characteristic curves for a FET at low drain voltages are straight lines passing through the origin, in this region the FET may be considered as a resistor whose magnitude is controlled by the gate voltage. We might thus expect that the channel noise would be given by a formula analogous to **11.2**

$$\overline{i^2} = \frac{4kTdf}{R_0} \qquad 11.9$$

where R_0 is the inverse of the slope of the line in question in the FET characteristics, and the detailed treatment shows that this is so. Although the FET is not normally operated in this region of its characteristics, it is found that the result given is a good approximation for the usual operating point also (reference 4).

To transform equation **11.9** into a corresponding mean square voltage at the gate, we divide by g_m^2 (since g_m relates drain current to gate voltage, and the square arises because we are talking of mean square voltages and currents), to obtain:

$$\overline{v^2} = \frac{4kTdf}{R_0 g_m^2}$$

Finally, putting $R_{eq} = 1/R_0 g_m^2$, we have for the channel thermal noise $\overline{v_C^2}$

$$\overline{v_C^2} = 4kTR_{eq}\,df = 2kTR_{eq}\frac{d\omega}{\pi} \qquad 11.10$$

where R_{eq} is the equivalent noise resistance of the channel. Incidentally, FET theory shows that $g_m = 1/R_0$, so R_{eq} is in fact equal to $1/g_m$, which gives us a useful feel for the magnitude of the equivalent noise resistance.

11·6·4 $1/f$ *noise in the channel*

The mean square $1/f$ voltage is given, as in equation **11.6** by

$$\overline{v_F^2} = A_f\frac{df}{f} = A_f\frac{d\omega}{\omega} \qquad 11.11$$

11·6·5 *Total noise at amplifier output*

To find the noise appearing at the output of the amplifier, each of the previous expressions for the mean square noise voltages must be multiplied by a factor $f(\omega)$, to take account of the dependence of the amplifier gain on frequency, integrated from $\omega = 0$ to $\omega = \infty$, and the integrated mean square values added. The factor $f(\omega)$ is given by the *square* of the value of $|A|$ in equation **2.26** (since we are dealing with mean *square* voltages). From section 3·6, $\omega_U = 1/T_R$, and $\omega_L = 1/T_F$: as, here $T_R = T_F = \tau$, we must put $\omega_U = \omega_L = 1/\tau$. Thus $f(\omega) = A_0^2\{\omega\tau/(1+\omega^2\tau^2)\}^2$. The integrations are straightforward, and the result is

$$\overline{v_{total}^2} = A_0^2\left(\frac{kT}{2R_g}\frac{\tau}{C_{in}^2} + \frac{eI_g}{4}\frac{\tau}{C_{in}^2} + \frac{kTR_{eq}}{2}\frac{1}{\tau} + \frac{A_f}{2}\right) \qquad 11.12$$

where the contributions appear in the same order discussed previously that is, gate-resistor noise, gate-current noise, channel thermal noise, and channel $1/f$ noise. (In obtaining the expression for the total noise given in **11.12**, the mean square voltages were added. This is appropriate for random, uncorrelated contributions, and contrasts with the direct addition of the voltages themselves for correlated effects.)

11·7 The equivalent noise charge equation

As a final method of stating the noise contribution, we introduce the idea of 'equivalent noise charge'. Let us consider a genuine signal from the detector which deposits a charge Q on the input capacity of the pre-amplifier. This is equivalent to a voltage step Q/C_{in} at the input, which would normally give an output signal of $A_0\,Q/C_{in}$. However, as shown in section 3·7, when the rise and fall times of the amplifier are made equal, the gain is reduced by a factor of exp(1) from the maximum obtainable. So the output pulse will be $A_0\,Q/C_{in}\,\epsilon$, where $\epsilon = \exp(1) = 2{\cdot}7$ approximately. The genuine charge signal which will produce the same output as the mean square noise – the 'equivalent noise charge' – is obtained by equating the quantity $(A_0\,Q/C_{in}\,\epsilon)^2$ to the value of $\overline{v^2_{total}}$ as given in equation **11.12.** Solving for Q^2:

$$Q^2 = (\text{equivalent noise charge})^2 = (\text{ENC})^2$$

$$= \epsilon^2\left(\frac{kT}{2R_g}\tau + \frac{eI_g}{4}\tau + \frac{kTR_{eq}}{2}\frac{C^2_{in}}{\tau} + \frac{A_f}{2}C^2_{in}\right) \qquad 11.13$$

Some of the terms in equation **11.13** depend on C_{in}, and some do not. This gives us another classification of noise sources – series and parallel. A parallel source, as its name implies, can be considered as in parallel with the input; consequently if the input capacity is changed, it affects both signal and noise together, and the ENC is not changed. Thus for a parallel source ENC is not a function of the input capacity. From equation **11.13** we see that gate-resistor noise, and gate-current shot noise can be considered as parallel noise sources. With series sources such as channel thermal and $1/f$ noise, when the input capacity is increased, the signal is reduced, while the noise remains the same. Thus the signal to noise ratio is worsened, and the equivalent noise charge increased.

A second point concerns the dependence of the noise on τ, the common rise and fall time, and the possibility of choosing a value for τ which will make the noise a minimum. Let us first write equation **11.13** as

$$(\text{ENC})^2 = L\tau + \frac{MC^2_{in}}{\tau} + NC^2_{in} \qquad 11.14$$

where we have simplified the equation by using constants L, M, and N, instead of the more complicated ones of equation **11.13.** Differentiating to find the minimum:

$$L - \frac{MC^2_{in}}{\tau^2} = 0$$

or

$$\tau_{min} = \left(\frac{M}{L}\right)^{\frac{1}{2}} C_{in} \qquad 11.15$$

Substituting back into equation **11.14** gives us the minimum value for the noise

$$(\text{ENC})^2_{min} = (LM)^{\frac{1}{2}}\,C_{in} + (LM)^{\frac{1}{2}}\,C_{in} + NC^2_{in} \qquad 11.16$$

We can see from equation **11.16** that the minimum is obtained by equalizing the contribution of the resistor and gate shot noise (which increases with τ) with that of the channel thermal noise (which decreases with increasing τ). When typical values for the quantities involved are inserted, τ_{min} comes out to be a few microseconds for values of C_{in} corresponding to the stray capacity of the f.e.t., and to correspondingly larger values when a detector with, say 50 pF capacity is added. This latter value of τ may be unacceptably large if the counting rate is high, and we may have to operate with values of τ considerably below the optimum. In this case the channel-thermal-noise contribution overshadows the resistor and gate shot noise, and if the $1/f$-noise contribution is small, the expression for the total noise, becomes

$$(\mathrm{ENC})^2 \approx \frac{MC_{in}^2}{\tau} = \frac{\epsilon^2 kTR_{eq} C_{in}^2}{2\tau}$$

or

$$(\mathrm{ENC}) \approx \left(\frac{M}{\tau}\right)^{\frac{1}{2}} C_{in}$$

Lastly, C_{in} can be split into two parts, the 'external' capacity C_{ext}, which depends on the detector being used, external wiring, etc., and C_0, the input capacity of the FET itself and its associated wiring, which represents the irreducible minimum of capacity. Thus,

$$\mathrm{ENC} = \left(\frac{M}{\tau}\right)^{\frac{1}{2}} (C_0 + C_{ext}) \qquad \mathbf{11.17}$$

$$= k_1 + k_2\, C_{ext} \quad \text{(say)} \qquad \mathbf{11.18}$$

For commercial pre-amplifiers, the manufacturers will quote these two constants, k_1 and k_2 (for a stated value of τ), in units discussed below.

11·8 Numerical units for noise

The units for $(\mathrm{ENC})^2$ in equation **11·13** are normally (coulombs)^2, that is ENC is in coulombs. ENC could also be expressed in terms of charge numbers at the input, by using the value of $e = 1{\cdot}6 \times 10^{-19}$ coulombs. But in the case of semiconductor detectors, towards which a great deal of recent work on low noise pre-amplifiers has been directed, it is preferable to use an alternative description. To understand it, we must briefly return to the action of a semiconductor detector. These devices as we saw in Chapter 1 are basically low-leakage junction diodes, operated under reverse bias. When an ionizing particle enters the depletion layer, a large number of hole–electron pairs are generated, and these are swept to their respective electrodes by the applied voltage, producing a signal charge on the capacity C_{in} of Figure 177. This capacity, as we know, represents the sum of the input capacity of the FET plus that of the detector. As there are competing processes, an energy loss of 3·66 eV (electron volts) is required to produce a hole–electron pair in a silicon detector, while for germanium with its smaller energy gap, only 2·96 eV is needed. A succession of pulses produced by incoming particles of the same energy will vary a little in size. One cause of this is the fact that the production of hole–electron pairs is a statistical process, subject to fluctuations; but the other, and often the main cause, is the fluctuations appearing simultaneously with the pulses due to the pre-amplifier noise. The energy spectrum recorded on an analyser will thus be broadened from an ideal vertical line, into what is shown in Figure 179. For the case where the pre-amplifier noise is predominant, the broadening may be calculated as follows.

The number of electrons corresponding to the equivalent noise charge is, as we have seen, $\text{ENC}/e = \text{ENC}/1{\cdot}6 \times 10^{-19}$. This in turn corresponds to the expenditure of energy in a silicon detector of $(\text{ENC}/1{\cdot}6 \times 10^{-19}) \times 3{\cdot}66$ eV, or one-thousandth of this if expressed in keV. This gives the standard deviation

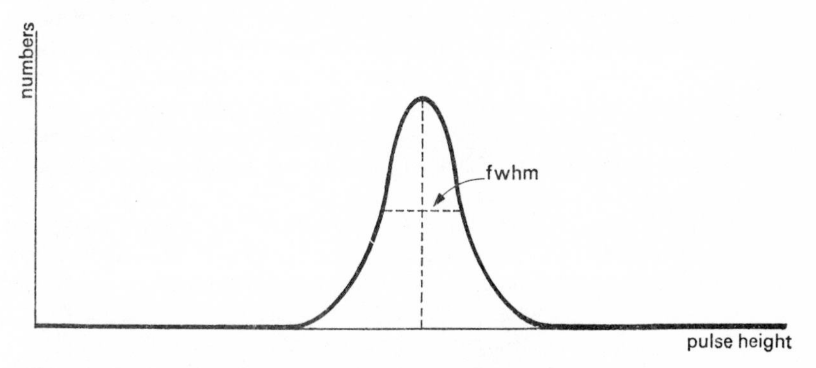

Figure 179. Noise broadening of an ideal line spectrum.

of the curve of Figure 179, which we consider to be Gaussian. The full width at half maximum (*fwhm* for short) that is, the width of the curve half way down from the peak, is obtained from the standard deviation by multiplying by 2·36. Hence

$$fwhm(\text{keV Silicon}) = \text{ENC}\left(\frac{3{\cdot}66 \times 10^{-3} \times 2{\cdot}36}{1{\cdot}6 \times 10^{-19}}\right)$$

$$= \text{ENC}(5{\cdot}4 \times 10^{16}) \qquad (\text{ENC in coulombs})$$

The full width at half maximum has the word 'silicon' added (or Si) because the use of the factor 3·66 makes the answer applicable only to a silicon device. It could equally well be quoted as 'keV Ge', for a germanium detector, if we substitute 2·96 for 3·66 in our calculations.

Referring back to equation **11.18**, and using now the new units for k_1 and k_2, the performance of an amplifier could be described by saying, for example, that its noise was 1·0 keV (Ge) *fwhm*, with zero external capacity (that is, giving the value of k_1), and its 'slope' (that is, the value of k_2) was 0·05 keV (Ge) per pF of added external capacity. One should also add the value of the time constant, τ, at which these results were obtained, say 2 μsec, as well as whether they were obtained with the FET at room temperature or cooled. Pre-amplifiers giving performance at room temperatures comparable with what has just been quoted, are commonly available, and considerably better figures have been reported in the research literature. The implication of such results is that spectroscopy of X- and gamma-ray photons of only a few kilovolts energy is possible with semiconductor detectors.

11·9 Further remarks on low-noise pre-amplifiers

We have concentrated until now on one particular type of pulse shaping, that is with one *RC*-differentiating and one *RC*-integrating network, whose time constants we found it advisable to make equal. The question arises whether some other type of shaping arrangement, say with additional *RC* networks, or by means of operational amplifiers might not produce better results from the point of view of noise. The pulse with the best possible signal to noise ratio can be shown to be a cusp shaped one, concave to the axis on both its leading and trailing edges. Such a pulse should provide an improvement of about 40% in signal to noise ratio over that produced by one *RC* integration and one *RC* differentiation. It is however a shape to which it is physically difficult even to generate an approximation. Triangular and Gaussian shaped pulses are easier to produce and produce appreciable improvement over simpler methods. However we will return to discuss this question in more detail in Chapter 13 when we have developed the necessary mathematical tools.

The other point we must underline is that all our discussion on noise so far has been in terms of the voltage sensitive amplifier, and since the charge sensitive amplifier is so widely used, particularly with semiconductor detectors, we must enquire whether the expression we have already obtained for the equivalent noise charge applies also to this latter type of amplifier. To do this we recall the comparison between the voltage sensitive and the charge sensitive amplifier in Section 6·5. For the former when a charge Q arrives at the input the output voltage is given by $-AQ/C_{in}$, while for the latter all the charge is pulled on to the feedback capacitor C_f producing an output voltage of $-Q/C_f$. We could thus have moved from the voltage sensitive to the charge sensitive case by replacing A/C_{in} by $1/C_f$, that is A by C_{in}/C_f. Returning to our present problem we deduce that the expression for $\overline{v^2_{total}}$ in equation **11.12** should be modified for the case of a charge sensitive amplifier by replacing the A_0^2/C_{in}^2 factor in the first two terms by $1/C_f^2$, and the A_0^2 factor in the last two terms by C_{in}^2/C_f^2. Similarly the quantity $(AQ/C_{in}\epsilon)^2$ in Section 11·7 becomes $(Q/C_f\epsilon)^2$ and when we equate it to our new value of $\overline{v^2_{total}}$ to obtain the equivalent noise charge we obtain the same result as we did in equation **11.13** for a voltage sensitive amplifier. Putting the matter even more informally, we can say that it is no surprise that the first two terms in equations **11.12** or **11.13** (the parallel noise source) contribute nothing new to the ENC when we change from the voltage to the charge sensitive configuration, since we know that they enter into our calculations in exactly the same way as the signal, and any reduction in signal output is compensated for by a corresponding reduction in this noise. The decrease of the output noise from the series sources (the other two terms) must be thought of in a different light - in fact as the result of the reduction of the gain of the amplifier when a feedback network consisting of C_f and C_{in} is placed across it. And since the noise source is inside the feedback loop we expect the feedback to affect signal and noise equally - something indeed which is true of feedback in general in such circumstances.

(Actually if we do our analysis a little more exactly we find that there is a small increase in ENC when we change to the charge sensitive configuration. This is because the capacitor C_f has its far side effectively connected to ground through the low output impedance of the amplifier. Its value, usually of the order of 1 pF, must therefore be included in the irremoveable stray capacity C_0 of equation **11.17**.)

11·10 Noise figures

Although we have made a very detailed study of noise sources, there is a more general criterion which is often used to describe the noise of say, a transistor, or even a complete amplifier – the noise figure. This is defined as follows. The available noise power produced by a resistor at temperature T is, from equation **11.3**

$$P = kTdf$$

If a transistor of power gain G is connected to this resistor, the power output, if the transistor were ideal, would be $G(kTdf)$. However, any real transistor will add noise to the signal, so the actual noise power output will be P^*, say, where $P^* > G(kTdf)$. The noise figure is defined as the ratio of P^* to $G(kTdf)$ and is usually expressed in decibels. Hence the noise figure F is given by

$$F = 10\log_{10}\frac{P^*}{GkTdf}$$

The case of an ideal transistor would be represented by $F = 0$.

While the practical methods of performing the idealized experiment described need not concern us (see reference 8) it is worth while quoting a typical result. For a noise source of 2000 ohms, and a bandwidth from 30 Hz to 15 kHz (a common test arrangement), a bipolar transistor would be considered to have good noise properties if its noise figure were 4 dB or less. Although a low noise figure as defined here is no guarantee that the transistor will meet the much stricter criteria discussed earlier, it will give an indication that such a transistor could be tested further as to its suitability for inclusion, say, as the second stage in a low-noise pre-amplifier for a semiconductor detector. As mentioned earlier, the noise performance of a bipolar transistor is normally not sufficiently good to allow its use in the first, critical stage.

References and further reading

1. H. NYQUIST, 'Thermal agitation of electronic charge in conductors', *Physical Review*, vol. 32, 1928, p. 110.
2. A. VAN DER ZIEL, *Noise: Sources, Characterization, Measurement*, Prentice Hall, 1970.
3. D. K. C. MCDONALD, *Noise and Fluctuations*, Wiley, 1962.
4. F. N. H. ROBINSON, 'Noise in field effect transistors at moderately high frequencies', *Electronic Engineering*, vol. 41, 1969, p. 353.
5. V. RADEKA, 'The field effect transistor – its characteristics and applications', *IEEE Transactions on Nuclear Science, NS-11*, no. 3, 1964, p. 358.
6. T. V. BLALOCK, 'A low noise charge sensitive pre-amplifier with a field effect transistor in the input stage', *IEEE Transactions on Nuclear Science, NS-11*, no. 3, 1964, p. 365.
7. F. S. GOULDING, 'Semiconductor detectors for nuclear spectrometry', *Nuclear Instruments and Methods*, vol. 43, 1966, p. 1: see also *Nuclear Instruments and Methods*, vol. 142, 1977, p. 212.
8. J. MILLMAN and C. C. HALKIAS, *Integrated Electronics*, McGraw-Hill, 1972.
9. A. B. GILLESPIE, *Signal, Noise and Resolution in Nuclear Counter Amplifiers*, Pergamon Press, 1953.
10. P. W. NICHOLSON, *Nuclear Electronics*, Wiley, 1974.

CHAPTER 12

The Laplace Transforms and its uses

12·1 Introduction

When discussing a non-sinusoidal repetitive waveform (like the train of pulses in Figure 40 for example) it is often convenient to think of it as broken down into a superposition of true sinusoidal waves. In it we have a fundamental wave and a series of harmonics of gradually decreasing amplitudes. This Fourier series method has been alluded to already in Section 9·2. Indeed we can go farther and think even of a single voltage step as analysed into a superposition of sinusoidal waves by the method of the Fourier transform. Here not just a fundamental wave and harmonics are present, but waves of all possible frequencies in various proportions. We have also briefly discussed this sort of picture intuitively in Section 3·6.

However such an analysis of a single isolated step into sinusoidals is rather artificial, and we tend to use a different technique for this and similar signals - the method of the Laplace transform - which is the subject of the present chapter. This is not to say that Laplace transforms can not also be applied to sinusoidal signals - indeed we shall have an example of this later - but it is true that the method is particularly suited for isolated steps and pulses generally.

12·2 Mathematical definition of the transform

We introduce the discussion by means of a very simple circuit which we met in Section 3·1. It is redrawn again here for convenience (Figure 180). A voltage v (of a form to be specified later) is applied to a capacitor and resistor in series. If v_1 and v_2 are the voltages across the resistor and capacitor respectively and i the instantaneous current at time t then we have

$$v = v_1 + v_2$$

also $\quad v_1 = iR$

and $\quad v_2 = q/C \quad$ where q is the instantaneous charge on the capacitor

$$= (1/C)\int_0^t i\,dt.$$

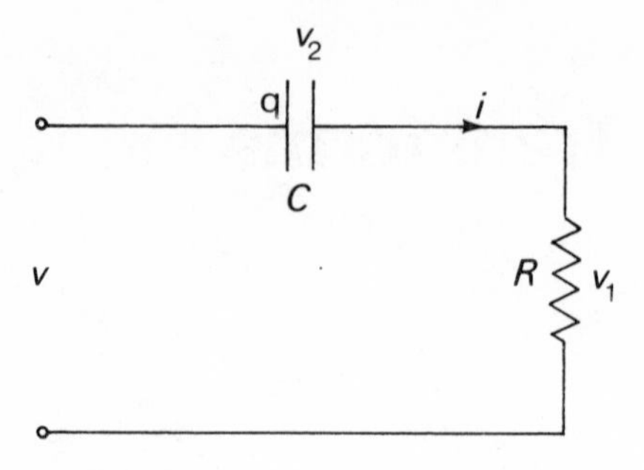

Figure 180. Circuit for analysis.

Hence eliminating v_1 and v_2 we obtain

$$v = iR + (1/C)\int_0^t i\,dt \qquad 12.1$$

We recognize equation **12.1** more easily by differentiating it to give:

$$\frac{dv}{dt} = R\frac{di}{dt} + (i/C)$$

or

$$\frac{di}{dt} + (1/RC)i = (1/R)\frac{dv}{dt} \qquad 12.2$$

This is just another form of the equation **3.4** which we obtained in Chapter 3 and which we solved there for the case v = constant = V_0. We can equally well solve equation **12.2** by elementary methods for the case $v = V_0$ or indeed more generally, if v varies with time in a fairly simple manner. Our purpose her is to show how Laplace transform methods may be used to solve this problem, and then to go on to discuss more complicated examples where the transform method is the only practical approach. We turn then to the general definition of the Laplace transform of some function $f(t)$ which varies with time (as the current did in our simple problem). This is given by;

$$\text{Laplace transform of } f(t) = \mathcal{L}[f(t)] = \int_{t=0}^{t=\infty} f(t)\exp(-st)\,dt \qquad 12.3$$

When we have performed the definite integration of $f(t)$ as required in equation **12.3** we naturally obtain some function of s only. We thus alternatively write $\mathcal{L}[f(t)]$ as $F(s)$. (Some authors use the symbol p instead of s.) Note also that since our integration starts at $t = 0$ we will be concerned only with signals which commence at this time. As a first example let us obtain the Laplace transform of a unit step function, that is a function which is zero up to time $t = 0$ and then suddenly rises after $t = 0$ to unit height (see Figure 181(a)). Here $f(t)$ is obviously unity and its Laplace transform is given by:

$$F(s) = \int_0^\infty 1.\exp(-st)\,dt$$

$$= [-(1/s)\exp(-st)]_0^\infty$$

$$\text{or } F(s) = 1/s \qquad \textbf{12.4}$$

The simplicity of this result is one of the reasons why the method of the Laplace transform is so useful for step functions of voltage. If the step is of size V_0 instead of unity the Laplace transform is clearly increased in the same ratio and becomes V_0/s.

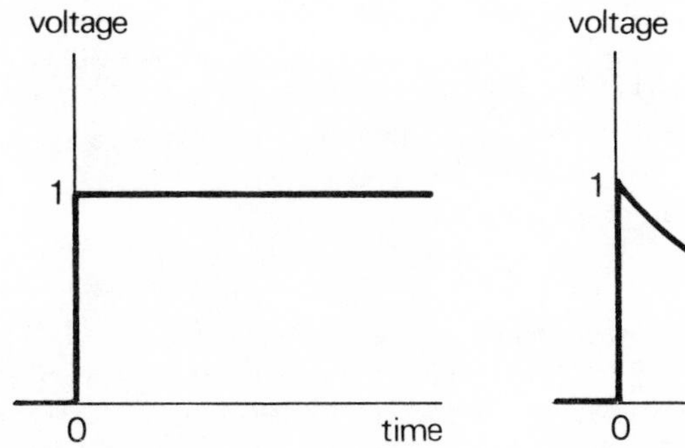

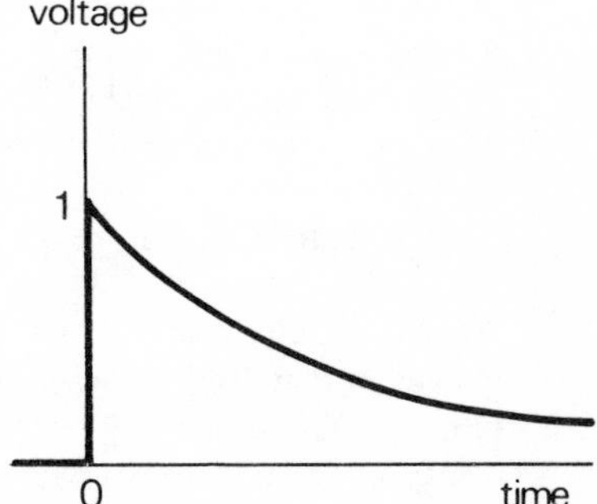

Figure 181.
(a) Step function of unit height. (b) Exponential function of unit height.

Similarly for the exponentially falling function $f(t) = \exp(-at)$ of Figure 181(b) the Laplace transform is given by

$$F(s) = \int_0^\infty \exp(-at)\exp(-st)\,dt$$

$$= \int_0^\infty \exp\{-(s+a)t\}\,dt$$

$$\text{or } F(s) = 1/(s+a) \qquad \textbf{12.5}$$

In fact our previous result for a step function is really a particular case of this latter answer with $a = 0$. We will find that these two Laplace transforms alone will be adequate for most of the problems involving non-repetitive signals with which we will have to deal. However one further transform will be of use in Chapter 13 and we discuss it now - the transform of a delta function. Consider

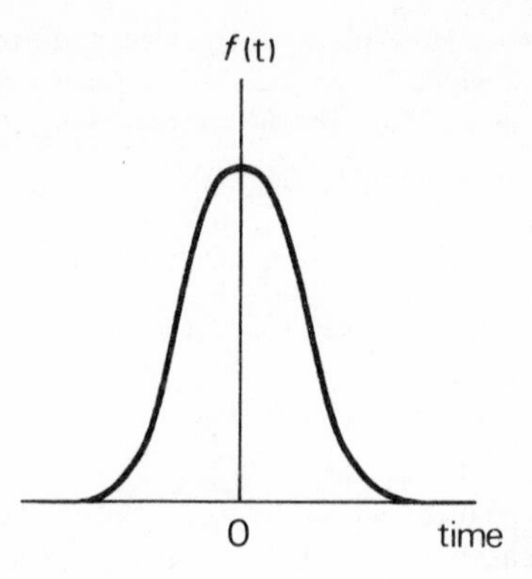

Figure 182. Introduction to the delta function.

a function of the general shape shown in Figure 182 and imagine that the height of the function continually increases while its width correspondingly decreases, its area remaining constant at a value of unity. In the limit we will arrive at a function of infinite height and zero width at $t = 0$, but still of unit area - the delta function, written $\delta(t)$. The Laplace transorm of this is given as usual by:

$$\mathcal{L}[\delta(t)] = \int_0^\infty \delta(t) \exp(-st)\, dt.$$

We can see informally what the answer to this integration is going to be if we remember that $\int \phi(t)\, dt$ (where $\phi(t)$ is any function) is defined as the limit of a sum $\Sigma\, \phi(t)\, \Delta t$ over the range in question. In our case $\phi(t) = \delta(t) \exp(-st)$ and it may be easier if we consider the case where the delta function is very tall and thin, but not quite infinitely so. Also near $t = 0$ we have $\exp(-st) \simeq 1$, therefore $\mathcal{L}[\delta(t)] \simeq \Sigma\, \delta(t)\, \Delta t$ where we now only need to sum around $t = 0$. The right-hand side of this last expression represents the area under our function, and when we go to the limit represents the area under the delta function, which is unity. Therefore $\mathcal{L}[\delta(t)] = 1$.

We may also have in the basic mathematical statements of the electronic problem we are dealing with, the operation of integration (see equation **12.1** for example) or differentiation. We thus require the Laplace transforms of $\int_0^t f(t)dt$ and $\frac{d}{dt} f(t)$ and we deal first with the case of the integral.

Here $$\mathcal{L}[\int_0^t f(t)\, dt] = \int_0^\infty [\int_0^t f(t)\, dt] \exp(-st)\, dt = F(s) \qquad \textbf{12.6}$$

In writing down this equation we have simply followed the same rule as in our basic definition given by equation **12.3**, that is, take the function whose Laplace transform is required, (in this case $\int_0^t f(t)\, dt$ multiply it by $\exp(-st)$ and integrate between $t = 0$ and $t = \infty$. We can evaluate the right-hand side of equation **12.6** using the method of integration by parts: $\int_a^b u\, dv = [uv]_a^b - \int_a^b v\, du$.

In our case we take $u = \int_0^t f(t)dt$ and $dv = \exp(-st)\,dt$, making $v = -\{\exp(-st)\}/s$.

Hence $F(s) = [\{\int_0^t f(t)\,dt\}\{\exp(-st)\}/s]_{t=0}^{t=\infty} - \int_0^\infty (\{-\exp(-st)\}/s) d(\int_0^t f(t)\,dt)$.

The quantity in the square brackets vanishes, because $(\exp(-st)$ is zero at $t = \infty$, and at $t = 0$ the term $\int_0^t f(t)\,dt$ becomes $\int_0^t f(t)\,dt$ which is also zero. Therefore

$$F(s) = \int_0^\infty \{\exp(-st)\}/s \; d(\int_0^t f(t)\,dt).$$

Now $\frac{d}{dt}(\int_0^\infty f(t)\,dt) = f(t)$ or $d(\int_0^t f(t)\,dt) = f(t)\,dt$, therefore

$$F(s) = (1/s)\int_0^\infty f(t)\exp(-st)\,dt$$

(since $1/s$ is not being integrated and can be taken outside the integral).

That is $\mathcal{L}[\int_0^t f(t)] = (1/s)\int_0^\infty f(t)\exp(-st)\,dt$.

But the integral on the right-hand side is nothing but $\mathcal{L}[f(t)]$ from our definition in equation **12.3**. Therefore we have that

$$\mathcal{L}[\int_0^t f(t)dt] = (1/s)\,\mathcal{L}[f(t)] \qquad \textbf{12.7}$$

That is, to obtain the Laplace transform of the integral of a function, we simply divide the Laplace transform of the function by s.

Similarly we can deal with the Laplace transform of the derivative of a function.

Here $\mathcal{L}[\frac{d}{dt}f(t)] = \int_0^\infty \{\frac{d}{dt}f(t)\}\exp(-st)\,dt$.

Evaluating this by parts, this time with $u = \exp(-st)$ and $dv = \{\frac{d}{dt}f(t)\}\,dt$, that is $v = f(t)$, we obtain

$$\mathcal{L}[\frac{d}{dt}f(t)] = [f(t)\exp(-st)]_{t=0}^{t=\infty} - \int_0^\infty f(t)\{-s\exp(-st)\}\,dt.$$

Once again the quantity in the square brackets vanishes at $t = \infty$ because there $\exp(-st) = 0$, but at $t = 0$ its value becomes $-f(0)$. The rest of the right-hand side can be written as $s\int_0^\infty f(t)\exp(-st)\,st$, and we recognize the integral as simply the Laplace transform of $f(t)$. Therefore finally we have

$$\mathcal{L}[\frac{d}{dt}f(t)] = s\mathcal{L}[f(t)] - f(0) \qquad \textbf{12.8}$$

Sometimes we write $f(0)$ as $f(0^+)$ to emphasise that we are concerned with signals which are zero before $t = 0$, and by $f(0^+)$ we mean the value of $f(t)$ just to the right of the origin. For example in the case of the step function of Figure 181(a), $f(0) = 1$, and similarly for the exponential function of Figure 181(b).

However in practice many signals do not have, even approximately, the very sharp initial rise shown in Figure 181 and for the sort of realistic pulse shown in Figure 38(c), for example, we have $f(0) = 0$. Consequently equation **12.8** takes, for these sorts of pulses, the simpler form:

$$\mathcal{L}[\frac{d}{dt}f(t)] = s\mathcal{L}[f(t)] \qquad \textbf{12.9}$$

We have been concerned up to now with finding the Laplace transforms of various functions. We shall find when applying the technique to real problems that we will also have to go back again from the s functions to the t functions, that is take the inverse Laplace transform. Usually this will be quite simple and we will be able to do it simply by recognizing the function. For example if we are faced with the function $1/(s + a)$ we recognize it as the Laplace transform of $\exp(-at)$. We write this inverse process as

$$\mathcal{L}^{-1}[1/(s + a)] = \exp(-at).$$

More complicated functions can often be broken up into the sum of simpler functions, using the method of partial fractions. For example if we wanted to obtain the inverse transform of $1/(s^2 + as)$ we would write it as $1/s(s + a)$ and then as $\frac{1}{a}(\frac{1}{s} - \frac{1}{s + a})$. We recognize this as the Laplace transform of $(1/a)\{1 - \exp(-at)\}$ since the transform of a unit step is $1/s$ and the transform of $\exp(-at)$ is $1/(s + a)$.

Thus $\quad \mathcal{L}^{-1}[1/(s^2 + as)] = \{1 - \exp(-at)\}/a.$

(Having written $1/(s^2 + as)$ as $1/s(s + a)$ or $\frac{1}{s}(\frac{1}{s + a})$ we could also have obtained our result from equation **12.7** - try it!)

Although these simple methods will suffice for our purposes, it is worth mentioning that for the inverse transform of more complicated products there exists a useful formula which we give here without proof:

$$\mathcal{L}^{-1}[F_1(s)F_2(s)] = \int_0^t f_1(\tau)f_2(t - \tau)\,d\tau \qquad \textbf{12.10}$$

This equation merely says that if we have a function of s which we recognize as the product of two functions $F_1(s)$ and $F_2(s)$ both of whose inverse transforms we know ($f_1(t)$ and $f_2(t)$) then the inverse transform of the product is given by the integral on the right. Here τ is a dummy variable. Although we would not normally use this complicated method to tackle our previous simpler function $1/(s^2 + as)$, let us do so now merely as an illustration of the method. $1/(s^2 + as) = (1/s)\{1/(s + a)\}$ and we put $F_1(s) = 1/s$ and $F_2(s) = 1/(s + a)$. Thus $f_1(t) = 1$ and $f_2(t) = \exp(-at)$. Now $f_1(\tau)$ is got by changing t everywhere into τ in $f_1(t)$ which still leaves it equal to unity: $f_2(t - \tau)$ is similarly obtained by changing t into $t - \tau$ in $\exp(-at)$ to give $\exp\{-a(t - \tau)\}$.

Hence using equation **12.10**

$$\mathcal{L}^{-1}[1/s^2 + as] = \int_0^t 1.\exp\{-a(t-\tau)\}\, d\tau$$

$$= \int_0^t \exp(-at)\exp(a\tau)\, d\tau$$

$$= \exp(-at)\int_0^t \exp(a\tau)\, d\tau$$

(since τ not t is being integrated)

$$= \exp(-at)[\exp(a\tau)]_0^t/a$$

$$= \exp(-at)\{\exp(at) - 1\}/a$$

$$= \{1 - \exp(-at)\}/a \quad \text{as before.}$$

The method will, of course work just as well if we put $1/(s+a)$ equal to $F_1(s)$ and $1/s$ equal to $F_2(s)$.

12·3 Simple examples of transform techniques

We are now in a position to discuss again our illustrative circuit of Figure 180. To underline the fact that the current i changes with time (that is, is a function of t) and that the voltage v may also do so, we rewrite equation **12.1** as:

$$v(t) = R\, i(t) + (1/C)\int_0^t i(t)\, dt \qquad \textbf{12.10}$$

Let us take the Laplace transform of both sides of this equation. We thus obtain:

$$v(s) = R\, i(s) + (1/C)\{i(s)\}/s. \qquad \textbf{12.11}$$

Here $v(s)$ represents the Laplace transform of $v(t)$ and similarly $i(s)$ represents the transform of $i(t)$, while we have treated the transform of $\int_0^t i(t)\, dt$ according to equation **12.7**, to give $i(s)/s$. Note the immense simplification we have achieved in going from equation **12.10** to equation **12.11**. We have changed an equation involving an integration into a simple algebraic equation. (The same would obviously have been true for an equation involving a differentiation.) This is the real power of the transform method in circuit theory. Equation **12.11** can now be solved for $i(s)$ to yield:

$$i(s) = \frac{v(s)}{R + (1/Cs)}. \qquad \textbf{12.12}$$

If we take the inverse transform of both sides we obtain:

$$\mathcal{L}^{-1}[i(s)] = i(t) = \mathcal{L}^{-1}\left[\frac{v(s)}{R + (1/Cs)}\right].$$

Thus $i(t)$ can in principle be found if we are told what $v(t)$, and hence what $v(s)$ is. Suppose $v(t)$ turns out to be a step function of height V_0: its transform is V_0/s. Thus

$$i(t) = \mathcal{L}^{-1}\left[\frac{V_0/s}{R + (1/Cs)}\right] = \mathcal{L}^{-1}\left[\frac{V_0}{R} \cdot \frac{1}{s + (1/RC)}\right]$$

where we have arranged matters so that the function of s on the right-hand side resembles one we already are familiar with. Clearly $i(t) = (V_0/R)\exp(-t/RC)$, and the voltage v_1 across R, if we require it, is $V_0 \exp(-t/RC)$.

We have taken a long time to obtain this elementary result, but in fairness it must be said that we can, once we know the rules, short circuit a lot of the working. To see this let us look again at equation **12.12** in conjunction with our original Figure 180. Equation **12.12** is really a statement, in the s notation, of Ohm's law, that is that the current $i(s)$ is given by the voltage $v(s)$ divided by the total impedance, R plus $1/Cs$. (It is no coincidence, as we shall see later, that the generalized impedance of the capacitor, $1/Cs$, resembles the quantity $1/jC\omega$ used in a.c. theory.) Conversely, when we have once explored the full mathematics of the problem, we can write down directly equation **12.12** from an examination of Figure 180. The rest of the solution then follows as before.

We will do one further simple example of the same type. Let us find the effect of applying a step function of voltage V_0 to the circuit of Figure 183. We can of course write down the complete mathematics of the problem, including our knowledge that the voltage across an inductor is given by $L\frac{di}{dt}$, and it is instructive to do this as a check. However we can, in line with out previous shortened approach, state immediately that $i(s)$ is given by:

$$i(s) = \frac{v(s)}{R + Ls}$$

where we have written the generalized impedance of the inductor as Ls from our previous discussion about the capacitor and our knowledge of the use of the quantity $jL\omega$ in a.c. theory. If we are interested in an output voltage across the resistor we have immediately

$$v_{out}(s) = i(s)R$$

or $$v_{out}(s) = \frac{R\,v(s)}{R + Ls}.$$

If $v(t)$ is a step function of size V_0 then $v(s) = V_0/s$ and

$$v_{out}(s) = \frac{R\,V_0}{s(R + Ls)} = V_0\left(\frac{1}{s} - \frac{1}{s + (R/L)}\right)$$

where as before we have used partial fractions to simplify the expression in s.

Remembering our results for the transform of a step function and an exponential, we see at once that:

$$v_{out}(t) = V_0\{1 - \exp(-R/L)t\}$$

in line with equation **3.11**.

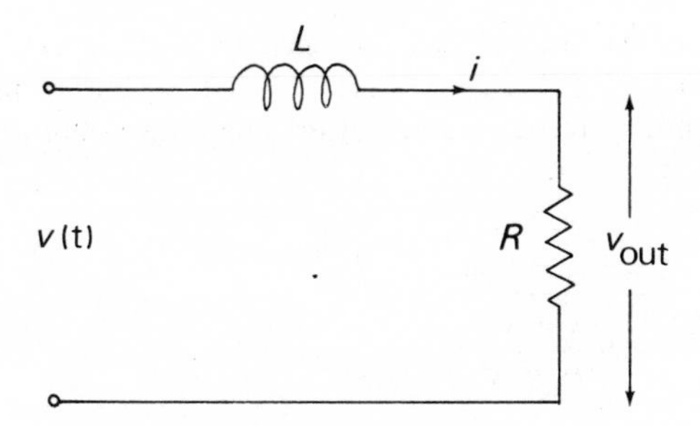

Figure 183. LR circuit for analysis.

12·4 Pole-zero cancellation

We give here one more example of a familiar circuit treated by the transform method, before turning to some newer topics. This is the pole-zero cancellation circuit of Figure 43. Figure 184 shows this redrawn in its *s*-notation form. If we write Z for the generalized impedance of R_1 and $1/C_1s$ in parallel, then clearly

$$v_{out}(s) = v(s).\frac{R_2}{R_2 + Z}.$$

Since $Z = (R_1^{-1} + C_1s)^{-1}$ we obtain:

$$v_{out} = v(s)\frac{R_2}{R_2 + (R_1^{-1} + C_1s)^{-1}}.$$

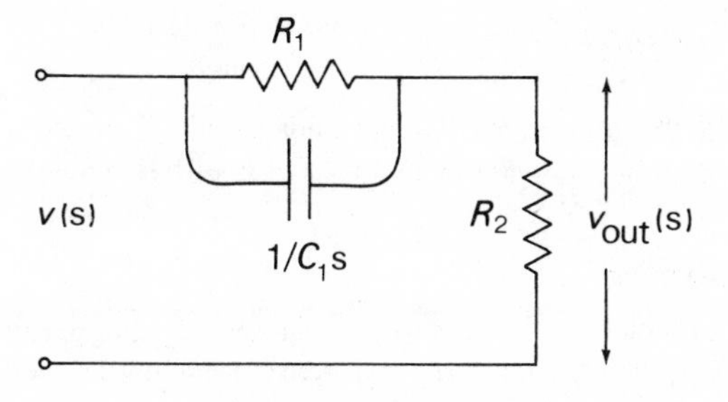

Figure 184. Circuit for pole-zero cancellation.

After tidying up this expression becomes:

$$v_{out} = v(s).\frac{s + (1/R_1C_1)}{s + \dfrac{R_1 + R_2}{R_1R_2C_1}}.$$

Recalling that the basic problem in pole-zero cancellation is to convert an exponentially decaying signal of size V_0 and time constant T_F into a more rapidly decaying signal without undershoot, we put the input signal $v(s)$ equal to $\frac{1}{s + (1/T_F)}$. Thus

$$v_{out}(s) = \frac{V_0}{s + (1/T_F)} \cdot \frac{s + (1/R_1C_1)}{s + \dfrac{R_1 + R_2}{R_1R_2C}}.$$

We have two terms of the form $s + \alpha$ on the bottom of the right-hand side and thus expect in general two exponential time constants in our answer. However if we arrange for the 'zero' term, $s + 1/(R_1C_1)$ to cancel with the 'pole' term $s + 1/T_F$ we will be left with the single decaying exponential we seek. Clearly this means choosing the components of our circuit so that $R_1C_1 = T_F$. Then

$$v_{out}(s) = \frac{V_0}{s + \dfrac{R_1 + R_2}{R_1R_2C_2}} = \frac{V_0}{s + (1/kR_1C_1)} \qquad \text{where } k = R_2/(R_1 + R_2)$$

and is less than one. Since we have previously put $R_1C_1 = T_F$ we finally have

$$v_{out}(s) = \frac{V_0}{s + (1/kT_F)}.$$

Taking the inverse transform we obtain our result

$$v_{out}(t) = V_0 \exp(-t/kT_F).$$

This is an exponentially decaying signal, falling more rapidly than the input signal ($k < 1$), and with no other exponential present, that is with no undershoot.

12·5 Current amplifier and integrator

Figure 185 shows an amplifier fed with a signal current i_{in} and producing an output voltage v_{out} - recalling in fact the transresistance amplifier we have met briefly before. Z_{in} is a generalized shunt impedance and would normally represent the parallel combination of the input resistance r_{in} and the input

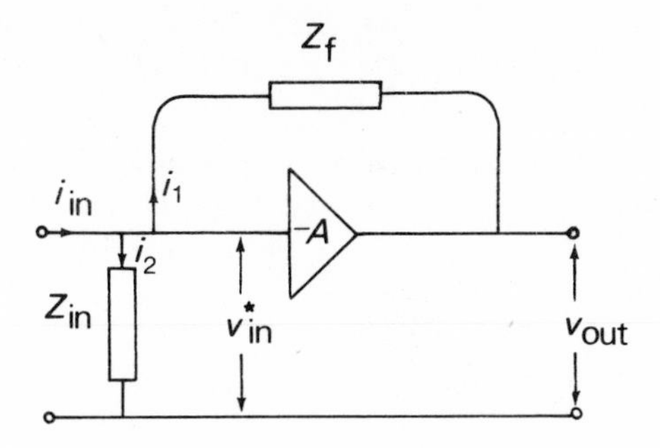

Figure 185. Current amplifier.

capacitance C_{in} of the complete amplifier. Z_f is a generalized feedback impedance, but in the present case we will be also taking it as a resistance R_f and capacitor C_f in parallel. The idealized part of the amplifier is assumed to have a voltage gain of $-A$ which is numerically very large and (since we have already included r_{in} in Z_{in}) to have an infinite input impedance. We can write down the relevant current and voltage relationships as follows:

$$i_{in} = i_1 + i_2$$

$$v^*_{in} - v_{out} = i_1 Z_f$$

$$v^*_{in} = i_2 Z_{in}$$

$$v_{out} = -A v^*_{in}.$$

(We remember we are working with the s terminology, and that we must use for Z_f and Z_{in} the appropriate functions of s.) Eliminating in turn i_1, i_2, and v^*_{in} we obtain for v_{out}

$$v_{out}(s) = \frac{-Z_f\, i_{in}(s)}{1 + \{1 + (Z_f/Z_{in})\}/A}.$$

Since A is assumed to be very large we have finally, to a close approximation:

$$v_{out}(s) = -Z_f\, i_{in}(s) \qquad \textbf{12.13}$$

(Indeed we could have easily guessed this result from our discussion in paragraph 6·5 about the action of high gain operational amplifiers in reducing v^*_{in} almost to zero.) Remembering that Z_f consists of a resistance R_f and a capacitor C_f in parallel we can write $Z_f = (R_f^{-1} + C_f s)^{-1}$, that is

$$v_{out}(s) = -(R_f^{-1} + C_f s)^{-1}\, i_{in}(s) = \frac{-i_{in}(s)}{C_f\{s + (1/R_f C_f)\}} \qquad \textbf{12.14}$$

If for simplicity we take the current i_{in} to have the exponentially decaying form $i_0 \exp(-t/T_F)$ as we might obtain from a scintillation counter, then $i_{in}(s) = i_0/(s + 1/T_F)$ and **12.14** becomes

$$v_{out}(s) = -\frac{i_0}{C_f}\left(\frac{1}{s + 1/T_F}\right)\left(\frac{1}{s + 1/R_f C_f}\right).$$

Using partial fractions we obtain

$$v_{out}(s) = -\frac{i_0}{C_f}\left(\frac{1}{R_f C_f} - \frac{1}{T_F}\right)^{-1}\left(\frac{1}{s + 1/T_F} - \frac{1}{s + 1/R_f C_f}\right) \qquad \text{and hence}$$

$$v_{out}(t) = -\frac{i_0}{C_f}\left(\frac{1}{R_f C_f} - \frac{1}{T_F}\right)^{-1}\{\exp(-t/T_F) - \exp(-t/R_f C_f)\} \qquad \textbf{12.15}$$

Just what this amplifier does depends very much on the relative values of R_f and C_f. Suppose first of all that C_f is small - perhaps just the stray capacity - then for reasonable values of R_f it is clear that $R_f C_f \ll T_F$. One exponential term in equation **12.15** is thus negligible and it reduces to

$$v_{out}(t) \simeq -i_0 R_f \exp(-t/T_F) \qquad \textbf{12.16}$$

The amplifier thus operates as a conventional feedback transresistance amplifier, converting an input current $i_0\exp(-t/T_F)$ into a proportional output voltage $-i_0 R_f \exp(-t/T_F)$ by means of the feedback resistor R_f. Suppose on the other hand that C_f and R_f are of such a size that $R_f C_f \gg T_F$ (for example C_f might be small but not negligible and R_f very large). In that case **12.14** reduces to

$$v_{out}(t) = -(i_0 T_F/C_f)\exp(-t/R_f C_f) \qquad \textbf{12.17}$$

Since C_f appears to be playing the dominant role here, we might suspect that we have some form of integrating device and this is indeed so. To see this let us remember that the total charge Q flowing into the amplifier from the signal $i_{in} = i_0 \exp(-t/T_F)$ is given by $Q = \int_0^\infty i\, dt = \int_0^\infty i_0 \exp(-t/T_F)\, dt = i_0/T_F$. Thus we can write **12.16** as

$$v_{out}(t) = -(Q/C_f)\exp(-t/R_f C_f)$$

and we recognize our device as the current integrator or charge sensitive amplifier that we discussed, without mathematical detail, at the very end of Chapter 6. This device produces a voltage proportional to the total charge in the pulse (and independent of the stray input capacity) and then allows it to decay away with time constant $R_f C_f$ in readiness for the arrival of the subsequent pulse.

It is possible to show that the circuit behaves in a similar fashion for any shape of pulse and not just an exponentially falling one. To do this however means combining equation **12.14** with the general relationship of **12.10**, and we will not deal with it here. At least we can see quickly that equation **12.14** reduces, for values of $R_f C_f$ large compared with the pulse lengths in question to

$$v_{out}(s) = (1/C_f)\frac{i_{in}(s)}{s}$$

and therefore from our knowledge of the Laplace transform of an integral, we can return to the time domain as follows

$$v_{out}(t) = (1/C_f)\int_0^t i_{in}(t)\,dt.$$

So we have shown (at least to the approximation we are now using) that our circuit does indeed integrate all varieties of pulses, though in fact we saw this just as simply in Section 6·5.

12·6 Gaussian shaping

We mentioned at the end of Chapter 11 that pulses of Gaussian shape are significantly better than simple *RC* shaped pulses from the point of view of noise. We indicate here, as a further example of the use of the Laplace transform, how pulses approximating to this shape can be generated. Figure 186 shows the circuit involved. It consists of a single *CR* differentiating circuit followed by n integrating circuits with the same values of R and C as in the differentiating circuit.

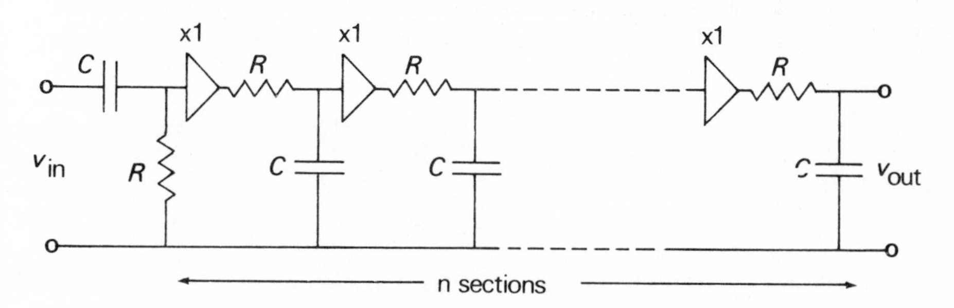

Figure 186. Circuit for Gaussian shaping.

Between each section there is an amplifier of unity gain, high impedance input and low output impedance, to prevent any section loading or interacting with the preceeding one. We shall first attempt to show that the output signal corresponding to an input step of voltage V_0 is $v_{out}(t) = V_0\{(t/RC)^n \exp(-t/RC)\}/n!$ This we do as follows. The 'transfer function' of the complete circuit, that is the ratio of v_{out} to v_{in} expressed in terms of s, is clearly

$$v_{out}(s)/v_{in}(s) = \left(\frac{R}{R + 1/Cs}\right)\left(\frac{1/Cs}{1/Cs + R}\right)^n .$$

Tidying up this equation we get

$$v_{out}(s)/v_{in}(s) = \left(\frac{RCs}{1 + RCs}\right)\left(\frac{1}{1 + RCs}\right)^n = \frac{RCs}{(1 + RCs)^{n+1}} .$$

Now if $v_{in}(t)$ is a step function of size V_0 then $v_{in}(s) = V_0/s$ and

$$v_{out}(s) = \frac{V_0 RC}{(1 + RCs)^{n+1}} = \frac{V_0}{(RC)^n(s + 1/RC)^{n+1}} .$$

We shall not try to perform an inverse Laplace transformation on this function: instead we shall attempt the simpler task of showing that the Laplace transform of the expression for $v_{out}(t)$ we quoted earlier gives indeed the value of $v_{out}(s)$ above. We do this as follows.

$$\mathcal{L}\{V_0(t/RC)^n \exp(-t/RC)\}/^n!$$

$$= \int_0^\infty \{V_0(t/RC)^n \exp(-t/RC)\}/n! \exp(-st)\, dt$$

$$= (V_0 a^n/n!) \int_0^\infty t^n \exp\{-(s + a)t\} dt$$

(where for convenience we have written $1/RC$ as a).

In order to integrate by parts we write this as follows:

$$-\{V_0 a^n/n!(s + a)\} \int_0^\infty t^n\, d \exp\{-(s + a)t\}$$ which of course equals

$$-\{V_0 a^n/n!(s+a)\} [t^n \exp\{-(s+a)t\}]_0^\infty$$

$$+ \{V_0 a^n/n!(s + a)\} \int_0^\infty \exp\{-(s + a)t\} nt^{n-1}\, dt.$$

The first term vanishes at $t = 0$ and $t = \infty$ and the second reduces to

$$\{V_0 a^n/n - 1!\, (s + a)\} \int_0^\infty t^{n-1} \exp\{-(s + a)t\}\, dt.$$

The integral resembles the one we started with, but with n changed to $n-1$, so by repeating the integration by parts n times in all we eventually arrive at the expression:

$$\{V_0 a^n/0!\,(s+a)^n\}\int_0^\infty t^0 \exp\{-(s+a)t\}\,dt \qquad \text{or,}$$

$$\{V_0 a^n/(s+a)^n\}\int_0^\infty \exp\{-(s+a)t\}\,dt.$$

Performing the final integration we obtain

$$\mathcal{L}\{V_0(t/RC)^n \exp(-t/RC)\}/n! = V_0 a^n/(s+a)^{n+1}$$

which on putting $a = 1/RC$ again, becomes

$$V_0/(RC)^n(s+1/RC)^{n+1} \quad \text{as required.}$$

The shape of this output pulse, $(V_0/n!)(t/RC)^n \exp(-t/RC)$ depends obviously on the value of n. We have already met some of these shapes. For $n = 0$ (that is one differentiating section and no integrating sections at all) we get $v_{out}(t) = (V_0/0!)(t/RC)^0 \exp(-t/RC)$ or $V_0\exp(-t/RC)$, a simple exponential. For $n = 1$ (that is one differentiating section and one integrating section) we get $(V_0/1!)(t/RC)^1\exp(-t/RC)$ or $v_{out}(t) = (V_0/RC)\, t \exp(-t/RC)$. This is a pulse we met previously in Section 3·7, and in contrast to the sharp rise of the pulse with $n = 0$ is one with a smooth rise and fall. Its maximum height occurs at $t = RC$ and is of size $V_0 \exp(-1)$ or only about 37 per cent of that for the case $n = 0$. As we increase n further the pulse becomes more symmetrical in its rise and fall about its peak position (which can easily be seen to occur at $t = nRC$). The peak height continues to reduce. Eventually for very large n the curve can be shown to approximate closely to a symmetrical Gaussian shape (see reference 4) and this is how the name for this type of shaping arises. Naturally in practice it is not feasible to go to very large values of n, but circuits with $n = 4$ are common, and offer some improvement from the point of view of signal to noise ratio over simple shaping systems.

12·7 Laplace transforms for sinusoidal signals

To indicate the use of transform methods with sinusoidal signals we will examine the result of applying such a signal to the simple circuit of Figure 187. First we must obtain the Laplace transform of $V_0\sin\omega t$, and although we can do this in a straightforward fashion, we can simplify matters by looking instead at the transform of $\exp(j\omega t) = \cos\omega t + j\sin\omega t$. The imaginary part of our result will then give the transform of $\sin\omega t$ and the real part will give (and we require it later) the transform of $\cos\omega t$. Thus

$$\begin{aligned}\mathcal{L}[\exp(j\omega t)] &= \int_0^\infty \exp(j\omega t)\exp(-st)\,dt \\ &= \int_0^\infty \exp(j\omega t - st)\,dt \\ &= \{1/(j\omega - s)\}[\exp(j\omega - s)t]_0^\infty .\end{aligned}$$

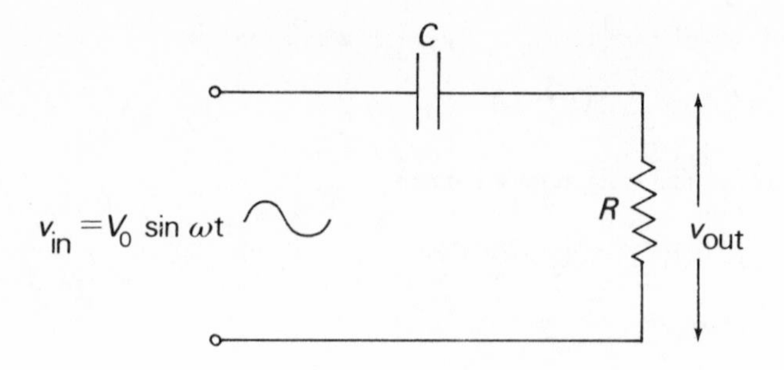

Figure 187. Circuit for analysis using sinusoidal signals.

The $\exp(-st)$ part of the term ensures that the quantity in the square brackets vanishes at $t = \infty$, and at $t = 0$ its value is unity. Therefore

$$\mathcal{L}[\exp(j\omega t)] = -1/(j\omega - s) = 1/(s - j\omega) = (s + j\omega)/(s^2 + \omega^2)$$

and thus

$$\mathcal{L}[\sin\omega t] = \omega/(s^2 + \omega^2) \quad \text{and} \quad \mathcal{L}[\cos\omega t] = s/(s^2 + \omega^2).$$

We can now tackle our problem. As usual

$$v_{out}(s) = v_{in}(s)R/(R + 1/Cs) = v_{in}(s)\{s/(s + 1/RC)\}$$

and putting in the value for $v_{in}(s)$

$$v_{out}(s) = \frac{V_0\omega}{s^2 + \omega^2} \cdot \frac{s}{s + 1/RC}.$$

Before we can put this into partial fractions we must write it as:

$$v_{out}(s) = \frac{V_0\omega}{(s + j\omega)(s - j\omega)} \cdot \frac{s}{s + a}$$

(where in addition we have written $1/RC$ as a for convenience).

This can be split into partial fractions as follows:

$$v_{out}(s) = \frac{A}{s + j\omega} + \frac{B}{s - j\omega} + \frac{C}{s + a} \qquad 12.18$$

and a simple calculation then gives

$$A = V_0\omega(a + j\omega)/2(\omega^2 + a^2) \quad B = V_0\omega(a - j\omega)/2(\omega^2 + a^2) \quad C = -V_0\omega a/(\omega^2 + a^2).$$

Clearly from **12.18**

$$v_{out}(t) = A \exp(-j\omega t) + B \exp(+j\omega t) + C \exp(-at)$$

or putting in the values for A, B and C

$$v_{out}(t) = \{V_0\omega/(\omega^2 + a^2)\} \times$$
$$\{\tfrac{1}{2}(a + j\omega)\exp(-j\omega t) + \tfrac{1}{2}(a - j\omega t)\exp(j\omega t) - a \exp(-at)\}$$

Now putting $\exp(j\omega t) = \cos\omega t + j \sin\omega t$ and similarly $\exp(-j\omega t) = \cos\omega t - j \sin\omega t$ and tidying up, we obtain:

$$v_{out}(t) = \{V_0\omega/(\omega^2 + a^2)\} \{a \cos\omega t + \omega \sin\omega t - a \exp(-at)\}.$$

We can finally write this as:

$$v_{out}(t) = \{V_0\omega/(\omega^2 + a^2)\} \{(\omega^2 + a^2)^{1/2} \sin(\omega t + \delta) - a \exp(-at)\} \qquad \textbf{12.19}$$

where $\delta = \tan^{-1} a/\omega$ (which can be checked by re-expanding $\sin(\omega t + \delta)$). Note incidentally that $v_{out}(t) = 0$ at $t = 0$.

Before we interpret this result let us look at what we would have got by the $j\omega$ method usually used in a.c. theory. Clearly here our voltage output would be:

$$v_{out} = V_0R/(R + 1/jC\omega) \quad \text{which can be written}$$

$$v_{out} = V_0/(1 - ja/\omega) \qquad \text{putting as usual } a = 1/RC$$

which in turn becomes

$$v_{out} = V_0(1 + ja/\omega)/\{1 + (a/\omega)^2\}.$$

The amplitude of the resultant signal is thus $\{1 + (a/\omega)^2\}^{-1/2}$ or $\omega/(\omega^2 + a^2)^{1/2}$ while its phase angle is given by $\tan \delta = a/\omega$.

If we now look back at equation **12.19** we see that the first term of this equation represents a sinusoidal signal with the same amplitude and phase as predicted by our $j\omega$ method. What does the second term in equation **12.19** represent? Clearly it represents a transient signal which dies away with a time constant of $1/a = RC$. After a time long compared with RC, we are left only with the steady sinusoidal solution given also by the $j\omega$ technique. We must now explain the presence of this second term. We obtain the Laplace transform of a function by integrating between zero and infinity, and we implicitly assume that our function is zero before $t = 0$. Thus what we impressed on our circuit was not a sinusoidal wave going from $-\infty$ to $+\infty$ in time, but one starting suddenly at $t = 0$, as indeed would occur in practice. The transient part of our answer thus corresponds to the effect of this sudden start of the signal. The $j\omega$ method can only predict the 'steady state' conditions which remain after a time long compared with the circuit time constant RC. Clearly more information is available from the Laplace transform method, and it is essential to use it if we are interested in the transient response of circuits to suddenly applied sinusoidal signals.

References and Further Reading

1. G. W. OWEN and P. W. KEATON, *Fundamentals of Electronics,* Vol. 1, Harper, 1966.
2. M. R. SPIEGEL, *Theory and Problems of Laplace Transforms,* Schaum, 1965.
3. R. L. KONIGSBERG, 'Operational Amplifiers', *Advances in electronics and electron physics,* vol. XI, 1959, p. 225.
4. E. FAIRSTEIN and J. HAHN, 'Nuclear pulse amplifiers – fundamentals and design practice', *Nucleonics,* vol. 23, no. 11, 1965, p. 50.

CHAPTER 13

Further topics in noise theory

13·1 Introduction

When we discussed shot noise in Chapter 11 we recognized that its presence was due to the fact that a current consists of the flow of a series of discrete charges, yet in the end our final expression, $\overline{i^2} = 2e\bar{I}df$, expressed as it was in terms of a mean square current fluctuation over a small frequency range, tended to obscure the basic processes at work. In this chapter we take a different approach (reference 1) and concentrate directly on estimating the effect at the output of an amplifier of the movement of the electrons at its input. Although the method will apply primarily to shot noise we will see how also to take account of thermal and other noise. In our attack on this problem we will make use of some of the Laplace transform techniques developed in the previous chapter, but in addition will need to apply some results which are contained in what are known as Campbell's theorems. We start, then, with a discussion of these.

13·2 Campbell's theorems

Campbell's theorems tell us how to calculate the response of a system (for example a voltage amplifier) to a train of events (for example a random stream of electrons arriving at its input) when we know the response of the system to a single event. Expressing this mathematically, if $f(t)$ is the output voltage of the system corresponding to a single event at the input, and if $\bar{n}$ is the mean counting rate (that is, the average number of events per second) then Campbell's theorem of the mean tells us that the average output voltage is given by

$$\bar{V} = \bar{n} \int_0^\infty f(t)\, dt \qquad \textbf{13.1}$$

and Campbell's theorem of the mean square tells us that the mean square output voltage is given by:

$$\overline{(V - \bar{V})^2} = \bar{n} \int_0^\infty \{f(t)\}^2\, dt \qquad \textbf{13.2}$$

(The quantity $\overline{(V - \bar{V})^2}$ is of course what we normally call $\overline{v^2}$.)
For a detailed discussion of these theorems reference 2 should be consulted. However in Appendix 1 to this chapter a plausible version of their proof is given. We now apply Campbell's theorems to the problem of optimum pulse shape in amplifiers.

13·3 Pulse shape for optimum ENC

In Chapter 11 we discussed the problem of noise in an amplifier with *RC–CR* shaping, indicated that the best situation was when the rise and fall times had a common value τ, and found a value for τ which gave the minimum equivalent noise charge (ENC). However we mentioned there the question of whether other forms of pulse shaping rather than the *RC–CR* type would give better results, and we now return to it. In fact we will attempt to deduce the pulse shape which gives the optimun value of ENC.

We start from the basic situation previously discussed in Section 11.6 and illustrated in Figure 177. An FET, used for the first stage of amplification of signals from a nuclear radiation detector, has a gate resistor of size R_g in parallel with a capacity C_{in}. As usual noise will arise from thermal effects in R_g, shot effect in the gate current I_g and thermal effects in the FET channel. (We will assume for simplicity that the contribution from $1/f$ noise is small.) In line with our discussion earlier in this chapter, we define $f(t)$ as the output voltage from the amplifier when unit charge arrives on the input. (We think in terms of charge on the input because ultimately we want to return to the concept of equivalent noise charge (ENC) which we introduced in Chapter 11.) The arrival of this charge on the input is mathematically equivalent to a unit step of charge placed on C_{in}, producing naturally a step of voltage there. Such a step (though not of unit size) will arise for example, when an electron of charge e from the gate current arrives on C_{in}.

We also want to discuss the effect of a delta function of charge on the input, for the following reason. We shall be considering the passage of a single unit charge through the channel of the FET. This corresponds to a very rapid rise in the current there, and its equally rapid decay after a short time. We may take as an approximation suitable for our calculations that this can be represented as an infinitely large current lasting for an infinitesimal time, such that $\int_0^\infty I\,dt = 1$ (for unit charge) – in fact a delta function. In line with our usual practice of referring everything to the input we can think of this effect in terms of a delta function voltage at the input, or equally well in terms of a delta function of charge on the input stray capacity. What will be the output of the amplifier corresponding to such a step? We can obtain the answer informally if we recognize that a delta function can be thought of as the derivative with respect to time of a unit step function. In the case of the latter the slope is zero before we come to the actual step: after we pass the step the slope is also zero, and at the actual step the slope is momentarily infinite. Thus a plot of the slope of a step function (that is, its derivative) gives us a delta function, as stated. Now we have defined $f(t)$ as the response of our system to a unit step of charge: therefore since we assume the system linear, we expect that it will respond to the derivative of a step function (that is a delta function) by the derivative of $f(t)$, that is $f'(t)$.

We can express the same idea more mathematically as follows. If $g(s)$ is the transfer function of our system, then since we know that the Laplace transform of unit step is $1/s$ and that of a delta function is unity we have:

$$v_{\text{out}}(s) \text{ for a step} = (1/s)\,g(s) \tag{13.3}$$

and $$v_{\text{out}}(s) \text{ for a delta function} = 1\cdot g(s). \tag{13.4}$$

But we have already defined $f(t)$ as the output voltage corresponding to a unit step input, therefore

$$\mathcal{L}[f(t)] = (1/s)\,g(s). \tag{13.5}$$

Now the Laplace transform of the derivative of a function is given by

$$\mathcal{L}[f'(t)] = s\mathcal{L}f(t) \quad \text{(provided } f(0) = 0\text{, as we will assume).}$$

Now in our particular case $\mathcal{L}[f(t)] = (1/s)\,g(s)$ from **13.5**, therefore

$$\mathcal{L}[f'(t)] = s \times (1/s)\,g(s) = g(s). \tag{13.6}$$

But from **13.4** $g(s)$ is just the output (in the s domain) from a delta function input, and equation **13.6** tells us that when we convert this back to the time domain we get $f'(t)$. Thus $f'(t)$ is the output from a delta function input.

13·3·1 *Shot noise from the gate current*

Let us now deal with the noise sources one by one, and we shall start on this occasion with the gate current. As we have defined $f(t)$ as the voltage at the output when unit charge arrives at the input, then for a single electron of charge e arriving there the output would be $ef(t)$. Using Campbell's theorem of the mean square we thus obtain

$$\overline{v^2} = \bar{n}\int_0^\infty \{ef(t)\}^2\,dt = \bar{n}\,e^2\int_0^\infty \{f(t)\}^2\,dt$$

where $\bar{n}$ is the average number of electrons arriving per second. But of course the current I flowing is simply $\bar{n}e$, and therefore

$$\overline{v^2} = Ie\int_0^\infty \{f(t)\}^2\,dt. \tag{13.7}$$

Remembering that in the present case we are talking of the gate current and gate noise we write this:

$$\overline{v_G^2} = I_g e\int_0^\infty \{f(t)\}^2\,dt \tag{13.8}$$

13·3·2 *Thermal noise in the gate resistor*

Our technique using Campbell's theorem can not be applied directly in this case because, of course, thermal noise is not due to a continuous unidirectional current as was the case with gate current noise. However a simple expedient will help us. Let us first turn the thermal noise into an equivalent shot noise to which we can apply our present methods. Now the thermal noise in the resistor can be written:

$$\overline{i^2} = 4kTdf/R$$

while the general equation for shot noise is given by

$$\overline{i^2} = 2eIdf.$$

Equating these two expressions we see that the thermal noise in a resistor of size R can be thought of in terms of the shot noise arising from an equivalent current $I = 2kT/eR$. In our case the equivalent current I_{eq} is equal to $2kT/eR_g$, where R_g is the gate resistor. Using then equation **13.7** we can write for the thermal noise in R_g

$$\overline{v_R^2} = I_{eq}e \int_0^\infty \{f(t)\}^2 \, dt \quad \text{or}$$

$$\overline{v_R^2} = (2kT/R_g) \int_0^\infty \{f(t)\}^2 \, dt \qquad \textbf{13.9}$$

13·3·3 *Thermal noise in the FET channel*

As in our earlier treatment of channel noise in Chapter 11, we regard the FET as generating noise like a resistor of size R_0, where R_0 is defined as in Section 11.6.3. Thus the noise is given by

$$\overline{i^2} = (4kT/R_0) \, df$$

As was the case for the thermal noise in R_g, we can not apply Campbell's theorem directly to this expression, but we can do so by again considering the noise as due to the shot noise in an equivalent current, given here by $I_{eq} = 2kT/eR_0$. We think of this equivalent current as a stream of electrons, each of charge e, passing through the FET channel at a rate of $\bar{n}$ per second. We wish to consider the passage of one of these electrons through the channel and transfer its effect to the input so as to be on an equal footing with the other noise contributions. To do this let us first consider the passage of a hypothetical unit charge. We have already discussed how we can approximate to this by a delta function of current. To convert this to an equivalent voltage at the gate we multiply by $1/g_m$, and to convert this to an equivalent charge we multiply by C_{in}, the input stray capacity. Remembering that the charge on the actual electrons in the channel is e and not unity, we finally think of the passage of a single electron through the channel as equivalent to a delta function of charge at the input, multiplied by the factor eC_{in}/g_m.

Now we have already seen that the effect of a delta function of charge at the input is to produce an output voltage of $f'(t)$. Therefore in our case the passage of an electron through the channel of the FET at the input of the amplifier will produce a voltage output of $(eC_{in}/g_m)\, f'(t)$. Applying Campbell's theorem we thus obtain for the mean square noise voltage the result:

$$\overline{v^2} = \overline{n} \int_0^\infty \{(eC_{in}/g_m)\, f'(t)\}^2\, dt \qquad \text{or since } I_{eq} = \overline{n}e$$

$$\overline{v^2} = I_{eq}(eC_{in}^2/g_m^2) \int_0^\infty \{f'(t)\}^2\, dt \quad \text{and putting in the value for } I_{eq}$$

$$\overline{v^2} = (2kTC_{in}^2/R_0\, g_m^2) \int_0^\infty \{f'(t)\}^2\, dt\,.$$

Finally writing $R_0 g_m^2$ as $1/R_{eq}$, where R_{eq}, as in Section 11.6.3 is the equivalent noise resistance of the channel, and identifying the noise as channel noise by using the appropriate suffix, we have:

$$\overline{v_C^2} = 2kTR_{eq}C_{in}^2 \int_0^\infty \{f'(t)\}^2\, dt\,.$$

13·3·4 *Minimization of the total noise*

Adding together the three noise contributions calculated in the previous paragraphs, we obtain for the total mean square noise voltage at the output:

$$\overline{v_{total}^2} = (I_g e + 2kT/R_g) \int_0^\infty \{f(t)\}^2\, dt + 2kTR_{eq}C_{in}^2 \int_0^\infty \{f'(t)\}^2\, dt$$

which we write for simplicity

$$\overline{v_{total}^2} = \mathcal{A}\int_0^\infty \{f(t)\}^2\, dt + \mathcal{B}\int_0^\infty \{f'(t)\}^2\, dt. \qquad \textbf{13.10}$$

Our task is now in some way to minimize the noise, by choosing the optimum form for $f(t)$, that is by selecting the most suitable type of pulse shaping in our amplifier. There are naturally some constraints which we must impose on $f(t)$ while attempting this minimization, and one of them appears when we remember the concept of equivalent noise charge (ENC), which is in fact the precise quantity we wish to minimize. The equivalent noise charge, as we defined it in Section 11.7, is the genuine charge signal on the input which will produce the same output as the mean square noise. If we call the equivalent noise charge Q, then the output voltage signal corresponding to it is $Qf(t)$, and if the peak value of $f(t)$, (occurring at a time t_P say), is written as $f(t_P)$, then our definition of ENC says that $\{Qf(t_p)\}^2 = \overline{v_{total}^2}$ or,

$$Q^2 = (\text{ENC})^2 = \overline{v_{total}^2}/\{f(t_p)^2\} \qquad \textbf{13.11}$$

In comparing various pulse shaping arrangements which not only shape the pulse but also change its height, we must in fairness alter the gain of the amplifier in each case until the peak pulse output reaches a standard height, and then compare the noise outputs. For simplicity we might arrange our output to be one volt, that is $f(t_P) = 1$. Equation **13.11** then tells us that if we calculate $\overline{v^2_{total}}$ we automatically obtain $(ENC)^2$, which is now numerically equal to it. This is what we will do, and so we impose the condition on our calculations that $f(t_P) = 1$. A second reasonable and obvious requirement is that the pulse must be zero for large values of t, that is, $f(\infty) = 0$, and we impose this condition also.

Furthermore we need only consider the integrals in equation **13.10** from t_P to ∞. This is because we can split the integrals in equation **13.10** into two parts, from 0 to t_P, and from t_P to ∞, with $f(t_P)$ held at unity. Now suppose as in Figure 118, that we had discovered that the solid curve to the right of t_P is the pulse shape that will minimize $\overline{v^2_{total}}$ from t_P to ∞. Then remembering that the two parts of the curve must meet at t_P and have the same value of unity there, and that integrals represent areas under curves, it should be clear that the mirror image of the right hand side of the curve is the appropriate shape for the left hand side. The overall shape must thus be symmetrical around t_P.

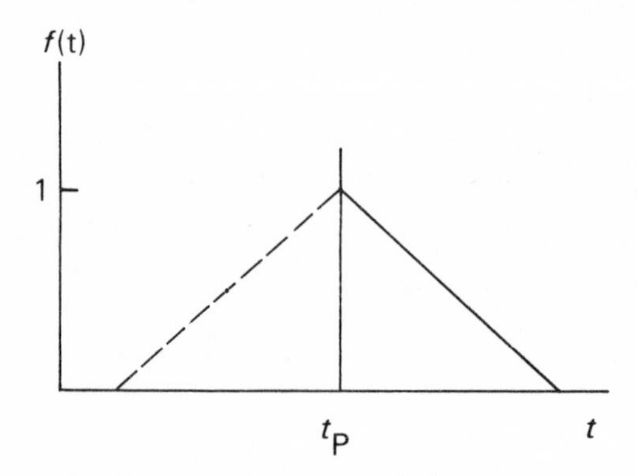

Figure 188. Symmetry of preferred pulse.

We are left then with the mathematical problem of minimizing the quantity

$$\phi = \mathcal{A}\int_{t_P}^{\infty}\{f(t)\}^2\,dt + \mathcal{B}\int_{t_P}^{\infty}\{f'(t)\}^2\,dt$$

subject to the end conditions $f(t_P) = 1$ and $f(\infty) = 0$. This is a problem in the calculus of variations, and for those familiar with this technique the answer can be written down almost directly:

$$\mathcal{A}f(t) = \mathcal{B}f''(t).$$

However we can obtain the same result in a plausible way as follows. Let us imagine that we are examining a number of pulse shapes (as in Figure 189) to try to find the optimum one from the point of view of noise: and let us further

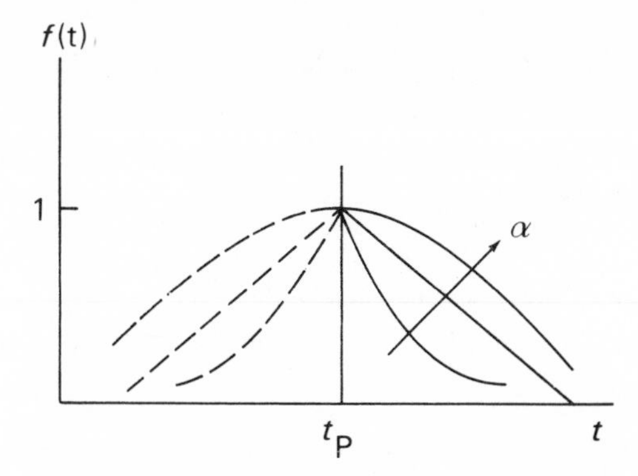

Figure 189. Possible pulse shapes.

imagine (and this is not in general true) that we can change from one to another by altering some parameter α as in the figure. Our problem then is to obtain the minimum of

$$\phi = \mathcal{A}\int_{t_P}^{\infty}\{f(t,\alpha)\}^2\,dt + \mathcal{B}\int_{t_P}^{\infty}\{f'(t,\alpha)\}^2\,dt$$

where we have written $f(t,\alpha)$ to indicate that the form of $f(t)$ may be varied by varying α. We do this by obtaining $\dfrac{d\phi}{d\alpha}$ and setting it equal to zero.

Now $$\frac{d\phi}{d\alpha} = \mathcal{A}\frac{d}{d\alpha}\int_{t_P}^{\infty}f^2\,dt + \mathcal{B}\frac{d}{d\alpha}\int_{t_P}^{\infty}f'^2\,dt$$

and since t and α can vary independently we can differentiate inside the integral to give

$$\frac{d\phi}{d\alpha} = \mathcal{A}\int_{t_P}^{\infty}2f\frac{df}{d\alpha}\,dt + \mathcal{B}\int_{t_P}^{\infty}2f'\frac{df'}{d\alpha}\,dt.$$

We can use the formula for integration by parts $u\,dv = uv - \int v\,du$ on the second integral, taking f' as u and $\dfrac{df'}{d\alpha}dt$ as dv, that is $\dfrac{dv}{dt} = \dfrac{df'}{d\alpha}$ or $v = \dfrac{df}{d\alpha}$ (by integrating with respect to t). Thus:

$$\frac{d\phi}{d\alpha} = 2\mathcal{A}\int_{t_P}^{\infty}f\frac{df}{d\alpha}dt + 2\mathcal{B}\left[f'\frac{df}{d\alpha}\right]_{t_P}^{\infty} - 2\mathcal{B}\int_{t_P}^{\infty}\frac{df}{d\alpha}f''\,dt. \qquad \textbf{13.12}$$

Now since, from our end conditions, all the possible curves for $f(t)$ come together at $t = t_P$ (where they are 1) and at $t = \infty$ (where they are zero) we must have $\dfrac{df}{d\alpha}$ vanishing at both places, and thus the quantity in square brackets in equation **13.12** disappears. So

$$\frac{d\phi}{d\alpha} = 2\int_{t_P}^{\infty}\frac{df}{d\alpha}(\mathcal{A}f - \mathcal{B}f'')\,dt.$$

And for $\dfrac{d\phi}{d\alpha}$ to be zero, as we require, we must have

$$\mathcal{A}f(t) = \mathcal{B}f''(t) \qquad \textbf{13.13}$$

as we previously stated.

13·3·5 *The optimum pulse shape*

From equation **13.13** it is clear that the pulse shape (from t_P to infinity) must be a decaying exponential in form, since this function has both the property that when differentiated twice it returns to its original form (apart from a constant) and will also satisfy the condition of vanishing after large values of the time. Thus $f(t) = C\exp(-t/\tau)$ where C and τ are constants. Now we have agreed to make $f(t) = 1$ at $t = t_P$, so $C\exp(-t_P/\tau) = 1$ and therefore

$$f(t) = \exp\{-(t - t_P)/\tau\}. \qquad \textbf{13.14}$$

We have not yet fully used the information in equation **13.13**. Substituting the value for $f(t)$ from **13.14** into **13.13** we obtain $\mathcal{A} = \mathcal{B}/\tau^2$ or

$$\tau = (\mathcal{B}/\mathcal{A})^{1/2}. \qquad \textbf{13.15}$$

Our optimum pulse shape has thus a cusp shaped form like the lower of the three curves in Figure 189, with the fall time given by equation **13.15**.

We can now calculate the minimum value for $\overline{v^2_{total}}$. This is obtained by substituting $f(t)$ as given by equations **13.14** and **13.15** into equation **13.10**. Thus:

$$\begin{aligned}
\overline{v^2_{min}} &= \mathcal{A}\int_0^{\infty}\{f(t)\}^2\,dt + \mathcal{B}\int_0^{\infty}\{f'(t)\}^2\,dt \\
&= 2\mathcal{A}\int_{t_P}^{\infty}\{f(t)\}^2\,dt + 2\mathcal{B}\int_{t_P}^{\infty}\{f'(t)\}^2\,dt \\
&= 2\mathcal{A}\int_{t_P}^{\infty}\exp\{-2(t - t_P)/\tau\}\,dt + (2\mathcal{B}/\tau^2)\int_{t_P}^{\infty}\exp\{-2(t - t_P)/\tau\}\,dt \\
&= 2(\mathcal{A} + \mathcal{B}/\tau^2)\int_{t_P}^{\infty}\exp\{-2(t - t_P)/\tau\}\,dt \\
&= 2(\mathcal{A} + \mathcal{B}/\tau^2)\tau\int_0^{\infty}\exp(-2u)\,du \qquad \text{(putting } (t - t_P)/\tau = u)
\end{aligned}$$

$$= \mathcal{A}\tau + \mathcal{B}/\tau$$

$$= \mathcal{A}(\mathcal{B}/\mathcal{A})^{1/2} + \mathcal{B}(\mathcal{A}/\mathcal{B})^{1/2} \quad \text{using the value of } \tau \text{ from equation } \mathbf{13.15}$$

$$= (\mathcal{A}\mathcal{B})^{1/2} + (\mathcal{A}\mathcal{B})^{1/2} \quad \text{(note how the two noise contributions have been equalized)}$$

$$= 2(\mathcal{A}\mathcal{B})^{1/2}$$

and from our previous discussion about the relationship between ENC^2 and $\overline{v^2_{\text{total}}}$ when $f(t_P)$ is made unity, we can also write

$$\text{ENC}^2_{\text{min}} = 2(\mathcal{A}\mathcal{B})^{1/2} \qquad \mathbf{13.16}$$

or $$\text{ENC}^2_{\text{min}} = 2\{(I_g e + 2kT/R_g)(2kTR_{\text{eq}}C^2_{\text{in}})\}^{1/2}. \qquad \mathbf{13.17}$$

13·3·6 *Noise with other pulse shaping*

We can now estimate how much improvement is obtained by using this ideal cusp shaped pulse, as compared with the *RC–CR* shaped pulse of Chapter 11. Equation **11.16** tells us that the minimum equivalent noise charge for that case is given by:

$$\text{ENC}^2_{\text{min}} = 2(LM)^{1/2}C_{\text{in}}$$

if we assume the contribution from $1/f$ noise negligible. Now the constants L and M used there are closely related to $\mathcal{A}$ and $\mathcal{B}$. In fact one can easily see that $L = \epsilon^2\mathcal{A}/4$ and $M = \epsilon^2\mathcal{B}/4C^2_{\text{in}}$. Therefore

$$\text{ENC}^2 = 2(\epsilon^2/4)(\mathcal{A}\mathcal{B})^{1/2} \quad \text{(for } RC\text{–}CR \text{ shaping)} \qquad \mathbf{13.16}$$

Thus $$\frac{\text{ENC}^2_{\text{min}}(RC\text{–}CR)}{\text{ENC}^2_{\text{min}}(\text{cusp})} = \epsilon^2/4$$

or $$\frac{\text{ENC}_{\text{min}}(RC\text{–}CR)}{\text{ENC}_{\text{min}}(\text{cusp})} = \epsilon/2 = 1.36$$

so there is a 36% increase in ENC when we change our ideal cusp shaped pulse to our previous *RC–CR* shaping.

Of course it is not physically possible to produce a perfect cusp shaped pulse, extending as it does to infinity on the left of the peak as well as on the right (nor indeed is it easy even to produce a good approximation to it). Consequently we will look briefly at two other possibilities, the triangular pulse and the Gaussian pulse. Figure 190 shows a triangular pulse (of unit peak height as we require for $f(t)$) and with its other dimension specified by the constant τ. (Such a triangular

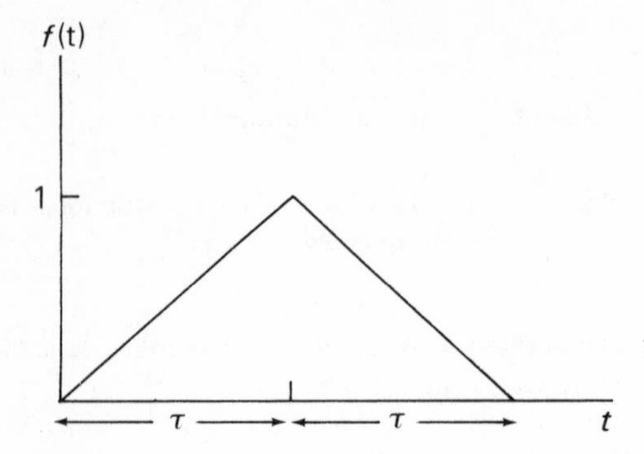

Figure 190. Triangular pulse.

pulse can be obtained in practice by integrating the double delay line shaped pulse shown in Figure 38(c).) As usual

$$\overline{v^2_{\text{total}}} = \mathcal{A}\int_0^\infty \{f(t)\}^2\,dt + \mathcal{B}\int_0^\infty \{f'(t)\}^2\,dt$$

where in this case $f(t) = t/\tau$ for $0 < t \leqslant \tau$

$$f(t) = 2 - t/\tau \quad \text{for} \quad \tau < t \leqslant 2\tau$$

and $f(t) = 0$ for $t > \tau$.

We are not, of course, seeking to minimize $\overline{v^2_{\text{total}}}$ here, but merely to deduce a value for it by substituting the expressions for $f(t)$ (and $f'(t)$) into the expression for $\overline{v^2_{\text{total}}}$, and working out the integrals. The integrations themselves are straightforward, and the result is:

$$\overline{v^2_{\text{total}}} = 2\mathcal{A}\tau/3 + 2\mathcal{B}/\tau.$$

We are now free to choose τ to obtain the smallest value of $\overline{v^2_{\text{total}}}$ and on differentiating the expression for it and putting the result equal to zero, we obtain:

$$2\mathcal{A}/3 - 2\mathcal{B}/\tau^2 = 0 \quad \text{or}$$

$$\tau = (3\mathcal{B}/\mathcal{A})^{1/2}.$$

Thus $$\overline{v^2_{\text{min}}} = (2\mathcal{A}/3)(3\mathcal{B}/\mathcal{A})^{1/2} + 2\mathcal{B}/(3\mathcal{B}/\mathcal{A})^{1/2}$$

$$= 2(\mathcal{A}\mathcal{B}/3)^{1/2} + 2(\mathcal{A}\mathcal{B}/3)^{1/2}$$

$$= 4(\mathcal{A}\mathcal{B}/3)^{1/2}$$

and ENC^2 has the same value $4(\mathcal{A}\mathcal{B}/3)^{1/2}$.

Comparing this with our ideal shaping we have:

$$\frac{\mathrm{ENC}^2_{\mathrm{min}}(\text{triangular})}{\mathrm{ENC}^2_{\mathrm{min}}(\text{cusp})} = \frac{4(\mathcal{A}\mathcal{B}/3)^{1/2}}{2(\mathcal{A}\mathcal{B})^{1/2}} = 2/\sqrt{3}$$

or
$$\frac{\mathrm{ENC}_{\mathrm{min}}(\text{triangular})}{\mathrm{ENC}_{\mathrm{min}}(\text{cusp})} = (2/\sqrt{3})^{1/2} = 1.075\,.$$

The triangular pulse is thus within 8% of the ideal case.

In a similar manner we can treat the shaping system we discussed in Section 12.6 which gave a pulse of the form $f(t) = V_0\{(t/RC)^n \exp(-t/RC)\}/n!$ (We noted incidentally that for large n this approximated to a symmetrical Gaussian shape.) The maximum value of this pulse occurs, as we saw in the earlier section, at $t = nRC$ and substituting back we obtain the peak height to be $f(t_\mathrm{P}) = V_0\{n^n \exp(-n)\}/n!$ Since we, as usual, put $f(t_\mathrm{P}) = 1$ we have $V_0\{n^n \exp(-n)\}/n! = 1$. Assuming we have a large number of stages in our shaping network, that is n large, we can write, using Stirling's formula $n! \sim (2\pi n)^{1/2} n^n \exp(-n)$. Hence

$$V_0/2\pi n)^{1/2} \sim 1. \qquad \mathbf{13.17}$$

We then proceed as usual to calculate

$$\overline{v^2_{\mathrm{total}}} = \mathcal{A}\int_0^\infty \{f(t)\}^2\,dt + \mathcal{B}\int_0^\infty \{f'(t)\}^2\,dt$$

and after some lengthy but not difficult integration, and the use of Stirling's formula again, one obtains:

$$\overline{v^2_{\mathrm{total}}} = \{V_0^2/(4\pi n)^{1/2}\}\{\mathcal{A}(RC) + \mathcal{B}/2nRC\}.$$

The quantity RC is at our disposal and by differentiation its optimum value is $RC = (\mathcal{B}/2n\mathcal{A})^{1/2}$ giving a minimum value for $\overline{v^2_{\mathrm{total}}}$ of

$$\overline{v^2_{\mathrm{min}}} = \{V_0^2/(2\pi n^2)^{1/2}\}(\mathcal{A}\mathcal{B})^{1/2}$$ and using equation **13.17** we obtain:

$$\overline{v^2_{\mathrm{min}}} = (2\pi\mathcal{A}\mathcal{B})^{1/2}.$$

Thus
$$\frac{\mathrm{ENC}_{\mathrm{min}}(\text{Gaussian})}{\mathrm{ENC}_{\mathrm{min}}(\text{cusp})} = \frac{(2\pi\mathcal{A}\mathcal{B})^{1/4}}{\{2(\mathcal{A}\mathcal{B})^{1/2}\}^{1/2}} = (\pi/2)^{1/4} = 1.12.$$

A Gaussian shaped pulse is thus somewhat worse for noise than a triangular shaped one, but is still considerably better than *RC–CR* shaping.

Finally we can calculate directly, using the methods of the present section, the value of the ENC for *RC–CR* shaping, even though of course we already have the answer using the methods of Chapter 11. Here $f(t) = \epsilon t \exp(-t/\tau)$ for a pulse with equal rise and fall times τ. The factor ϵ in front ensures that $f(t_P) = 1$. Applying the now familiar procedures for calculating $\overline{v^2_{total}}$, one obtains:

$$\overline{v^2_{total}} = (\epsilon^2/4)(\mathcal{A}\tau + \mathcal{B}/\tau)$$

and choosing a best value for τ of $(\mathcal{B}/\mathcal{A})^{1/2}$ one has

$$\overline{v^2_{min}} = (\epsilon^2/2)(\mathcal{A}\mathcal{B})^{1/2}$$

which agrees with our previous value for $\mathrm{ENC}^2_{min}(RC\text{–}CR)$ from Chapter 11.

13·4 Optimum *RC–CR* shaping

In the preceeding paragraph, and indeed in our original discussion of the matter in Section 11.5 we assumed intuitively that in using *RC–CR* shaping the best arrangement from the point of view of noise is one in which the rise and fall times of the amplifier are made equal. We indicate now how a rigorous proof of this may be obtained.

We know from our discussion in Section 3.5 that the output from an amplifier with fall time T_F (= T_1, say) and rise time T_R (= T_2, say) can be written:

$$f(t) = C\{\exp(-t/T_1) - \exp(-t/T_2)\}. \qquad \textbf{13.18}$$

Here the constant C in front (which includes among other things the amplifier gain) will later be chosen to fulfill our usual condition of $f(t_P) = 1$. Now

$$f'(t) = C\{(-1/T_1)\exp(-t/T_1) + (1/T_2)\exp(-t/T_2)\} \qquad \textbf{13.19}$$

and to obtain the peak position one puts $f'(t) = 0$. Hence

$$t_P = \{T_1T_2/(T_1 - T_2)\}\ln(T_1/T_2)$$

and substituting back into $f(t)$ one obtains for the peak height (after simplification)

$$f(t_P) = C\left[(T_2/T_1)^{\frac{T_2}{T_1 - T_2}} - (T_2/T_1)^{\frac{T_1}{T_1 - T_2}}\right].$$

The subsequent calculations are made easier if, instead of using T_1 and T_2 we use T_1 and λ (= T_2/T_1). Using these parameters we then obtain for $f(t_P)$, again after simplification:

$$f(t_P) = C\lambda^{\frac{\lambda}{1-\lambda}}(1 - \lambda)$$

and as we are putting $f(t_P) = 1$

$$C = \left(\lambda^{\frac{\lambda}{1-\lambda}}\right)^{-1} /(1 - \lambda) \qquad 13.20$$

which we will use later. Now

$$\overline{v^2_{\text{total}}} = \mathcal{A}\int_0^\infty f(t)^2 dt + \mathcal{B}\int_0^\infty f'(t)^2 dt.$$

Putting in the values for $f(t)$ and $f'(t)$ respectively from equations **13.18** and **13.19** and performing the integrations gives:

$$\overline{v^2_{\text{total}}}/C^2 = \mathcal{A}\left(\frac{T_1}{2} - \frac{2T_1T_2}{T_1 + T_2} + \frac{T_2}{2}\right) + \mathcal{B}\left(\frac{1}{2T_1} - \frac{2}{T_1 + T_2} + \frac{1}{2T_2}\right)$$

or using T_1 and λ instead of T_1 and T_2,

$$\overline{v^2_{\text{total}}}/C^2 = \frac{\mathcal{A}T_1(1 - \lambda)^2}{2(1 + \lambda)} + \frac{\mathcal{B}(1 - \lambda)^2}{2T_1\lambda(1 + \lambda)}$$

or obtaining a value for C^2 from equation **13.20**

$$\overline{v^2_{\text{total}}} = \left(\lambda^{\frac{\lambda}{1-\lambda}}\right)^{-2} \left[\frac{\mathcal{A}T_1}{2(1 + \lambda)} + \frac{\mathcal{B}}{2T_1\lambda(1 + \lambda)}\right]. \qquad 13.21$$

To obtain the condition for minimum noise one differentiates $\overline{v^2_{\text{total}}}$ in turn with respect to the two variables T_1 and λ, and sets the results equal to zero. The first of these gives:

$$\frac{\mathcal{A}}{2(1 + \lambda)} - \frac{\mathcal{B}}{2T_1^2\lambda(1 + \lambda)} = 0$$

or $\qquad T_1^2 = \mathcal{B}/\mathcal{A}\lambda. \qquad 13.22$

Differentiating $\overline{v^2_{\text{total}}}$ with respect to λ, putting the result equal to zero and using **13.22** should then give us the optimum value for λ. We can do things in a slightly easier fashion by obtaining $\frac{\partial \overline{v^2_{\text{total}}}}{\partial \lambda}$ and showing that $\lambda = 1$ (that is $T_1 = T_2$) makes it equal to zero. We thus proceed to differentiate each term in turn in equation **13.21** (and in doing so it is helpful to take logs and use the result $\frac{du}{d\lambda} = u\frac{d}{d\lambda}(ln\ u)$) and put in the value of $\lambda = 1$ in the result. One finds that singularities occur in some of the terms and there it is necessary to put $\lambda = 1 + x$ where x is a small quantity, which can later be allowed to go to zero, and also to use the fact that $ln(1 + x) = x - x^2/2$ approximately. One finally obtains:

$$\left(\frac{\partial \overline{v^2_{\text{total}}}}{\partial \lambda}\right)_{\lambda=1} = (\epsilon^2/8)(\mathcal{A}T_1 - \mathcal{B}/T_1)$$

where $\epsilon = \exp(1) = 2{\cdot}7$ approximately.

The right-hand side of this equation can be seen to be zero, using equation **13.22** with $\lambda = 1$. Thus $\frac{\partial \overline{v^2_{\text{total}}}}{\partial \lambda}$ is zero for $\lambda = 1$ and hence the optimum situation for noise is when $\lambda = 1$ that is $T_1 = T_2$. The rise and fall times should thus be equal for best results, as we intuitively assumed previously.

Appendix

13·5 Plausible proof of Campbell's theorems

Our aim here is to demonstrate in a plausible way the correctness of Campbell's theorems of the mean and the mean square, that is

$$\overline{V} = \bar{n} \int_0^\infty f(t)\, dt \qquad \textbf{13.23a}$$

$$\overline{(V - \overline{V})^2} = \bar{n} \int_0^\infty \{f(t)\}^2\, dt \qquad \textbf{13.23b}$$

where $\overline{V}$ and $\overline{(V - \overline{V})^2}$ are the mean output voltage and mean square output voltage respectively from our system, $f(t)$ is the response of the system to a single event, and $\bar{n}$ the average number of events per second.

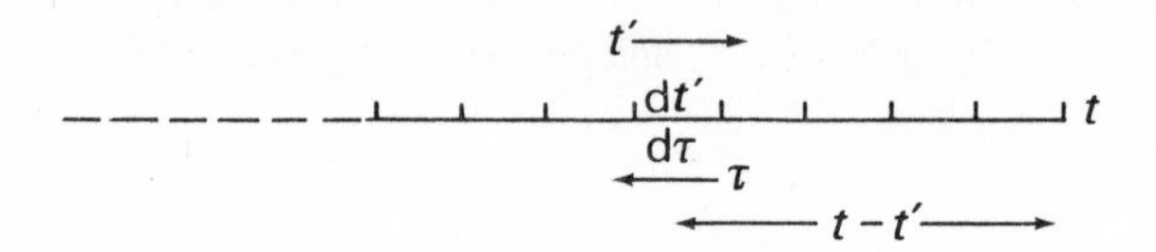

Figure 191. Definition of times for Campbell's theorems.

Let us imagine that we are trying to estimate the effect at the output at some particular time t, and let us divide the time leading up to t into small intervals dt' as shown in Figure 191. If $n(t')$ is the number of events per unit time then the number occurring in dt' is $n(t')\, dt'$ (or more correctly if dt' is small enough, $n(t')\, dt'$ is the probability of an event in dt'). The effect at time t of each event is given by $f(t - t')$ and the effect at time t from the interval dt' is thus $f(t - t')\, n(t')\, dt'$. The total effect is then

$$V(t) = \int_{t'=-\infty}^{t'=t} n(t')\, f(t - t')\, dt'$$

where we have taken the lower limit at $-\infty$ because we assume we have left the system for a long time after switching on in order for it to settle down. Changing the variable in the previous integral by putting $t - t' = \tau$ we have:

$$V(t) = \int_{\infty}^{0} n(t-\tau) f(\tau)(-d\tau) \quad \text{or}$$

$$V(t) = \int_{0}^{\infty} n(t-\tau) f(\tau)\, d\tau \qquad 13.24$$

The mean value of the output is got by averaging over various values of t and is given by:

$$\overline{V(t)} = \overline{\int_{0}^{\infty} n(t-\tau) f(\tau)\, d\tau}$$

where a bar above a term indicates an average with respect to t. Remembering that an integral is just the limit of a sum of terms, what we have to do is to average a large number of these sums. Let us organize this in a different way. Suppose we fix our attention on a particular $d\tau$ and average the $n(t-\tau)d\tau$ over the various t's. The result is $\bar{n}\, d\tau$ where $\bar{n}$ is the mean value of n. This corresponds to a particular $d\tau$ and a corresponding $f(\tau)$: we then sum over all the $d\tau$'s to get:

$$\overline{V(t)} = \int_{0}^{\infty} \bar{n}\, f(\tau)\, d\tau. \qquad 13.25$$

Now $\bar{n}$ is a constant and can come outside the integral. Also since τ is just a dummy variable we can finally write:

$$\overline{V(t)} = \bar{n} \int_{0}^{\infty} f(t)\, dt \quad \text{as required}$$

We now calculate $\overline{\{V(t) - \overline{V(t)}\}^2}$. From equations **13.24** and **13.25** we have:

$$\overline{\{V(t) - \overline{V(t)}\}^2} = \overline{[\int_{0}^{\infty} n(t-\tau) f(\tau) d\tau - \bar{n} \int_{0}^{\infty} f(\tau) d\tau]^2}$$

$$= \overline{[\int_{0}^{\infty} \{n(t-\tau) - \bar{n}\} f(\tau) d\tau]^2}$$

$$= \overline{\int_{0}^{\infty} \{n(t-\tau) - \bar{n}\} f(\tau) d\tau \times \int_{0}^{\infty} \{n(t-\tau) - \bar{n}\} f(\tau) d\tau}.$$

Here we have two sums of terms to multiply and average. Again let us start with a particular $d\tau$ in the first sum. We must multiply the term corresponding to it by all the terms in the second sum (see schematic diagram in Figure 192). But first let us concentrate on the result of multiplying the $d\tau$ term by the corresponding factor in the other sum (solid arrow). The result, apart from the factor $\{f(\tau)\}^2$ is $\{n(t-\tau) - \bar{n}\}^2 (d\tau)^2$. (Note that this is always positive regardless of the sign of $n(t-\tau) - \bar{n}$.) Proceeding to average this over all the various t's we obtain: $\overline{\{n(t-\tau) - \bar{n}\}^2}\, (d\tau)^2$. Now we know that poissonian statistics (which we believe our events obey) predict that if $\bar{N}$ is the mean number of events in a given time then the standard deviation is given by $(\bar{N})^{1/2}$ and the variance, that is $\overline{(N - \bar{N})^2}$ is given by $\bar{N}$. Thus in our case the quantity $\overline{\{n(t-\tau) - \bar{n}\}^2}\, (d\tau)^2$ reduces to $\bar{N} = \bar{n}\, d\tau$. Multiplying by $\{f(\tau)\}^2$ and summing over all the $d\tau$'s we obtain for this part of the contribution

$$\overline{\{V(t) - \overline{V(t)}\}^2} = \int_0^\infty \{f(\tau)\}^2 \, \bar{n} \, d\tau$$ or since $\bar{n}$ is a constant

$$\overline{\{V(t) - \overline{V(t)}\}^2} = \bar{n} \int_0^\infty \{f(\tau)\}^2 \, d\tau. \qquad 13.26$$

We still have to go back and calculate the result of multiplying the term corresponding to our particular $d\tau$ by a non-corresponding term in the other sum (one of the dotted arrows in Figure 192) and averaging for various values of t. However in this case we expect the result to average to zero since we are multiplying and averaging two uncorrelated terms, and we are liable to get as many positive values of $\{n(t - \tau) - \bar{n}\} d\tau$ as negative. Clearly no net contribution arises from this part of the calculations, and the result we have already obtained in **13.26** stands. Since τ is only a dummy variable we can, as before, change it to t to obtain:

$$\overline{\{V(t) - \overline{V(t)}\}^2} = \bar{n} \int_0^\infty \{f(t)\}^2 \, dt$$

which is the result previously quoted.

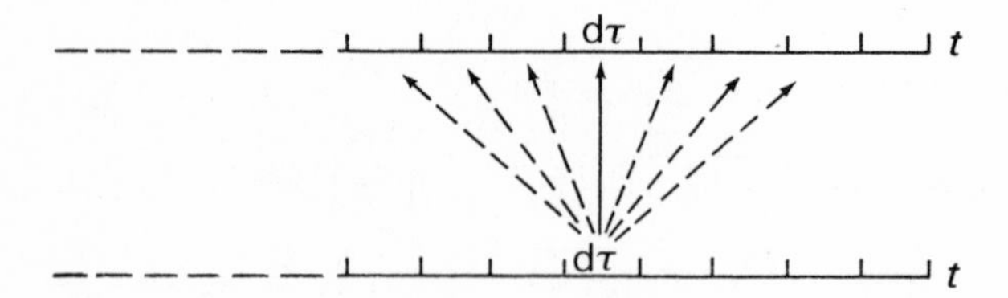

Figure 192. Cross multiplication of terms.

References and further reading

1. R. WILSON, 'Noise in ionisation chamber pulse amplifiers', *Philosophical Magazine*, vol. 41, Series 7, 1950, p. 66.
2. N. R. CAMPBELL and V. J. FRANCIS, 'A Theory of valve and circuit Noise', *Journal of the Institution of Electrical Engineers*, vol. 93(3), 1946, p. 45.
3. E. FAIRSTEIN and J. HAHN, 'Nuclear pulse amplifiers – fundamentals and design practice', *Nucleonics*, vol. 23, no. 11, 1965, p. 50.
4. P. W. NICHOLSON, *Nuclear Electronics*, Wiley, 1974.

Index

A page reference in bold type indicates main discussion of topic

MANSFIELD COLLEGE
LIBRARY
WITHDRAWN
OXFORD